621.312 G 1

PASADENA CITY COLLEGE
LIBRARY
PASADENA, CALIFORNIA

Geothermal Energy

Geothermal Energy

Resources, Production, Stimulation

Edited by Paul Kruger and Carel Otte

Stanford University Press, Stanford, California 1973

Stanford University Press
Stanford, California
© 1973 by the Board of Trustees of the
Leland Stanford Junior University
Printed in the United States of America
ISBN 0-8047-0822-3
LC 72-85700

Contents

732766 WITHDRAWN

Foreword

The prospects for geothermal power have been one of my special interests during my years as a member of the Senate Interior Committee and the Joint Congressional Committee on Atomic Energy. It is therefore a pleasure for me to say a word about this book, which resulted from the American Nuclear Society's Special Session on Geothermal Energy, held in Las Vegas, Nevada, in June 1972.

Although Italy has been generating electricity from natural-steam energy since 1904, and its present capacity at Larderello totals about 390 megawatts, the growth of The Geysers steam plant in California already makes the U.S. an important nation in this field. We are now producing about 300 megawatts of electricity from energy supplied by the field, and expansion to some 600 megawatts by the end of 1975 will make the U.S. the largest producer of electricity from geothermal energy. Other countries, notably New Zealand, Mexico, Japan, Iceland, and the Soviet Union, are all increasing their attention to geothermal power. And geothermal-energy potentials are being assessed in many other lands. In 1970 the United Nations responded to this growing attention by convening an international symposium on the development and utilization of geothermal resources.

Geothermal energy is known to exist in many other areas of the U.S., and it could indeed be a sleeping giant among the nation's energy resources. I am convinced it can be successfully and safely developed to play an important role in overcoming our burgeoning power problems. Since geothermal energy is a relatively clean form of energy, it could play an even greater role in meeting the nation's mandate for a halt to the environmental degradation resulting from power production. Moreover, geothermal energy may be put to a variety of other uses, such as hot-water heating and industrial processes.

What are the prospects for geothermal power? The answer depends on the willingness of the Federal Government and the willingness and capacity of private industry and the scientific community to get on with the task of ascertaining the technical and economic feasibility and environmental acceptability of geothermal-resources development. I plan to do my share in the Federal Government through the Senate Interior Committee and the Joint Committee on Atomic Energy. I hope this book acts as a stimulus to each of its readers to do his share in bringing the full array of the nation's scientific and technological expertise to bear on the development of environmentally acceptable geothermal-energy resources.

Geothermal energy is at once an ageless phenomenon and a challenging new frontier. I commend it to you as a national resource meriting your expert attention. It is time to awaken this sleeping giant and put it to work!

Alan Bible

Las Vegas
November 30, 1972

ALAN BIBLE
Senator from Nevada

Preface

This book originated as the proceedings of the Special Session on Geothermal Energy held at the American Nuclear Society Annual Meeting June 19–20, 1972, in Las Vegas, Nevada. The Special Session was organized initially to review the progress achieved in the technology for stimulating geothermal fluids by explosive means. It was sponsored by the Civil Explosives Application Division of the Society. As the plans for the Special Session developed, it became apparent that a broader view of the state of the art of geothermal-energy development was required before we could reasonably examine the potential role of stimulation technology. Thus the Special Session on Geothermal Energy became a general conference covering three major aspects of this emerging energy field: the resource, the production technology, and potential stimulation methods.

The program committee called upon experts in these fields to present state of the art papers, with the request that manuscripts be submitted at the meeting to be included in a suitable published proceedings. The tremendous interest in this promising energy resource was reflected in the strong attendance at the Special Session; and most of those attending were not members of the Society. With the evident need for a serious study and review of the field of geothermal energy, the program committee agreed to have the proceedings published as a scholarly work, by Stanford University Press. Accordingly, each of the papers presented at the meeting received a thorough technical review and was returned to its author for preparation of final manuscript. Each paper, and its illustrative materials, then received a careful editing at the hands of the publisher. Although the decision to build in these technical and editorial reviews substantially deferred the publication date of the proceedings, the consequent investment in time and effort should prove to have been justified: a more worthy and more broadly useful volume will have been added to the sparse literature on geothermal energy.

The editors acknowledge with gratitude the assistance of their colleagues on the program committee, Anthony H. Ewing of the U.S. Atomic Energy Commission and Donald E. White of the U.S. Geological Survey. The editors also acknowledge with appreciation the technical reviews prepared by Richard G. Bowen, State of Oregon; Chester F. Budd, Jr., Union Oil Company; Jim Combs, University of California, Riverside; Walter C. Day, U.S. Army Corps of Engineers; Richard Dondanville, Consultant; John P. Finney, Pacific Gas and Electric Company; John W. Harbaugh, Stanford University; Harvey Hennig, Union Oil Company; Jack Howard, Lawrence Livermore Laboratory; Tsvi Meidav, United Nations; L. J. P. Muffler, U.S. Geological Survey; Henry J. Ramey, Jr., Stanford University; Robert W. Rex, University of California, Riverside; Herbert Rogers, Rogers Engineering Company; Morton C. Smith, University of California, Los Alamos; and Alan K. Stoker, Stanford University.

Acknowledgments are also due the American Nuclear Society, for permission to publish the manuscripts, and the U.S. Atomic Energy Commission, for support in preparation of the final manuscripts.

PAUL KRUGER
Stanford, California

CAREL OTTE
Los Angeles, California

January 15, 1973

Geothermal Energy

1. Introduction: The Energy Outlook

CAREL OTTE AND PAUL KRUGER

Energy is the driving force of industry—and, indeed, of civilization. Hubbert (1969) noted that until recent centuries one of the factors limiting population growth was the constancy of the energy supply per capita—only slightly more than the food supply. It was only where fossil fuels began to be employed that civilization was released from the constraints of subsistence agriculture and home industry and began its breathtaking growth in both population and per capita standard of living. Coal mining began about eight centuries ago, the production of petroleum as an important fuel just over one century ago, and the use of natural gas on a large scale even more recently. To be sure, other forms of energy contribute to the world's energy budget— hydroelectric and nuclear power come readily to mind—but it is a fact of life in the 1970s that fossil fuels are paramount.

From its modest beginnings, world consumption of energy from fossil fuels has increased at about 4 percent per year. With the population of the world now increasing at a rate of just under 2 percent per year, the surplus of energy that is available contributes to a steady improvement in living standards in virtually all corners of the globe. But average figures do not adequately portray the *pattern* of energy consumption. The United States leads the world not only in per capita income but also in per capita consumption of energy. With 6 percent of the world's population, we consume 33 percent of the world's energy production (United Nations, 1971). And human labor in the United States

---

Carel Otte is a Vice President and Manager of the Geothermal Division of the Union Oil Company of California, Los Angeles; and Paul Kruger is Professor of Nuclear Civil Engineering, Stanford University, Stanford, California.

constitutes far less than 1 percent of the work performed in factories, refineries, and mills. The general worldwide correspondence between per capita income and energy use is shown in Fig. 1. The apparent anomalies of Canada and Sweden can be understood by considering their particular circumstances: the greater use of energy by the former relative to income is owing to its vastness, harsh climate, and energy-consuming extractive industries; the contrasting relationship in Sweden reflects its greater compactness and greater reliance on light industries. More typical are the United States and West Germany; for the former, both measures are roughly double the corresponding values for the latter.

The graph reflects an important point. It is becoming increasingly evident that the United States has supported its population expansion, its economic growth, and its high standard of living by means of an accelerating utilization of low-cost energy—energy derived chiefly from a bountiful endowment of domestic fossil-fuel resources. But fossil fuels are finite in amount and nonrenewable in periods less than millions of years. Sooner or later the limits of supply of all of our energy supplies are going to be reached.

In petroleum products the United States indeed became a net importing nation in the late 1960s and is now importing 30 percent of its petroleum liquids. Moreover, the nation was unprepared for the abruptness of the turnover: in the past our vast pipeline system had been able to meet immediate demand and our excess producing capacity had served as reserve "storage"; but with the immense increase in crude-oil imports came a shortage of storage capacity for fossil fuels. At the same time, serious environmental concerns have slowed and delayed the installation of nuclear-reactor electric-generating capacity in the United States. The net effect has been an inability to match new capacity to the growth in energy consumption. And the result has been power shortages, particularly in periods of extreme weather conditions. In the heat of summer, with air conditioners producing peak demand, many population centers are threatened with brownouts, and some have already experienced them. And in the dead of winter, with transportation and supply curtailed in the colder regions, fuel shortages have come as a severe shock.

Unlimited supplies of low-cost energy have been taken largely for granted in the United States, and long-range projections of increases

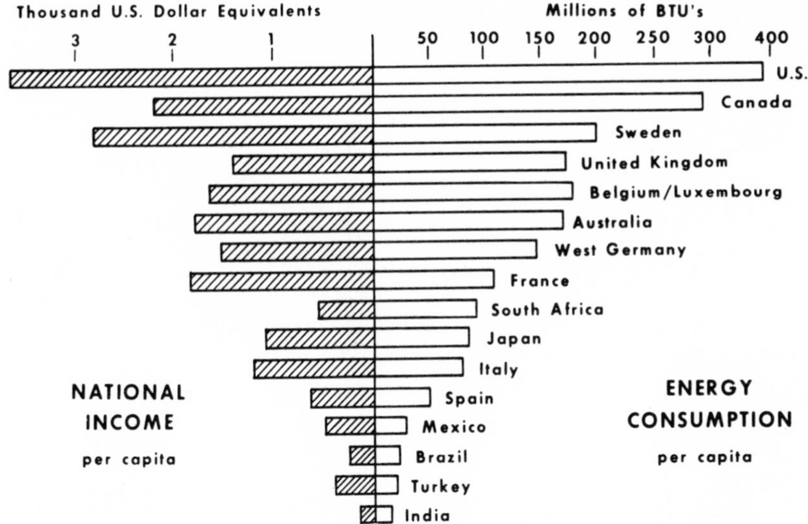

Fig. 1. 1968 per capita income and energy consumption for seventeen selected countries. (Source: U.S. Department of the Interior, "United States Energy: A Summary Review," 1972.)

in the real gross national product and improvements in the quality of life have been based on assumptions of a continued abundance of energy. Yet it is a fact that the United States is falling from a position of relative energy self-sufficiency into dependence on other countries to meet its requirements.

Meanwhile, the rate of increase of energy consumption in the rest of the world is half again that in the United States. While U.S. energy requirements are projected to increase on the order of 4 percent per year in the 1970s, free-world requirements outside the United States will probably grow at a rate in excess of 6 percent per year. None of the industrialized nations are self-supporting in energy, and together with the developing nations they will compete vigorously with the United States for available supplies (U.S. Department of the Interior, 1972).

The increasing awareness of the exponential growth of energy consumption, the rapid depletion of our natural resources, the lagging development of new energy resources and technologies, the strain on the dollar in world markets, and the growing public and institutional demands for energy and materials conservation and environmental protec-

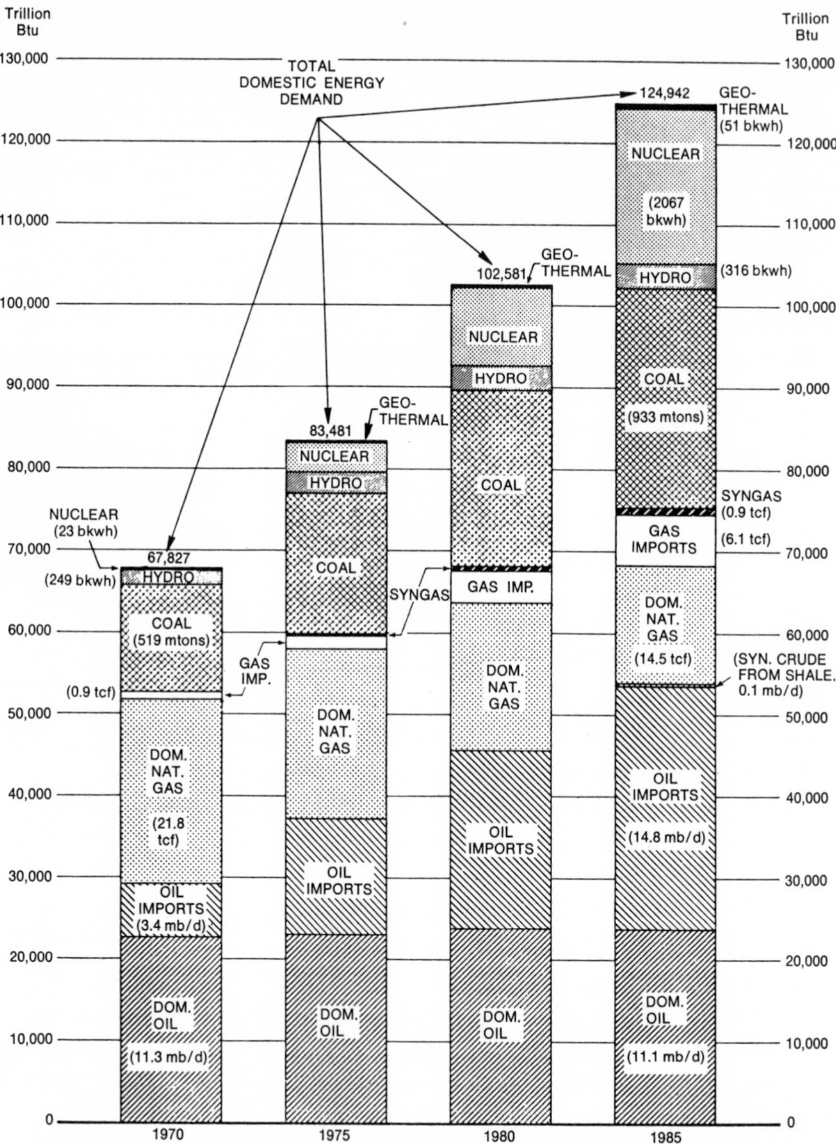

Fig. 2. United States energy balance, 1972, by types and origins of resources. Categories are domestic oil (millions of barrels per day), oil imports (same units), synthetic crude oil from domestic shale (same units), domestic natural gas (trillions of cubic feet), natural gas imports (same units), syngas (natural gas produced by coal gasification, same units), domestic coal (millions of tons), hydroelectric power (billions of kilowatt-hours), nuclear power (same units), and geothermal power (same units). (Source: National Petroleum Council, "U.S. Energy Outlook: An Interim Report," November 1971.)

tion have forced the Federal Government to look toward a more rational plan for the nation's energy economy. As a result, many Government and private agencies have initiated broad studies of projected energy consumption and the possibilities for conserving energy while enhancing the quality of the environment. Examples of these studies are the reports published by the National Petroleum Council (1971; 1972). Figure 2, from the 1971 report, shows the U.S. energy balance for the years 1970 to 1985 (the 1972 report supplies comparable figures in a different form). Projections are based on a general continuation of current trends. The values are given in British thermal units (Btu) so that the relative contributions of the various energy resources can be compared. The data for the total domestic energy supply from the 1971 report are given in Table 1 in conventional units. The conclusions reached in these reports may be summarized as follows:

1. *Energy consumption.* During the period 1971–85, energy consumption in the United States is forecasted to grow at an average rate of 4.2 percent per year. The annual growth rates by market sectors would be as follows: electric utilities, 6.7 percent; raw-material uses, 5.4 percent; transportation, 3.7 percent; residential and commercial, 2.5 percent; and industrial, 2.2 percent.

2. *U.S. energy supplies in relation to consumption.* In 1970, domestic energy supplies satisfied 88 percent of U.S. energy consumption. Domestic supplies will grow at an average rate of perhaps 2.6 percent per year during the 1971–85 period. Since domestic demand will increase at a greater rate, the nation will become increasingly dependent on imported supplies. By 1985, domestic supplies will satisfy only about 70 percent of U.S. consumption.

3. *Petroleum liquids.* Domestic supplies, consisting chiefly of crude oil but including condensate and natural-gas liquids (butane, propane, etc.) totaled 11.3 million barrels per day in 1970, which was 31 percent of total energy consumption. Total U.S. production in 1985 is estimated at only 11.1 million barrels per day. In order to meet growing demands for petroleum liquids, imports would have to increase more than fourfold by 1985, reaching a rate of 14.8 million barrels per day, and accounting for 57 percent of total petroleum supplies or 25 percent of total energy consumption.

4. *Natural gas.* In the absence of supply limitations (an assumption that may prove to be unwarranted), potential gas demand would ap-

TABLE 1
U.S. Energy Balance, 1970–85, from Domestic Sources
(In conventional physical units)

Energy source	1970	1975	1980	1985
Projected Domestic Supply				
Oil (millions of barrels per day)				
Conventional	11.3	11.1	11.8	11.1
Synthetic (from shale)	—	—	—	0.1
TOTAL	11.3	11.1	11.8	11.2
Gas (trillions of cubic feet)				
Conventional	21.82	19.80	17.47	14.50
Synthetic	—	0.37	0.55	0.91
TOTAL	21.82	20.17	18.02	15.41
Coal (millions of short tons)				
For domestic use	519	651	799	933
For export	71	92	111	138
TOTAL	590	743	910	1,071
Other (billions of kilowatt-hours)				
Hydroelectric	249	271	296	316
Nuclear	23	326	926	2,067
Geothermal	0.7	12	34	51
TOTAL	272.7	609	1,256	2,434
Imports Required to Balance				
Oil (millions of barrels per day)	3.4	7.3	10.7	14.8
Gas (trillions of cubic feet)	0.92	1.55	3.75	6.08

SOURCE: National Petroleum Council, "U.S. Energy Outlook: An Interim Report," November 1971.

proximately double between 1970 and 1985, reaching a level of about 39 trillion cubic feet per year. Dependence on imports would rise from 4 percent of gas supplies in 1970 to more than 28 percent in 1985, assuming the availability of foreign supply. (Under current federal policies, domestic natural-gas supplies could be expected to fall by about 40 percent by 1985; but if gas from Alaska's north slope, gas imports from Canada, synthetic gas from domestic coal and naphtha, and imports of liquefied natural gas (LNG) and liquefied petroleum gas (LPG) make up for the decline, as expected, total gas supplies in 1985 would be 21.5 trillion cubic feet per year, slightly less than the 1970 supplies.)

5. *Coal.* The supply of domestic coal, including exports, will increase from 590 million tons in 1970 to some 1,071 million tons in 1985. Coal reserves are ample and could support a faster growth rate in production.

6. *Nuclear power.* The supply of nuclear power will increase from 23 million megawatt-hours (Mwh) in 1970 to perhaps 2 billion Mwh in 1985. No shortage of domestic fuels is foreseen, if prices for U_3O_8 hold at up to $10 per pound. By 1985, nuclear energy will be supplying 48 percent of total electric-power requirements.

7. *Other fuels.* The remaining fuels—hydroelectric power, geothermal power, and crude oil from shale—will together contribute only 3 percent of energy requirements in 1985. Not considered as available in 1985 are the emerging technologies of thermonuclear fusion reactors and the harnessing of solar energy in meaningful quantities.

The 1972 National Petroleum Council report examines the potential of geothermal energy in some detail. Growth from the 1970 electric-generating capacity of 82 Mw is estimated to 1985 for four cases: the most optimistic, 19,000 Mw, assumes maximum technological progress with no impediments to development. Case II, which projects 9,000 Mw installed capacity by 1985, assumes that large areas of land will be available for prospecting with little delay owing to ecological questions. Case III is considered relatively optimistic, 7,000 Mw or about 1 percent of U.S. electric-energy requirements in 1985. And Case IV, the least optimistic, assumes 3,500 Mw by 1985.

Another source of energy forecasts is the reports published by the U.S. Department of the Interior (1971; Dupree and West, 1972). Table 2 summarizes by type the energy sources that will provide the gross energy input. Gross energy production is forecasted to increase by 69 percent from the actual output in 1971 to the predicted output in 1985. There is general agreement between these forecasts and the estimates of the National Petroleum Council (Fig. 2).

TABLE 2

Energy Sources for the U.S. Economy

(In trillions of Btu)

Energy source	Actual 1971	Forecasted 1985
Coal	12.6	21.5
Oil	30.5	50.7
Natural gas	22.7	28.4
Nuclear power	0.4	11.7
Hydroelectric power	2.8	4.3
TOTAL	69.0	116.6

SOURCE: U.S. Department of the Interior, "United States Energy: A Summary Review," 1972.

TABLE 3
Nuclear Power Forecast for the United States

Power category	Installed 1970	Forecasted 1985
U.S. electric-generating capacity (thousands of megawatts)	309	995
Capacity of domestic nuclear power plants (thousands of megawatts)	5	300
Fraction of total capacity from nuclear plants (percent)	1.6	30
Fraction of new electric-power capacity from nuclear plants (percent)	15	45

SOURCE: U.S. Atomic Energy Commission, "Forecast of growth of nuclear power," Report WASH-1139.

The U.S. Atomic Energy Commission (1971) reported a broad forecast of electric-energy growth only (i.e. excluding energy consumed by transportation, industry, etc.), and its conclusions (Table 3) are in general agreement with those of the National Petroleum Council reports (Table 4). From a total of 309,000 Mw installed generating capacity in the United States at the end of 1970, the AEC expects a growth to 995,000 Mw by 1985. Of these totals, the amounts derived from domestic nuclear power plants are, respectively, 5,000 Mw in 1970 and 300,000 Mw by 1985, representing 1.6 percent of installed capacity in 1970 and 30 percent by 1985.

All these studies indicate a marked increase in the demand for energy in the United States by 1985. With the production of domestic petroleum products likely to remain roughly constant, the growth is expected to be met from three sources: oil importation, representing 57 percent of the needed oil supply and contributing perhaps as much as 25 percent of total energy consumption; an almost doubled output from the nation's coal supply, from approximately 600 million tons to about 1,100 million tons; and the installation of some 250 nuclear power plants of 1,000-Mw generating capacity, which would provide from 10 to 13 percent of the national energy output.

But although each major source of energy is capable of meeting the projected supply requirements, each has its own unique problems. For example, domestic petroleum production is now actually declining, and prices are likely to escalate rapidly as a result of supply shortages and competition from other industrial nations for the available reserves.

TABLE 4
Fuel Mix and Total Power Production for U.S. Electric Utilities
(In trillions of Btu)

Power source	Power produced 1970	1985
Oil	2,050	4,530
Natural gas	3,900	3,900
Coal	7,800	13,900
Nuclear[a]	240	18,713
Hydroelectric	2,677	3,320
TOTAL	16,667	44,363

SOURCE: National Petroleum Council, "U.S. Energy Outlook: A Summary Report," December 1972.
[a] Includes minor contribution from geothermal energy (500×10^{12} Btu in 1985).

Dependence on large volumes of imported petroleum products will constitute a serious threat to our national security and to the economy by increasing the imbalance in our foreign payments. Oil imports already constitute a major component of our payments imbalance. And the report by Winger et al. (1972), another broad energy forecast to 1985, prepared for the Chase Manhattan Bank, estimates an annual deficit in payments for petroleum supplies alone of as much as 25 to 30 billion dollars by 1985. Such a deficit the nation cannot tolerate.

Our coal reserves are vast—even at present mining rates they may last for centuries—but in the crucial near term the availability of coal supplies may be constrained by lack of manpower, inadequate transportation facilities (coal deposits tend to be concentrated geographically), and reasons of miner health and safety. Moreover, the imposition of environmental controls on mining, particularly stripmining, as well as on the emission of such products as sulfur dioxide and fly ash from coal-fired plants, may further impede the growth of coal production as an energy source. All of these problems warrant increased attention to technologies permitting the recovery of coal energy in situ, such as coal gasification of underground strata (syngas).

Domestic natural-gas reserves, like those of petroleum, are being depleted rapidly, and few new major fields are being developed. Moreover, artificial restraints on the price of natural gas encourage its use in grossly unsuitable applications, such as boiler fuel for industry and utilities. The importation of natural gas requires that the gas be liquefied for shipment and storage, a costly process reflected in much higher prices to the consumer. Thus natural gas, which should be conserved essentially for chemical raw-material, light-industrial, and domestic use, is in serious jeopardy as a contributor to the national energy budget.

If nuclear power's contribution to the nation's energy economy is to be realized, a greater public acceptance must somehow be fostered. The forecasted growth of nuclear power production will be achieved only if the delays currently caused by considerations of reactor safety, siting (e.g. away from fault lines), disposal of radioactive and thermal wastes, and other environmental impacts (particularly in the mining and refining of reactor material) are resolved. In addition, the Department of the Interior forecast notes that the attainment of substantial growth in nuclear power is predicated on the introduction of both the high-temperature gas reactor and the liquid-metal fast breeder reactor, both of which are still in the developmental stage. Any significant problems encountered in the realization of either of these reactor types, or in the continued installation of conventional reactors, would yield a corresponding increase in consumption of fossil fuels. And although nuclear power is expected to grow rapidly in the next few decades, it is essentially limited to large central-station generation and cannot be offered as a substitute for fossil fuels in the transportation industry or in various other forms of energy utilization.

As for other sources of power, the Dupree and West (1972) forecast suggests that after 1980 our energy sources may be supplemented not only from increased oil imports and new reserve discoveries, but also from shale oil, coal liquefaction, and tar sands, each dependent on commercial development of new technologies. In this regard it is interesting to note, in contrast to the 1972 report of the National Petroleum Council, that the Dupree and West report, prepared for the Department of the Interior, does not even consider geothermal energy.

The problems associated with the nation's future energy needs and the major sources of supply are painfully complex; many conflicting forces are at work. It is in fact likely that some of the forecasts will not be met; deviations are likely to occur, particularly in light of our rapidly changing economic and institutional climate. A trend toward reduced energy consumption and more efficient production and utilization of energy would help. Such a trend would correspondingly reduce the forecasted levels of energy supply. Policies tending to reduce consumption would include encouraging the use of mass transportation and smaller, less powerful automobiles, or establishing building codes that provide for improved insulation of homes and other heated or air-conditioned structures. Policies tending to increase utilization efficiency would include discouraging the use of natural gas as boiler fuel and encouraging its use in more appropriate applications.

Energy, moreover, is consumed in a great many ways, and some sources of energy (hydroelectric, nuclear, geothermal, and, for the most part, coal) are exploitable only in central-station electric-power generation; if these sources were more fully exploited, other sources, such as oil and natural gas, could be freed to a greater extent for energy needs that cannot be supplied from central-station power, such as transportation. Considering, then, the vast reserves of coal and oil shale available to us, and the high consumption of energy encouraged by low prices, we could postulate not so much an energy crisis as a crisis of misallocation. But in the near term, over the next decade, we may indeed be facing a supply crisis.

Studies with the objective of decreasing the exponential demand for power have already begun at the Federal level. In its report, the Office of Emergency Preparedness (1972) suggested that energy-conservation measures can reduce U.S. energy demand by 1980 by as much as the equivalent of 7.3 million barrels of oil per day. This value is equivalent to about two-thirds of the oil imports projected for that year. Clearly, conservation of energy will be a necessary part of future plans for a comprehensive national energy program. But just as clearly, programs to improve the quality of the environment will themselves consume energy; all such efforts are laudable, so only that the finiteness of available energy is reckoned with in a comprehensive national plan.

Therefore, it seems prudent, with the realization that serious energy shortages may occur between now and 1985, that the United States attempt to utilize *all* of the indigenous energy supplies that can be developed with modern and future technologies. Among the "miscellaneous" sources of energy, a likely candidate for early development in large quantities is geothermal energy. This volume attempts an assessment of the potential of geothermal energy in meeting the increasing energy demand within the United States during the next few decades and in conserving our fossil fuels. The book, in fulfilling the objectives of the Special Session on Geothermal Energy of the American Nuclear Society meeting in June 1972, consists of three major parts: (1) a review of the worldwide status, principles of occurrence, resource potential, and exploration techniques of geothermal energy; (2) a discussion of resource production, forms of utilization, and the environmental impact of geothermal resource development; and (3) an examination of potential stimulation concepts for resource development by nuclear, chemical, and mechanical techniques.

Several conclusions became apparent during the symposium. Some of them may be summarized as follows:

1. The utilization of geothermal energy will be largely for the generation of electric power. Mineral extraction, fresh-water production, or the use of the resources for process or space heating, although of potential value, will generally be limited to special occurrences or circumstances.

2. The nation's geothermal-energy resource is potentially vast, particularly the portion that may be tapped by deeper drilling technology.

3. Most of the resources will be of the liquid-dominated type, rather than the vapor-dominated type exemplified by the field at The Geysers, in California.

4. Development of the technology for the vapor-turbine cycle will permit the development of liquid-dominated resources of intermediate-range temperatures (in excess of 300°F), and thus vastly increase the exploitable portion of the nation's geothermal resources.

5. Electric power produced from a geothermal field appears to be economically competitive with that from conventional fossil-fuel stations.

6. The environmental impact of geothermal power development appears to be more acceptable than that of fossil- or nuclear-fuel power development.

7. Successful development of any of the stimulation techniques would substantially enlarge the economically exploitable geothermal reserves; stimulation techniques permitting development of the deep, hot, dry rock masses would enormously expand the geothermal energy reserves.

Other assessments of the potential for geothermal energy are appearing. For example, the National Science Foundation sponsored a formal review of geothermal energy for the purpose of evaluating the need for research support. In the report, Hickel (1972) forecasts that the United States could be producing at least 132,000 Mw of electric power from its geothermal resources in 1985. Whether geothermal energy can supply about 13 percent of the nation's power requirements by that time remains an open question. But with improvements in utilization technology and the removal of certain institutional barriers (legal and regulatory, Federal and local), it could become a major element of central-station electric-power generation, and in the process make up for delays in the growth of the other sources of supply, reduce the need for additional importation of petroleum products, and release petroleum and natural gas for more appropriate uses. If the recommendations made in this book are acted upon, the promise of geothermal energy becomes bright, and the forecast of the National Petroleum Council could be far exceeded.

REFERENCES

Dupree, W. G., and J. A. West. 1972. United States energy through the year 2000. U.S. Department of the Interior.

Hickel, W. J. 1972. Geothermal energy: A special report. University of Alaska, Fairbanks.

Hubbert, M. K. 1969. Resources and man. San Francisco: Freeman.

National Petroleum Council. 1971. U.S. energy outlook: An interim report, November 1971.

National Petroleum Council. 1972. U.S. energy outlook: A summary report, December 1972.

Office of Emergency Preparedness, Executive Office of the President. 1972. The potential for energy conservation, a staff study. Washington, D.C.: U.S. Government Printing Office.

United Nations. 1971. World energy statistics.

U.S. Department of the Interior. 1972. United States energy: A summary review.

U.S. Atomic Energy Commission. 1971. Forecast of growth of nuclear power. WASH-1139.

Winger, J. G., J. D. Emerson, G. D. Gunning, R. C. Sparling, and A. J. Zraly. 1972. Outlook for energy in the United States to 1985. New York: Chase Manhattan Bank.

2. Worldwide Status of Geothermal Resources Development

JAMES B. KOENIG

The Earth is a great reservoir of heat energy, but most of its heat is buried too deeply or spread too diffusely to be felt tangibly at the surface, especially as it is masked by incoming solar radiation. We become aware that the Earth is a great heat engine during episodes of volcanic eruption, when material at temperatures upwards of 800°C is emitted at the surface. Hot springs, geysers, and fumaroles are other surface manifestations of the Earth's heat content.

When recoverable, the heat of the Earth can be made to do useful work. The uses depend upon the enthalpy (heat content), physical state, and chemistry of the transporting medium. Despite very high enthalpy, molten lava has found no direct use. Similarly, hot, dry rock has found only the most limited and local uses. High-enthalpy aqueous fluids (above about 200 cal/g) are of proved use in the generation of electricity and in industrial processes, and may find application in desalination. Fluids of lower enthalpy have many uses as process heat in industry, and perhaps can be used also in desalination. Low-enthalpy fluids (above about 100 cal/g) are used widely in space heating and in agriculture. Additionally, certain mineralized waters may yield industrially valuable chemicals as a by-product of heat extraction or desalination. Developments in the technology of heat-exchanging and desalination may extend the range of uses of low-enthalpy fluids significantly in the next few decades.

High-enthalpy geothermal systems are known only in regions of youthful volcanism, crustal rifting, and recent mountain building. The

---

James B. Koenig is with the California Division of Mines and Geology, Sacramento, California.

major geothermal and volcanic belts are the circum-Pacific margin; island groups of the mid-Atlantic rift; the rift zones of east Africa and the adjacent Middle East; and the irregular belt of mountains and basins extending from the Mediterranean basin of Europe and north Africa across Asia to the Pacific (see Fig. 1).

Lower-enthalpy fluids are far more abundant in volcanic zones and elsewhere, and may represent a greater reserve of useful energy by an order of magnitude or more. Significant areas of lower-enthalpy geothermal fluids include the Gulf Coast of the United States, an extensive region in western Siberia, and portions of central Europe just north of the Alps and the Carpathian Mountains. Geologically, these are subsiding sedimentary basins at the margins of folded mountain ranges. Water encountered during exploration for oil in sedimentary basins has usually been considered a nuisance. But in the future these hot waters of relatively low enthalpy may represent an energy source as valuable as oil, and perhaps as widespread. Other low-enthalpy waters have been encountered in wells and mines or as occasional warm springs in older folded mountains, but rarely in the ancient stable platforms and shields of continental interiors.

This paper summarizes the distribution of geothermal resources on a

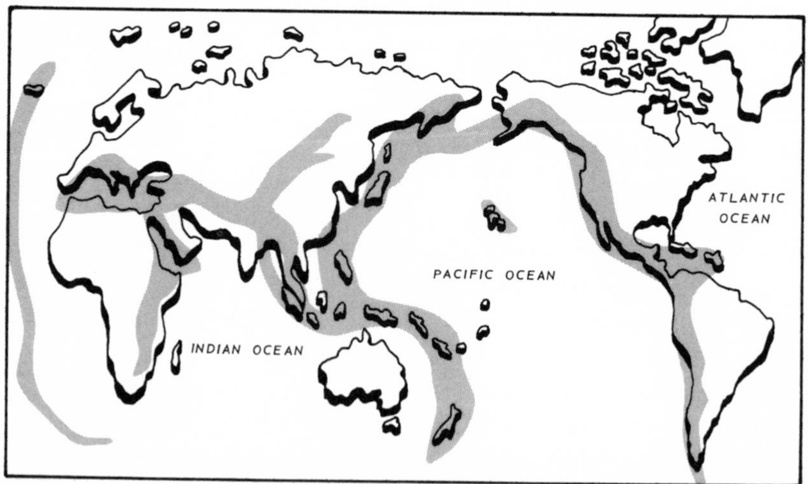

Fig. 1. Regions of intense geothermal manifestations; the distribution accords generally with recent volcanism and youthful mountain-building, and in part outlines boundaries of mobile crustal plates.

worldwide scale, reviews development and exploration activities, and evaluates the potential for development over the next decade.

The Framework for Utilization of Geothermal Energy

Because heat dissipates rapidly, geothermal fluids cannot be transported far from their point of recovery without suffering diminished quality. The maximum transport distance depends upon the initial enthalpy (heat content) of the fluid and the use for which the fluid is intended. For steam used in electric-power generation, the probable maximum distance is on the order of 2 to 3 km. Hot water for agricultural use or space heating can be transported farther, the greatest reported distance being 20 km for municipal hot water in Iceland.

Where heat energy is converted to electric power, the energy can then be distributed throughout the power-transmission grid. Conversion to electric power represents a practical means of transferring heat energy from what are often remote geothermal fields to population centers. However, the price of conversion is a substantial loss of energy, and further losses occur in transmission and in the subsequent use of electricity to do mechanical or thermal work. Additionally, electricity cannot be stored easily or efficiently, which means that a power system must be built to meet peak demand, almost without regard for base load. In terms of capital investment, this can be expensive and wasteful. Because of these unfavorable factors, direct utilization (i.e. without conversion) is likely to increase in such applications as space heating, agriculture, industrial processing, and perhaps desalination, especially as costs rise for other fuels. Energy for all of these processes can, of course, be supplied by electricity; but the conversion and reconversion are wasteful, especially with reconversion to heat. If energy costs rise sufficiently over the decades, new communities may develop near the geothermal-energy source, much as cities were once established to take advantage of running water to power mills. This concept is under active consideration in Hungary.

The countries that have successfully developed geothermal electricity to date are among the wealthiest and most industrially developed in the world. These include the United States, Italy, Japan, the Soviet Union, New Zealand, and Iceland. Even Mexico, where a plant is to begin operation in early 1973, is among the relatively more developed nations. This pattern appears to indicate the need for an adequate economic and technological infrastructure. However, the majority of countries in which exploration is under way, and where opportunities for future development are greatest, are generally among the least developed of nations, lacking the economic and technological base. But even within most highly developed countries there are regions that are lacking in energy resources, low in population, or remote from the industrialized heartland, where costs of electric power often are higher. Examples are Hokkaido, the northern island of Japan, the Kamchatka Peninsula of the Soviet Union, and parts of the Great Basin of the United States, all of which exhibit geothermal potential.

Many of the industrialized nations are poor in reserves of fossil fuels, and some are chronically short of foreign exchange. The United States now typifies the latter. Fuel-short countries include Italy, Japan, and, until very recently, New Zealand. In Japan, 75 percent of all energy is imported, and this figure is expected to increase over the next 10 years to over 85 percent. Even a wealthy nation like Japan cannot afford to import four-fifths of its energy indefinitely. Finally, environmental considerations have begun to limit other modes of energy conversion in the industrialized nations. Nuclear power, although technologically feasible, has not developed as predicted—for economic and environmental reasons, and because of a deep-rooted public resistance.

Conversely, in New Zealand, where geothermal exploration has been under way since the end of World War II, recent discoveries of sizable reserves of natural gas have brought the development of geothermal-electric plants to a halt. However, the Government of New Zealand will continue to encourage the direct utilization of geothermal energy in industry, agriculture, and municipal heating. If the demand for electricity continues to increase, or if foreign markets are obtained for New Zealand natural gas, this decision may be reversed. Geothermal-energy development thus rests in an uneasy relationship with local political, economic, and technological conditions.

In summary, we can expect increased use of geothermal energy, certainly in direct utilization. Despite certain disadvantages, such as inability to sustain transport over long distances, restriction to base-load power applications, relatively low efficiency, and transmission losses, the doubling of demand for electricity every 7 to 12 years and the accelerating demand for other forms of energy throughout the world will require us to develop our geothermal-energy resources. Especially in countries plagued by a shortage of fossil-fuel reserves or an unfavorable balance of payments, there is an incentive to develop indigenous energy sources.

Costs

The production costs of various modes of energy generation are difficult to compare. Available data (see Table 1) are not always accurate, and may reflect different interest and taxation rates, amortization periods, special allowances, or other hidden costs. In any case, it has been shown—in Hungary, Iceland, New Zealand, and the Soviet Union—that direct utilization of geothermal energy in industry, agriculture, and space heating is appreciably less expensive than the use of crude oil, gasoline, or diesel fuel for the same purposes. Natural gas, where obtainable, and coal are more nearly competitive, though still more expensive than hot water for heating purposes.

In the generation of electricity, only hydroelectric power has been found to be cheaper, and only in certain situations. In Iceland, for example, hydroelectricity has been shown to be less expensive than geothermal power in most circumstances. But once the more ideal hydroelectric sites were developed, geothermally generated electricity became economically competitive. Moreover, in Iceland the direct utilization of hot water for municipal heating is far cheaper than heating via hydroelectricity. At The Geysers in the United States, geothermal-

TABLE 1
Selected Comparative Cost Data for Geothermal Energy

Geothermal field	Geothermal production	Local average, other fuel
Electricity, U.S. mills/kwh		
Namafjall, Iceland	2.5 – 3.5	—
Larderello, Italy	4.8 – 6.0	~ 7.5
Matsukawa, Japan	4.6	~ 6.0
Cerro Prieto, Mexico	4.1 – 4.9	~ 8.0
Pauzhetsk, U.S.S.R.	7.2	~10.0
The Geysers, United States	5.0	7.0
Space heating, U.S.$/Gcal energy		
Reykjavik, Iceland	4.0	6.7
Szeged, Hungary	3.0	11.0
Refrigeration, U.S.$/Gcal energy		
Rotorua, New Zealand	0.12	2.40
Drying diatomite, U.S.$/ton		
Namafjall, Iceland	~2	~12

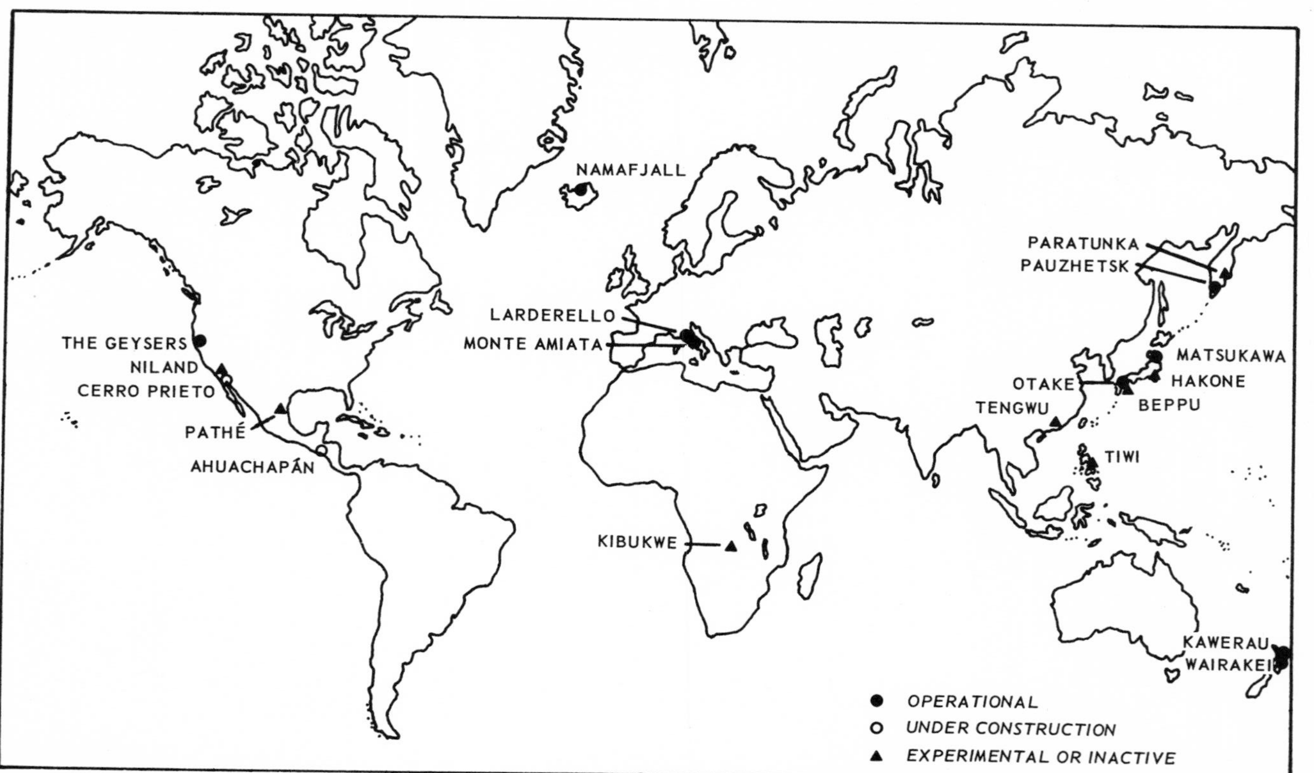

Fig. 2. Geothermal electric power stations.

electric power has proved to be cheaper than power from other fuel sources, regardless of plant size. Even in developed countries, then, geothermal power compares favorably with power from very large generating stations that have the advantage of economy of scale. And indeed, the ability of geothermal-generating systems to be developed economically in relatively small power-unit increments—say, 25 to 50 Mw —is a major consideration for underdeveloped countries, where the load and the load growth are commonly small. But the New Zealand decision must remind us that wherever it is politically or economically more advantageous to develop other energy sources, geothermal-power development may be retarded.

History of Development and Exploration

The early history of geothermal development saw the utilization of thermal springs as baths and health resorts, and the occasional use of thermal waters to heat buildings. Primitive peoples had already used the heat of fumaroles for cooking food and, in arid lands, steam condensate for drinking water. Sulfur deposited from the steam of fumaroles, kaolinitic clays formed by the decomposition of rocks in fumarole zones, and to a lesser extent fumarolic mercury and alum were utilized for centuries. But it was the recovery of boric acid from the fumaroles of Larderello, Italy, that marked the beginning of modern geothermal development. Starting in 1812, mineralized hot-spring waters were boiled to dryness in cauldrons heated by wood fires, and boric acid recovered from the residue. In 1827, fumarolic steam was substituted for wood as a fuel for this operation. Shortly thereafter, the first borings were made for steam at Larderello, both as fuel and to increase the flow of borate source material.

The first experimental generation of electricity from natural steam was undertaken at Larderello in 1904. In 1913 a 250-kw generating station came into service, marking the beginning of continuous generation of geothermal electricity (see Fig. 2).

After World War I, the concept of geothermal energy was carried to the ends of the world. Experimental borings at Beppu, Japan, began in 1919, and in 1924 a 1-kw generator was installed and operated experimentally. In the United States, test borings were drilled at The Geysers and Niland, California, in the 1920s. Although low-pressure steam was found in abundance, the projects were abandoned for lack

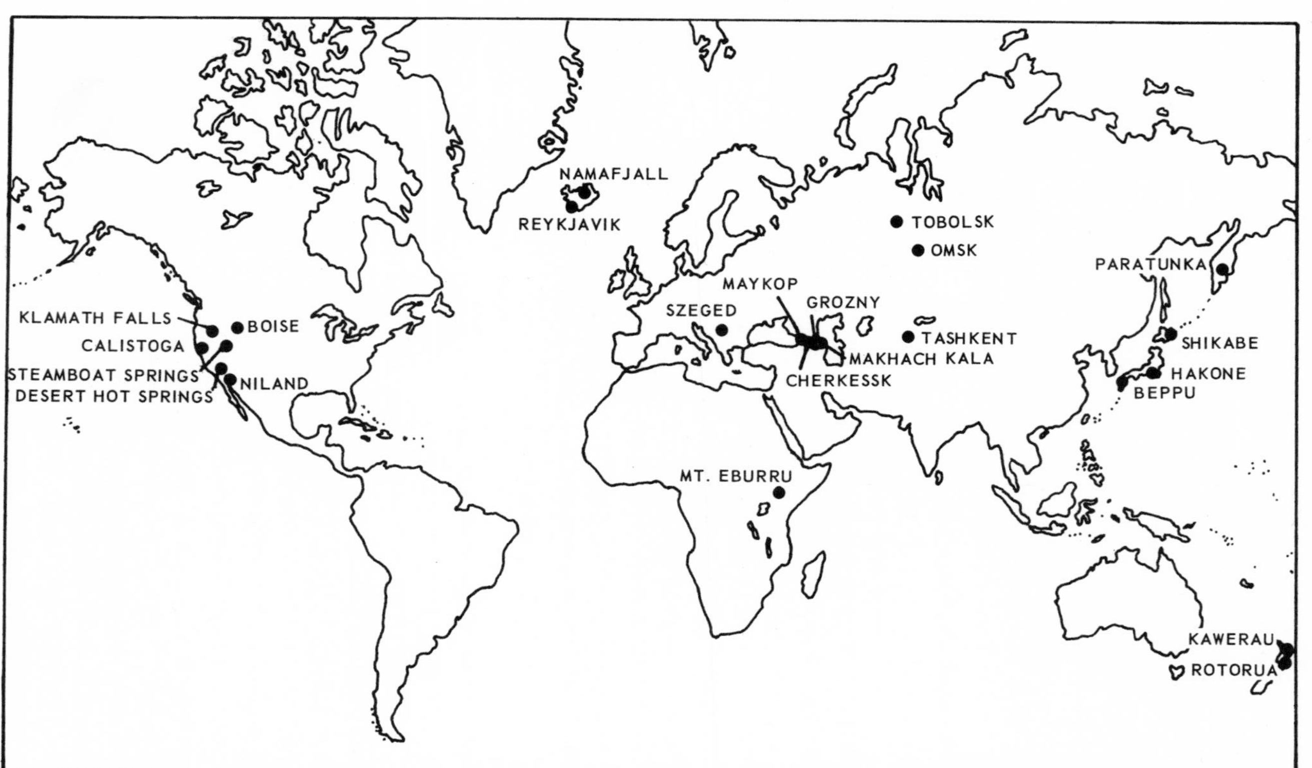

Fig. 3. Areas of significant use of geothermal energy for purposes other than generation of electric power.

of a market for electricity. Holes were drilled at other fumarole areas in the United States in the 20s and early 30s, most notably in Yellowstone National Park. A test hole was drilled in Java in 1928, but no development followed.

In Iceland, the exploration of hot-water aquifers by drilling began in 1928 at Reykjavik and in 1933 at Reykir, a few kilometers to the east (see Fig. 3). Hot water from these systems was distributed to consumers by the Reykjavik Municipal District Heating Service. Before 1940, hot-water wells had been drilled for heating purposes at Rotorua, New Zealand. In that year a great many wells were drilled for domestic use in Rotorua and in towns south of Lake Taupo.

World War II disrupted traditional patterns of living; in the reconstruction of war-devastated economies, attention focused anew on geothermal energy. This was especially true in Italy, Japan, and New Zealand: all three were short of fossil fuels for power generation, and generation and transmission facilities had been largely destroyed in Italy and Japan.

The remainder of this paper undertakes to recount, country by country, the history of geothermal-energy exploration and exploitation, for electricity generation and for industrial-process heating, space heating, chemical extraction, or the like. Table 2 gives pertinent data on thirteen important geothermal fields in eight countries; Table 3 indicates the current status of geothermal exploration and development in some fifty countries.

Italy

From the simple 250-kw generator of 1913, the Italian geothermal-generating complex has grown to over 390,000-kw capacity, and is currently the world's largest. (The original boric acid recovery works, established in 1812, was shut down in 1969 because of inability to compete economically with other sources of borax.) Over 365 Mw of the electrical capacity is produced from the thirteen plants comprising the Larderello field. Some 25 Mw are supplied by the four plants at Monte Amiata, 75 km to the southeast. Individual turbine size is small, ranging from the 900-kw noncondensing unit at San Ippolito to the 26,000-kw plants at Castelnuovo and Larderello. Approximately 43,000 kw are generated from a series of noncondensing turbines at the Larderello and Monte Amiata fields. Noncondensing turbines were chosen

TABLE 2

Characteristics of Selected Geothermal Fields

Field	Reservoir temperature, °C	Reservoir fluid	Enthalpy, cal/g	Average well depth, meters	Fluid salinity, ppm	Mass flow per well, kg/hr	Non-condensable gases, %
Larderello	245	Steam	690	1,000	<1,000	23,000	5
The Geysers	245	Steam	670	2,500	<1,000	70,000	1
Matsukawa	230	Mostly steam	550	1,100	<1,000	50,000	<1
Otake	200+	Water	~400	500	~4,000	100,000	<1
Wairakei	270	Water	280	1,000	12,000	—	<1
Broadlands	280	Water	400+	1,300	—	150,000	~6
Pauzhetsk	200	Water	195	600	3,000	60,000	—
Cerro Prieto	300+	Water	265	1,500	~15,000	230,000	~1
Niland	300+	Brine	240	1,300	260,000	~200,000	<1
Ahuachapán	230	Water	235	1,000	10,000	320,000	~1
Hveragerdi	260	Water	220	800	~1,000	250,000	~1
Reykjanes	280	Brine	275	1,750	~40,000	~400,000	~1
Namafjall	280	Water	260	900	~4,000	400,000	6

TABLE 3
Status of Geothermal Exploration and Development, 1972

Nation	Electric-power generation/ construction	Experimental power stations	Significant direct utilization	Other geothermal-field discoveries	Additional exploration under way[c]
Chile				x	
China		x			
Ethiopia					x
Guadeloupe (Fr. W. Indies)				x	
Hungary			x		
Iceland	x		x	x	
Indonesia					x
Italy	x				x
Japan	x	x[b]	x	x	x
Kenya			x		x
Mexico	x[a]	x	x	x	x
New Zealand	x		x	x	
Nicaragua				x	
Philippines		x		x	
El Salvador	x[a]			x	
Taiwan				x	
Turkey				x	
U.S.S.R.	x	x	x	x	x
United States	x	x[b]	x	x	x
Zaire		x[b]			

[a] Under construction. [b] Inactive.
[c] Other geothermal exploration/interest: Algeria, Argentina, Bulgaria, Burundi, Colombia, Costa Rica, Czechoslovakia, Ecuador, Fiji Islands, Greece, Guatemala, India, Israel, Malawi, Mali, Morocco, New Britain, New Hebrides, Peru, Poland, Portugal (Azores Is.), Rwanda, Spain (Canary Is.), Tanzania, Tunisia, TFAI (French Somaliland), Uganda, Venezuela, Yugoslavia, Zambia.

originally because of their simplicity, lower capital cost, and ease of construction. However, they consume approximately twice as much steam per kwh as condensing turbines, 20 kg vs. 10 kg. In a situation where steam is abundant, capital scarce, and time short, noncondensing turbines appear to be most favorable. But with the expansion of the Larderello complex to the point where many geologists believe that field capacity has been reached, or nearly so, effective utilization of steam becomes the more critical variable. For this reason, conversion to condensing turbogenerators is planned. This could add upwards of 40,000 kw of generating capacity at Monte Amiata and Larderello without any increase in steam production.

Producing pools of the Larderello field are at Capriola, Castelnuovo, Gabbro, Lago, Lagoni Rossi, Larderello, Montecerboli, Monterotondo,

San Ippolito, Sasso Pisano, and Serrazzano. From Monterotondo on the south to the northern end of the field at Gabbro is a distance of some 20 km. Total field area is probably in excess of 250 km^2. In addition, steam has been discovered at Travale, Boccheggiano, and Roccastrada, east and southeast of the main Larderello field. Roccastrada is nearly halfway between the center of the Larderello field and the Bagnore and Piancastagnaio pools of the Monte Amiata field. Radicofani may represent an eastern extension of the Monte Amiata field.

Approximately 500 wells have been drilled across the Larderello and Monte Amiata fields, of which nearly 200 were in production in 1971. The reservoir fluid is steam, with a variable content of noncondensable gases, averaging (see Table 2) about 5 percent, although the initial gas content of each well is much higher; e.g., the gas content at Piancastagnaio had decreased from almost 90 percent initially to 20 percent by 1970, and was still decreasing. Reservoir temperatures reach a maximum of about 250°C.

Average well yield is about 23,000 kg/hr of steam at Larderello and perhaps 36,000 at Monte Amiata. The greater average yield at Monte Amiata reflects the shorter production history, as both mass and pressure declines are reported with time. Individual wells may deviate greatly from these averages: mass flows as large as 270,000 kg/hr have been reported. Average well depth at Larderello is slightly over 1,000 m. Wells are completed with 34-cm-diameter casing, which is the largest production diameter at any active geothermal development in the world.

There is evidence of interference between wells in certain pools, and it is reported that maximum sustainable mass flow has been attained for several parts of the field. This is reflected in the low average yield per well, which in turn reflects declines in yield per well with time. Many geologists have thus been led to state that field capacity has been reached. However, successful exploration and development is continuing at the previous margins of the field, most notably at Travale, where, beginning in 1951 with the drilling of five successful steam wells, a geothermal field was developed at the site of an old boric acid works. Two 3,500-kw turbines were installed in 1952 and operated until 1962, when decreases in well yield required the plant be dismantled. In February 1972 it was reported that several new wells had been completed, and that they were capable of producing at least 100,000 kg/hr each of dry steam. This may significantly extend field capacity.

The Larderello–Monte Amiata region is being explored by various methods, including infrared imaging, to determine if significant data have been overlooked in past work. Drilling is continuing in several pools of the Larderello field and at Travale. The volcanic centers at Roccastrada and Radicofani are also under study. Work includes geochemistry, temperature-gradient drilling, geologic mapping, and drilling of exploratory holes. As a result of this continued exploration and development effort, costs of steam have stayed constant, instead of declining as originally predicted by the operating company. Also because of the capital cost of conversion from noncondensing to condensing turbines, net costs per kwh have increased slightly over the decade.

As more condensing turbines are installed, more water effluent will require disposal. Since the cessation of boric acid recovery in 1969, borated waters have been discharged to the natural river drainage. No adverse effects upon agriculture are reported.

Steam at Larderello is produced from permeable to cavernous limestone, dolomite, and anhydrite of Upper Triassic to Upper Jurassic age. Field depth is controlled by a decrease in permeability with penetration into the carbonate sequence and underlying crystalline basement. The reservoir is capped by a thrust sheet comprising impermeable carbonates, argillites, and ophiolites of Jurassic to Eocene age. Surface leakage of steam occurs along faults extending to the carbonate-anhydrite reservoir beneath the thrust plate. There is no obvious source of heat in the immediate vicinity, although the presence of a deep pluton has been suggested. Tertiary granitic rocks are exposed on the island of Elba, some 80 km to the southwest. The closest Late Tertiary volcanic rocks are exposed at Roccastrada.

Monte Amiata consists of Pliocene and Pleistocene acidic and alkaline volcanic rocks extruded through a sequence of shales, marls, limestones, and sandstones similar to those at Larderello. The steam reservoir is also beneath an impermeable thrust sheet. Post-volcanic collapse is believed to have occurred, fragmenting the reservoir and controlling mercury mineralization and weak hot-springs activity. More basic volcanic rocks occur at Radicofani and at Monte Volsini to the south. The relationship between acidic and alkaline volcanism, post-volcanic collapse, mercury mineralization, and hot-springs activity is observed at many geothermal fields throughout the world.

A series of seven target areas has been chosen for further geothermal

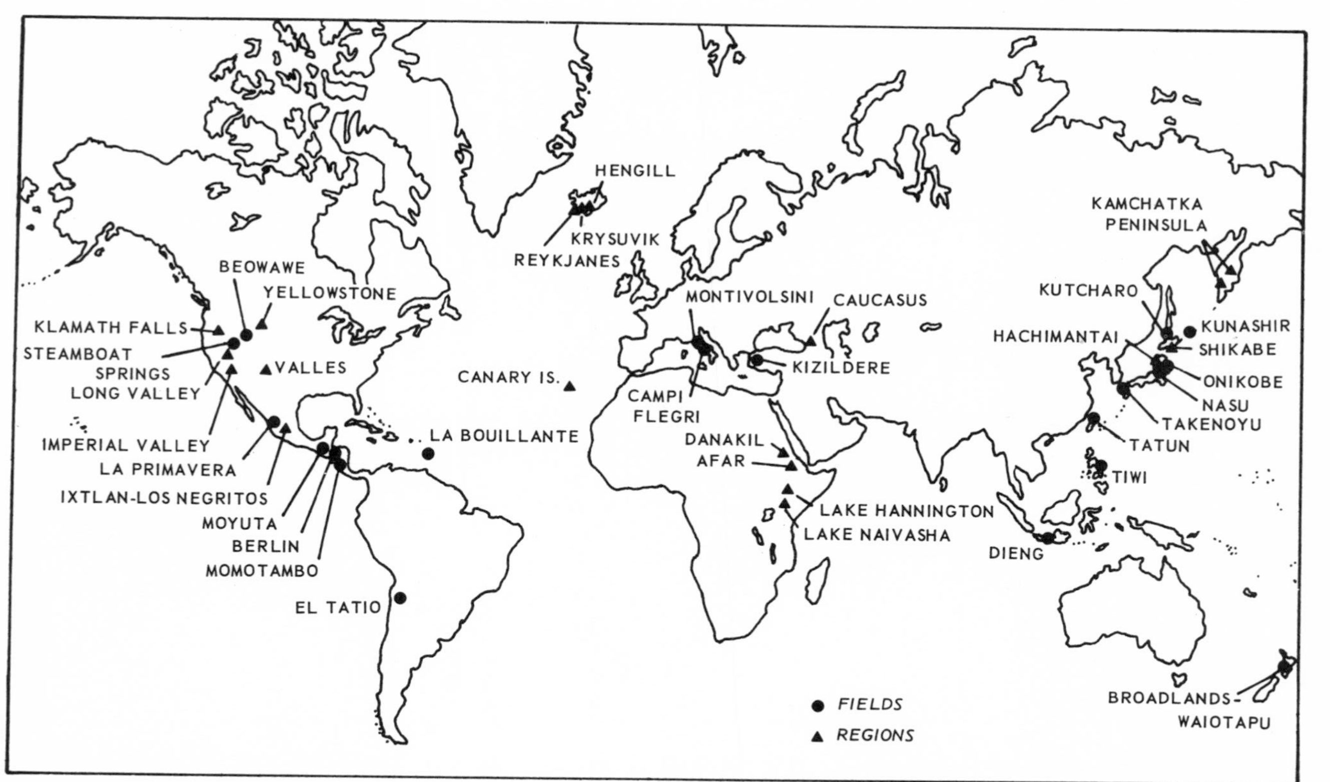

Fig. 4. Major undeveloped geothermal fields and regions under exploration.

exploration in the Apennine mountain chain (see Fig. 4). South from Monte Amiata these are: Monte Volsini; Monte Cimini, including the Viterbo area; Monte Sabatini; Colli Albani, to the southeast of Rome; the region about Naples, including Pozzuoli and the Campi Flegri solfatara fields of classical fame; and Monte Vulture, in south-central Italy. Other areas in northern Italy have been investigated and rejected. These include Monte Berici, near Padua, where there are Late Tertiary silicic volcanic rocks, and Montecatini, between Larderello and Pisa. Drilling has also been done at La Tolfa, a volcanic center southwest of Viterbo.

In the 1920s and 30s, drilling was carried out on the island of Ischia and in the vicinity of the city of Pozzuoli. Low-pressure steam was reported from certain of these tests. A project to study the feasibility of using a freon-based heat exchanger to generate electricity at the Campi Flegri was never carried through.

New Zealand

The development of geothermal energy in New Zealand has been confined largely to the Taupo volcanic depression of the North Island. Utilization has centered at Wairakei, where a 160-Mw power station operates; at Kawerau, where some 180,000 kg/hr of steam are used to produce newsprint and sawn lumber and to generate 10 Mw of electricity; and at Rotorua, where steam and hot water are used extensively for heating purposes. Natural hot water is used on a small scale at Ngawha, also on North Island, and elsewhere in the Taupo depression.

The Taupo volcanic depression extends for over 200 km in a north-northeasterly direction, parallel to the main structural grain, and culminates in the north at the active volcano White Island, in the Bay of Plenty. The zone is some 25 to 30 km broad at its widest. Fumaroles and hot springs are abundant in the central 100-km-long portion of the depression. At least a dozen areas have been explored by drilling.

As a result of drilling, hot-water reservoirs of sodium-chloride composition have been found, at temperatures commonly as high as 270°C, the highest being the 306° at Rotokawa. Steam is formed during the upward flow of water in wells, giving a mixture of steam and water at the wellhead. The fraction of steam varies, averaging 20 percent at Wairakei. Only rarely is dry steam produced in wells, and this is believed to reflect a flashing effect in the producing formation brought on by lowered fluid pressure near the wellbore. Gas content varies with the area. At Rotorua, hydrogen sulfide gas evolves from the superheated waters and is a nuisance and potential health hazard.

Declines in mass flow and pressure are observed commonly in wells of the Wairakei and Kawerau fields, necessitating redrilling and deepening of individual wells. These declines have also made it necessary to set intake pressures for turbines lower than was originally planned. Ground subsidence also has occurred at Wairakei. Deposition of calcite and silica, and corrosion at the points of oxidation of reservoir fluid, have created further problems in handling of the fluid. Saline effluent, representing about 80 percent of the original geothermal fluid, is disposed of to rivers, apparently without harmful effect.

Despite these problems, there was a rapid growth of interest in geothermal development through the late 1940s, culminating in the completion of a 160-kw generating station at Wairakei. The first experimental plant had to be abandoned in 1964, after a year of operation, because of insufficient yield from wells. Subsequent construction was centralized at one facility, with extensive steam-collection lines, to utilize a local river for cooling purposes. Increases in capacity were planned to 250 Mw at Wairakei and to between 90 and 120 Mw at the Broadlands field, just to the northeast, by 1976. But the discovery of natural gas in New Zealand changed these plans, and no geothermal power development is planned for the 1970s. However, the government is encouraging direct municipal and industrial utilization of these hot-water resources.

The most significant examples of direct utilization are at Kawerau and Rotorua. The New Zealand Department of Scientific and Industrial Research and the Ministry of Works, in conjunction with the Tasman Pulp and Paper Company, began exploration at Kawerau in 1952. Over a dozen wells have since been drilled or redrilled at Kawerau, and steam from these wells is used for heat exchanging with boiler-quality water for the generation of high-quality steam for mill processes. Additional natural steam is used for timber drying, to operate log-handling equipment, and for generation of 10,000 kw of electricity. At Rotorua, a city of some 30,000 people, over 1,000 hot-water wells supply thermal energy to individual houses, schools, hospitals, hotels, and commercial and industrial establishments. A geothermal air-conditioning scheme is in operation at a 100-room hotel at Rotorua. The construction costs were said to be as low as those of conventional cooling systems, and the operating costs are perhaps only 5 percent as high. At Rotorua, the average well depth is 100 to 150 m, with a few as deep as 250 m. Temperatures commonly are above 120°C, and reach as high as 175° at these shallow depths. When allowed to flow, these wells produce a mixture of steam, hot water above 100°, and noncondensable gases. Because of corrosion and pollution problems associated with this fluid, heat exchanging often is accomplished with municipal water, which is circulated to consumers. Unlike the geothermal-heating systems in Iceland, which are operated by the municipality, space heating in New Zealand is not operated by government agencies. It is, however, sharply regulated, especially concerning corrosion, hydrogen-sulfide emissions, disposal, contamination of other water supplies, and effects of production upon neighboring wells.

Many hot-water wells in the Lake Taupo area supply heat to farms. Examples of agricultural applications include steaming of raw garbage as swill for hogs, heating of stock pens, cleaning of runs, and sterilization of various equipment. Also, an experimental forestry station uses steam to dry seeds and lumber, as well as to heat seed beds and buildings.

Total heat consumed for direct applications of geothermal fluid in New Zealand is probably greater than that used to generate electricity at Wairakei.

United States

Development of electric power has proceeded rapidly at The Geysers, California, since the first turbine of 12,500 kw was installed in 1960. Pacific Gas and Electric Company purchases steam from producing companies and converts it to electricity through condensing turbines at a rate of 9 kg/kwh. Late in 1972, some 2.75 million kg of dry steam were consumed hourly to produce 302,000 kw of electricity. Another 110,000-kw plant is nearing completion, and is due for testing late in 1973. And construction has begun on still another plant of the same size, scheduled for completion in 1974. It is estimated that 110 Mw will be installed annually through the 1970s, which by 1980 will result in approximately 1,180,000 kw of generating capacity. This would be the largest producing geothermal complex in the world at that time.

The capacity of The Geysers field is a matter of speculation. On the basis of different estimates of field extent, well yield, well spacing, and total fluid in storage in the reservoir, estimates of maximum sustainable production range upward from 1,200 Mw to 4,800 Mw and even more. To date, some 110 wells have been drilled across a 12-km zone with maximum width of about 3 km. Some 85 of these were completed as producers of dry steam. Field boundaries still are known only approximately, and exploration is continuing at distances of many kilometers from produceable wells.

A shallow and a deeper reservoir have been encountered, the former apparently fed from the latter along faults. Average well yield from the deeper (greater than 500 m) reservoir is on the order of 70,000 kg/hr. The production declines that have been documented for wells in the shallow reservoir have recently been demonstrated to occur also in the deeper reservoir (Budd, this volume).

The reservoir consists of highly fractured, slightly metamorphosed, sedimentary and igneous rocks of Cretaceous and Upper Jurassic age. The deepest wells have surpassed 2,500 m in depth without notable reduction in fracture permeability. Recharge is unknown, and the subject of study. Low-grade metamorphic reactions, including the deposition of silica and calcite, may have contributed to a lateral decrease in permeability, but this is not clearly documented. Reservoir temperatures reach about 250°C. The heat source is apparently an igneous mass at a depth of perhaps 5 to 8 km, which has yielded a series of alkaline and acidic volcanic rocks at the surface. The age of these rocks is probably Pleistocene, and they cover an area of perhaps 300 to 400 km^2.

Of the few unproductive deep boreholes, almost all encountered high temperatures. Permeability remains the critical variable; and it is still a question whether this is controlled by local or regional fracture patterns or by mineral solution and deposition activity. If The Geysers represents a water-deficient system, as has been postulated, continued production may result in increasing fluid enthalpy over time.

Potentially countering this is the limited reinjection of condensate from the condensing turbines. After evaporative losses in cooling, only about 20 percent of the fluid remains for disposal. The high content of boron and ammonia in the condensate have precluded disposal to local streams, whose volume of flow varies sharply with the season. Injection of cooled condensate in the rock reservoir at approximately 2,000 m depth may serve as recharge allowing secondary recovery of heat from the reservoir. But it may also locally decrease enthalpy, permitting a water phase to predominate. Some calculations of enthalpy from deep wells have suggested that steam and liquid water do coexist in the reservoir. Therefore, indications of long-term trends in enthalpy, mass flow, and field pressure are eagerly awaited.

Significant discoveries have been made in about half a dozen localities elsewhere in the western United States. Most attention is given currently to the Imperial Valley of California, where several companies and Government agencies are active. The U.S. Bureau of Reclamation and the Office of Saline Water are sponsoring an exploration program aimed at joint production of electricity and desalination of waters for agricultural use. Magma Energy Company and its associates are also drilling in the Imperial Valley, searching for geothermal brine suitable for use in generating electricity by heat-exchanging. A completely closed system is envisaged, with heat-depleted fluid reinjected back into the reservoir rocks. An agreement has been signed with San Diego Gas and Electric Company to construct and operate a pilot plant of this type at Niland in the Imperial Valley, with the first plant to operate in 1974.

In 1963, Standard Oil Company of California drilled a 4,000-m oil-exploration hole west of Brawley. It encountered sodium-chloride brine at a temperature of about 260°C. This brine is somewhat similar to the reservoir fluid at Cerro Prieto, Mexico, and is presumably what is sought elsewhere in the Imperial Valley.

Exploration at Niland revealed an extremely hot (to 370°C), highly saline (to 260,000 ppm total dissolved solids) brine. Niland appears to be the northernmost and most saline field of a 150-km-long geothermal province extending northward from the Gulf of California, Mexico. Data from electrical-resistivity surveys support the concept of a salinity gradient across the Valley, increasing to the northwest. An experimental generator installed at Niland in the early 1960s was never operated at its rated 3,000-kw capacity. It is now inactive. At Niland, recovery of various chlorides was attempted, most notably KCl. Market conditions and problems of corrosion, scaling, and waste disposal prevented commercial development of either electricity or a major chemical-recovery works. However, on an intermittent basis, calcium-chloride solution was marketed for a few years in the late 1960s. Also, for over 30 years carbon dioxide gas was produced from shallow wells at the northeast end of the Niland field. Over 65 wells were drilled, and approximately 100 million m^3 of gas were produced between 1933 and 1954. Average well life was about 2 years. Well depth averaged 150 m. Temperatures of the CO_2 gas ranged between 50° and 75°. The entire gas production was converted to dry ice for use in refrigeration or railroad cars.

Space heating with geothermal fluids has succeeded most effectively in south-central Oregon. At Klamath Falls some 350 wells supply heat to buildings via heat-exchanging with pure municipal water. Elsewhere in Oregon, greenhouses, resorts, baths, farm buildings, and schools are heated by geothermal waters. Similar projects have been undertaken in California (Calistoga and Desert Hot Springs) and Idaho (Boise), and in farms and villages across the western states. Several state and national parks, such as Yellowstone, Lassen, and Mount Katmai, have been established around major geothermal fields. In a few desert locations, fumarolic steam is condensed for livestock. An unusual application at Steamboat Springs, Nevada, is the use of hot well-water as the energy source in the preparation of plastic explosive.

At Yellowstone National Park, the U.S. Geological Survey drilled thirteen shallow test holes between 1967 and 1969, the deepest being slightly over 300 m. Maximum temperature was about 240°C. One hole produced essentially dry steam, whereas the others yielded a variable ratio of steam and hot water. Because this famous area of geysers, fumaroles, and hot springs is a National Park, no development is anticipated.

Union Oil Company is the latest to undertake exploration of the Valles Caldera of north-central New Mexico. Five exploration holes have been drilled, to depths of at least 1,600 m. Temperatures to 250° are reported. Essentially dry steam is said to have been produced from a hole drilled in 1970. The others have yielded hot water with variable percentages of steam flashover.

Rock temperatures above 200° are reported from holes at Brady's Hot Springs and Beowawe, Nevada. Temperatures of at least 175° have been measured in wells at Clear Lake, California; Steamboat Springs, Nevada; Casa Diablo (Long Valley), California; and Surprise Valley, California.

Along the Gulf Coast, high-temperature and high-pressure waters have been encountered in the search for oil: a maximum temperature of 273° was measured at 5,859 m. The great depth to productive reservoirs (a minimum of 2,500 m to the 120°C isotherm), as well as problems of production, utilization, and disposal, have forestalled any attempts at commercial development.

Altogether, over 35 areas have been explored by drilling. In many of these, however, results are inconclusive because of insufficient or misdirected efforts. Exploration activity continues in California, Nevada, and Oregon, as in earlier years, but the emphasis has shifted to include Idaho, Utah, New Mexico, Arizona, and adjacent areas.

Iceland

Iceland pioneered in the use of hot water for municipal heating in the 1930s. Approximately 50 percent of the 200,000 population receives geothermal heating, and this is to rise to over 60 percent in this decade. Nine out of ten homes in Reykjavik, the nation's capital, receive geothermal water for home heating, distributed by the Reykjavik Municipal District Heating Service. Low-enthalpy hot-water fields at Reykir and within Reykjavik (Laugarnes field) supply this energy from reservoirs at base temperatures of 98° and 146°C. Over 100 wells have been drilled into these fields, the deepest being 2,200 m, at Reykjavik. Another low-enthalpy field is under development at Ellidaar, 3 km from Reykjavik. It is expected that this field will supply energy for expansion of geothermal heating in the Reykjavik area. A concrete conduit with twin 35-cm steel pipelines carries hot water for distances up to 18 km and with temperature losses on the order of 5°. Distribution to users is at about 80°. Because the mineral content of the water is remarkably low for geothermal fields (less than 400 ppm average), no processing other than separation of contained gases is needed.

Space-heating systems in operation or under construction elsewhere in Iceland will serve the needs of an additional 25,000 persons, mostly in small towns and villages on the southwest coast and in the north-central part of the island. Geothermal greenhouse operations in southwestern Iceland supply much of the fresh vegetables for the Reykjavik market. Additionally, heated baths and pools are found all across the western and northern parts of the country.

At Namafjall, near Lake Myvatn in northern Iceland, rich sublacustrine deposits of diatomite are dried with geothermal steam. Nearby is a 3,000-kw geothermal power station, opened in 1969. Exploration began at Namafjall around 1947, in connection with attempts to mine sulfur from the fumarole field. Drilling revealed a high-enthalpy, hot-water reservoir with a base temperature above 260°C and a maximum measured temperature of 286°. Seven wells were drilled to an average depth of 700 m, the deepest in excess of 1,380 m. Four of these wells are operated, yielding 1.8 million kg/hr of superheated water. Some 240,000 kg/hr of steam separates from the water and is used to operate the power plant and diatomite plant. The water, which contains mostly sodium bicarbonate and silica, is not strongly mineralized. Gas content is similar to that at Larderello. Some shallow wells have produced almost-dry steam.

In 1967, the Johns Manville Corporation, in conjunction with the Icelandic firm Kisilidjan h/f, began mining diatomite at Namafjall and purchasing steam from the government of Iceland. The initial production rate for refined diatomite was doubled to 24,000 tons per year by 1970. Simultaneously, the Laxa Power Works, which produces 22 Mw mostly by hydroelectric power, arranged for the construction of a non-densing 2.5-Mw turbine. This turbine soon is to be reconditioned to produce 3.4 Mw. A district heating system is under design to utilize the residual hot water.

Other uses for geothermal fluids are under consideration in Iceland. An experimental salt-recovery scheme was discontinued for economic reasons. There is a pilot operation under way to dry seaweed with natural heat (water at 100°C) for recovery for alginates. A geothermal brine with the approximate composition of seawater is under study for the recovery of various chemicals, including magnesia, table salt, and bromine. Freeze drying of fish, Iceland's main export, is also being evaluated. In the 1950s, when the present generation of nuclear reactors was under design, the potential for recovery of heavy water, D_2O, from geothermal fluids was evaluated. The process appeared economic but the market for D_2O did not materialize.

The conditions for using noncondensing turbines in Iceland appear favorable: heat energy is cheap and abundant, and no advantage accrues from large plant size, since a small population is distributed in villages and on farms across a large area. Multiple, small-capacity plants appear to be the most effective scheme. This suggests a bright future for low-cost geothermal electric power. The savings in fuel oil for heating purposes in the Reykjavik district amounts to some 200,000 tons per year. An additional 20,000 tons of fuel oil would be required for existing greenhouse operations. This is an appreciable savings for a small nation; if this heating were done entirely by electricity, which is not the cheapest alternative and therefore not an ideal comparison, some 200,000 kw of generating capacity would be required for peak demand.

Four high-temperature regions are known, three in the southwest and one in the north-central part of the country. Myvatn thermal region, in the north-central part, has a base temperature of approximately 280°C and constitutes an area of 50 km^2. It includes the Namafjall field.

In southwest Iceland, aligned from southwest to northwest at intervals of 30 km, Reykjanes, Krysuvik, and Hengill have base temperatures of 280°, 220°, and 260°, respectively. Hengill, the largest, has an area of 70 km^2 and includes the important field of Hveragerdi at its southern end. At least eight deep and a great many shallow holes have been drilled in this area. The deepest hole at Hveragerdi is about 1,200 m, and produces dilute superheated water. Up to 25 percent flashes to steam, and between 13 and 32 Mw of electric power are planned. Reykjanes has had seven holes drilled, to a maximum depth of 1,750 m. These wells yield brine with chloride concentration up to 29,000 ppm. This brine is considered to derive from seawater by steam separation in a thermal regime of 270°. A system for recovery of various salts from the fluid, with concurrent generation of electricity, is under evaluation. At Krysuvik, about 30 km south of Reykjavik, plans have been made to transport superheated water to the capital for space heating in the event that growth of the system exceeds the capacity of the Reykir and Reykjavik (Laugarnes) fields. Here, too, large flows of dilute superheated waters are obtained from more than a dozen wells as deep as 1,200 m.

These and other high-temperature fields occupy the zones of crustal rifting and active volcanism in Iceland. Outward east and west from this zone are extensive sheets of Late Tertiary basalts through which abundant warm and hot springs issue. The high-temperature reservoirs, however, are associated with the main Quaternary rift. Further, many are closely associated with Late Quaternary centers of dacitic and rhyolitic volcanism rather than with plateau basalts. Significant associations of high-enthalpy reservoirs and acidic volcanism occur at Myvatn, Askja, Geysir, Hengill, and at least half a dozen other areas. Only in the zone from Reykjanes to Krysuvik, some 30 km long, is silicic volcanism absent. Permeability within a reservoir may be controlled by lithology or by fractures. Thermal emissions to the surface are generally fracture-controlled.

Low-temperature systems have been explored at several parts of the southwest, west, and northwest coasts of Iceland. At Hlidardalur, 35 km southeast of Reykjavik, a hole drilled 1,500 m deep into an area of high temperature gradient failed to yield water. Temperature measurements of 150°C at 700 to 800 m indicated a potential reservoir formation. This formation was fractured artificially and induced to flow, representing perhaps the world's first geothermal-fracturing experiment. At more than a dozen localities in western and northern Iceland, hot water is produced from wells at temperatures between 90° and 180°. These, where not already in use, have potential for use in agriculture and space heating. In most cases, water quality is high; however, one or two instances of geothermally heated seawater are reported, less saline than that at Reykjanes and thus representing lower-temperature systems. These lower-enthalpy fields are almost exclusively associated with centers of basaltic volcanism.

Japan

Geothermal exploration began in 1919 at Beppu, on Kyushu Island, where 1 kw of electricity was generated in 1924. Somewhat earlier, the first geothermal heating of greenhouses began. However, it was not until the years immediately after World War II, with Japan's industrial base in ruins, that serious development began. The Agency of Industrial Science and Technology, which includes the Geological Survey, and several private electric utilities, drilling companies, mining companies, and prefectural governments, independently and jointly began exploration of Japan's geothermal resources.

At Beppu, experimentation resulted in the generation of 30 kw of electricity in 1951. Even earlier, the Tone Boring Company had begun a drilling program in several areas. At Yunosawa, on the south-central coast of Honshu, geothermal steam was used in a demonstration plant to generate 8 kw in 1948. At Hakone, also on the south coast of central Honshu, 30 kw of electricity has been generated since 1960 from steam produced in a shallow well, for use at a hotel. Other experimental generation of electricity has taken place on Hokkaido Island (Atagawa) and in northern Honshu (Narugo).

Japan's hot springs have been used for therapy and recreation for centuries. Japan probably leads the world in the use of natural hot water in baths, therapeutic spas, and resorts, there being literally thousands of these. More recently, volcanic parks such as Hachimantai have been established for the preservation of the geothermal phenomena. However, space-heating systems have not developed as widely as in Iceland or the Soviet Union. Four district heating systems were in operation in 1969, using hot water (not greater than 70°C) equivalent to about 5,000 tons of fuel oil per year.

Greenhouse heating with geothermal waters is common in farms and experimental agricultural stations, chiefly in south-central Honshu. Some are near Beppu, and a few are in the Shikabe geothermal area of Hokkaido Island. Commonly, garden vegetables, tropical fruits, and other plants are cultivated. At least two farms raise chickens. At one pond, eels and carp are raised for consumption, and at another, alligators.

There are several more specialized industrial applications. One, on Kyushu Island, involves the recovery of sulfur from deposits at fumaroles. At Beppu, 98° water is used to process about 30 tons of rice annually for use by a bakery. In the Shikabe area of Hokkaido, salt is recovered from seawater by evaporation, using geothermal hot water and steam from a 70-m well. Though the operation produces only 150 tons of salt per year, a combined geothermal power plant and salt-recovery works planned at Shikabe could produce 7,000 kw of electricity, up to 100,000 tons of salt annually, and some amount of fresh water. The proposed plant would use multi-stage vacuum distillation.

Exploration began in 1952 at Matsukawa, in northern Honshu, with the drilling of wells for steam to supply bathhouses. Japan Metals and Chemicals Company became active in exploration there in 1956, and was joined in 1958 by the Geological Survey and later by other organizations. At least nineteen wells have been drilled since 1952. Of these, six are classed as production wells, having an average depth of about 1,200 m. Five wells used for power generation yield an average of 110,000 kg/hr of dry steam. Two wells yield both steam and water in the ratio of about 4:1. Reservoir temperature, about 240° to 250°C, is much like that at The Geysers, California. The reservoir formation is a fractured series of welded dacite tuffs and lavas of probable Pliocene age. These overlie Miocene sands and shales, Lower Tertiary "green tuffs," and Paleozoic slate and chert. Above the reservoir rocks are Quaternary andesites and constructional debris of very youthful volcanoes. The area about Marumori volcano comprises a caldera of perhaps 10 km^2.

Construction of the Matsukawa power plant began in 1961. Initial operation was at 9,000 kw. The facility has been expanded to 20,000 kw, and plans call for expansion to 27,000 kw in the 1970s, with future expansion to at least 60 Mw. Noncondensable gases, which average about 0.5 percent, are expelled to the air. Condensate from the condensing turbine is discharged to the natural drainage, without known harmful effect.

Exploration begun in 1953 at Otake, on Kyushu Island, culminated in 1967 with the completion of a 13,000-kw power station by a group headed by Kyushu Electric Power, Inc. Ten wells were drilled at Otake, and two at Hachobaru to the south; five are connected to the condensing turbine. One well produces about 37,000 kg/hr of dry steam; the others produce larger amounts (up to 540,000 kg/hr) of steam and water, with steam making up 10 to 25 percent of flow. Noncondensable gases total less than 0.5 percent of the steam by weight. The reservoir fluid is of sodium-chloride composition, but relatively dilute, averaging about 4,000 ppm. Maximum reservoir temperature is about 200°C. Maximum well depth at Otake is 900 m; at Hachobaru, one well is 785 m deep.

The geology consists of Pleistocene andesite lava and tuff-breccia, and occasional pumice beds, overlaid by Holocene sediments and volcanic ash. Structure is largely obscured by the constructional features of Quaternary volcanoes. However, faulting is believed to be significant, forming down-dropped blocks. Hydrothermal alteration and thermal emissions are intense and widespread, suggesting fracture permeability to the surface. The wide range of flow rate and steam percentage from well to well suggests that permeability and stored fluid are limited in portions of the field. But no deep wells have been drilled, and the main reservoir may not yet have been encountered.

The Otake field is located in Aso National Park, thus requiring special permission for construction of plant facilities. To avoid pollution of natural streams, water discharges are pumped to a nearby hydroelectric reservoir. Limited life is predicted for the present wells because of scaling and fluctuations in pressure. Silica deposition in the discharge pipeline caused a restriction of power production in 1968 and 1969. Studies are still under way to control scaling. Calcium-carbonate scale, requiring acidizing, has been found in at least one well. Despite these problems, additional wells are planned and production is to increase to perhaps 180,000 kw over the next decade or so.

At least 23 areas have been explored by drilling in Japan. Several are being evaluated for development, the most significant being at Shikabe, on Hokkaido Island; at several places near Matsukawa; at Onikobe, south of Matsukawa; and at Takenoyu and Hachobaru, near Otake.

In the Hachimantai volcanic area, northwest of Matsukawa, extensive geophysical work culminated in the drilling of at least nine holes to average depths of about 750 m. Temperatures reached 200°C at the south end of the drilled area, where hot water was encountered with up to 30 percent steam fraction. Exploration is continuing there, and the construction of a 10,000-kw power plant is being considered. The geology at Takinokami, southwest of Matsukawa, is similar. Temperature surveys to 50 m (maximum temperature, about 200°) were followed by a 400-m well, which produced a steam and water mixture.

Onikobe has been the scene of exploration since 1962. It is a structural basin of about 60 km^2, in which there are numerous fumaroles and hot springs. Exploration has included geologic mapping, geophysical surveys, and test drilling. On the basis of ten holes, with maximum depth over 700 m, and temperatures to 190°, drilling of production wells is under way. One hole is reported to produce nearly 30,000 kg/hr of superheated steam.

Takenoyu has been investigated since the early 1950s. Two wells were drilled to about 250 m in 1962. Bottom-hole temperatures were approximately 200°. One well ceased to produce after a short period, indicating limited permeability in the reservoir. The other, however, continues to yield 3,000 kg/hr of saturated steam. This well has been studied extensively; additional drilling was in progress in 1970.

Geothermal exploration has been carried out also at Kucharo and Showa-shinzan volcanic centers on Hokkaido Island; Nasu in central Honshu; Hakone and Atagawa, on the south coast of Honshu; Oshirakawa, in west-central Honshu; and several places in the south of Kyushu Island. Many of the drill holes were to shallow depths, and exploration cannot be considered conclusive at most locations. High-temperature reservoirs were located at Nasu (194°) and on southern Kyushu Island (over 170°).

A relationship has been noted between high-temperature reservoirs and dacitic or dacitic-rhyolitic-andesitic volcanism in Japan. This observation is similar to findings in Iceland, New Zealand, the United States, and elsewhere. The Japanese fields, however, have not been marked by the high temperatures encountered at centers of silicic volcanism in New Zealand or Iceland. Well flow has varied widely, and probably reflects fracture rather than intergranular permeability in most Japanese fields. Indeed, restricted permeability may represent a constraint upon development. But in power-short Japan, exploration is being pursued vigorously on all fronts.

Soviet Union

Eleven geothermal developments were reported in the Soviet Union in 1969. Most significant among these were the generation of electricity at Pauzhetsk; the experimental, freon-based heat exchanger at Paratunka; and the space-heating operations at Makhach Kala. There were other space-heating projects at places in the Caucasus, in western Siberia, in central Asia, and in northeastern Siberia.

Pauzhetsk is at the southern tip of the Kamchatka Peninsula, an area of active volcanism. At Pauzhetsk there are several large, boiling springs, fumaroles, and geysers; the fumarolic volcano Koshelev is nearby to the southwest. Drilling began in 1957; by 1964, 21 holes had been drilled. Eighteen of these were considered to be production wells, ranging in depth from 200 to 1,200 m. They revealed a succession of nearly horizontal, Quaternary dacitic and andesitic tuffs and agglomerates, overlying Tertiary tuffaceous sediments, the whole cut by faults into horsts and grabens. This structure had collapsed into a caldera. Heat emissions are localized on the caldera rim and along pre-caldera faults.

Reservoir fluid is hot water, containing abundant sodium and potassium chloride. It is not as saline as fluids produced at Ahuachapán, El Salvador, or Cerro Prieto, Mexico, but is similar in composition. Maximum field temperature is about 200°C, and enthalpy, even in deeper wells, is probably less than 195 cal/g. Steam flashover is about 15 to 20 percent.

Construction of a 5,000-kw power plant began in 1964, and full operation was achieved in 1967, utilizing flashed steam from three to seven wells. Plans have been laid to increase electricity production in two stages. The first involves reconstructing the two original turbines, the second installing additional low-pressure turbines, to make total installed capacity 22 to 25 Mw by 1980. Electricity from Pauzhetsk is transported 27 km to consumers at costs 30 percent less than that of electricity generated using fuel oil.

The most significant experiment to date in heat-exchanging with geothermal fluids has been undertaken at Paratunka, near the city of Petropavlovsk, on the east coast of Kamchatka. A 680-kw freon-based generating plant was installed in 1967, and after extensive testing began routine operation in 1970, supplying power to the Paratunka State Farms. The plant comprises two 340-kw turbines and utilizes water from shallow wells at 81°C. Excess heat is used to heat greenhouses and seedbeds, and warm water is used for irrigating crops. Other space-heating projects in the vicinity use water at under 80° to heat homes, baths, and swimming pools.

There has been extensive development of geothermal-heating systems at several points across a 500-km zone on the north slope of the Caucasus. Beginning in 1947 at Makhach Kala, on the Caspian Sea, after the accidental discovery of 60° to 70° waters in the course of prospecting for oil, space heating has developed rapidly. At Makhach Kala, 15,000 people receive home heat and hot water from a municipal geothermal system; and geothermal water is used as boiler feed in an oil refinery and as heating for greenhouses, seedbeds, and baths. Trending northwest, there are similar activities at Grozny; in the Kabardin-Balkaria district; at Cherkessk, where 18,000 people are served; and in the vicinity of Maykop. In the Kabardin-Balkaria district alone, site of one of the more modest projects, an annual saving of 2,500 tons of fuel oil results. Water temperatures are between 47°, near Cherkessk, and 86°, near Maykop, though the reservoirs supplying these wells may be at temperatures of 100° to 150°C. Many of these waters are but lightly mineralized, allowing direct utilization. In some cases, however, mineralization is strong, and heat is exchanged to fresh, cool water.

Other geothermal-heating projects are near the Black Sea coast of Georgia; on the Caspian Sea near the border with Iran; in several places in northeast Siberia; in the cities of Omsk and Tobolsk in western Siberia; at several points near Tashkent in central Asia; and on the shore of Lake Baikal. Figures for the total consumption of geothermal energy in the U.S.S.R. are fragmentary and contradictory. However, the total energy consumed in these several space-heating systems is believed to be the equivalent of 1 million tons of fuel oil per year.

Exploration of high-temperature reservoirs has continued on Kamchatka Peninsula and on the island of Kunashir' in the Kurile chain. In addition to the developed fields of Paratunka and Pauzhetsk, there is geological, geophysical, and geochemical exploration under way at Bolshe-Banny, Uzon Geyzermy, Nalychersky, and six other areas on Kamchatka. At Bolshe-Banny, located about 30 km west of Paratunka, a maximum temperature of 171° was measured in drill holes.

The search for low-enthalpy waters for use in heating and in agriculture is accelerating in Georgia, the west Siberian lowlands, and areas on the northern slopes of the Pamir, Tien Shan, and Altai ranges in central Asia. Vast basins containing waters at temperatures up to 120° have been outlined in western Siberia and along the northern foothills of the Caucasus. Temperature-gradient drill holes, geochemistry, and seismic-refraction surveys are commonly used in these studies.

Hungary

Extensive hot-water aquifers have been found to underlie the southwestern part of the Hungarian basin, near the Yugoslav border. Although reservoir temperatures are not high enough for the generation of electricity with present-day technology, the hot-water resource has been extensively developed for space heating. Eighty wells had been completed by early 1970, to depths of 1,800 to 2,000 m. Water temperatures from these depths are 85° to 110°C. Temperature gradients of 35° to 55° per km occur, which is not unusual for deep sedimentary basins. Like other high-gradient basins, such as the Gulf Coast of the United States, the Hungarian basin is devoid of Quaternary volcanic rocks or high-temperature surface waters. Basin fill comprises several thousand meters of Tertiary and Quaternary sediments. Reservoir formations in the Hungarian basin are highly permeable lower Pliocene sands. Average well yield is 80 to 90 m^3/hr. Since the water contains calcium carbonate it must be allowed to stand, and thus to precipitate within tanks rather than within the heating system. Wells are reconditioned periodically.

The Hungarian government is developing an extensive geothermal-heating system in the province of Csongrad. Ironically, plans to heat the entire city of Szeged with natural hot water were abandoned when oil—ultimately the largest oil and gas field in Hungary—was discovered during the drilling of a hot-water well. Even so, some 1,200 housing units and associated municipal and commercial buildings are heated by hot water in Szeged. Geothermal fluids are used also to heat greenhouses, animal pens, runs, and other farm buildings, and to dry crops. In all, well over 1 million m^3 of space is heated geothermally, at costs well below those of conventional fuels (see Table 2). Data are incomplete and perhaps inconsistent, but the heating done by hot water in Hungary would require about 80,000 tons of fuel oil annually.

Geothermal development is expected to increase at about 15 percent per year through the decade, mainly in agricultural application.

Mexico

A 75-Mw power plant is to begin operation at Cerro Prieto, Baja California, early in 1973. This will not be the first Mexican geothermal station—an experimental facility has operated at Pathé, Hidalgo, since 1960. At Pathé at least fourteen wells have been drilled, and 500 kw of electricity is produced from a generator rated at 3,500 kw and now supplied by a single well. Pathé is located in the extensive east-west volcanic belt that traverses central Mexico from the Pacific to the Gulf coastal plain. Exploration at Pathé began in 1955. The reservoir comprises a thick sequence of altered and fractured Tertiary andesites and basalts. These overlie Cretaceous carbonates and are in turn overlaid by Quaternary and Late Tertiary tuffs and sediments. Wells produce steam or steam and water, although only four can sustain flow. Production is from variable depths, since permeability is believed to be controlled by fractures. The low productivity per well and the tendency of wells to produce dry steam indicate a low permeability and limited fluid content in the reservoir. The presence of a deeper reservoir in the Cretaceous carbonate rocks has been postulated, but has not been explored by drilling.

Cerro Prieto, 30 km south of Mexicali on the Mexican-United States border, is the southernmost explored portion of the structural rift extending from north of the Salton Sea in the United States into the Gulf of California, a total distance of several hundred kilometers. Exploration is occurring in several portions of this rift in the United States. At Cerro Prieto, fumaroles and mud volcanoes in the vicinity of a Quaternary dacite-basalt volcano, aligned along a probable extension of the San Jacinto fault, attracted exploration in the late 1950s. Twenty-three deep holes have been drilled, of which fifteen have been completed as production wells. Average yield is about 230,000 kg/hr of superheated brine. Approximately 20 percent flashes to steam, giving an average electric-power yield of some 5 Mw per well, although the steam percentage varies from 13 to 25 percent in individual wells. The strongest well of the field yields nearly 700,000 kg/hr of brine, which is equivalent to 15 Mw of electricity. Though average well depth is about 1,500 m, one deep well reached 2,600 m, penetrating crystalline basement. Permeable sands are encountered between 600 to 1,200 m and at depths below

2,400 m. Only the shallower reservoirs are to be used for production. Reservoir temperatures are over 300°C, with a maximum recorded temperature of 388°C.

As elsewhere in this structural rift, several thousand meters of Late Tertiary and Quaternary sediments have filled the opening trench caused by the separation of Baja California and adjacent California from the continental mainland. Volcanism is exhibited only locally, and the heat source is believed to be molten rock of the upper mantle, which is here at depths of only 15 to 20 km. Water in these sediments may be derived from the ancestral Colorado River system.

The reservoir fluid contains between 13,000 and 25,000 ppm total dissolved solids, mostly chlorides of sodium and calcium. The feasibility of commercial extraction has been studied for such substances as potassium chloride, lithium, and boron, but no recovery operations are planned for the first plant. Disposal will be via ditches to the Rio Hardy, a distributary of the Colorado River, and thence to the Gulf of California. Some concern has been expressed over long-term consequences, and the alternatives of reinjection and ponding are under consideration. Other problems include brine corrosivity and the possibility of induced ground subsidence.

However, a second set of wells is to be drilled beginning in 1972, to supply steam for a second 75-Mw station. This would go on line in about 1980. A steam-powered drilling rig has been brought to Cerro Prieto, to operate on natural steam. This should have the effect of reducing drilling costs slightly. Water condensed from steam is used in construction and maintenance operations in this extremely arid region. Only a small portion of the Cerro Prieto geothermal zone has been drilled, and it is considered likely that total potential is many times that due for production in this decade. Exploration and development are carried out by an arm of the Comisión Federal de Electricidad, the Mexican national electricity agency.

Several high-temperature fumarole and hot-springs systems have been under exploration in the east-west volcanic belt of Mexico. This zone contains many active volcanoes and extensive evidence of Quaternary volcanism. Five systems have been explored in some detail and at a sixth, San Marcos, exploration is just under way. The five are, from west to east, La Primavera, in Jalisco; Los Negritos, Ixtlán, and Los Azufres, in Michoacán; and Los Humeros, in Pueblo. This zone of volcanoes and interspersed heat emissions is over 700 km long. La Primavera occupies the margins of a graben down-dropped within a Quaternary caldera. Superheated steam flows from fumaroles in altered rhyolites at temperatures to 100°C. Geophysical and geochemical work was in progress in 1970. Los Negritos is perhaps the best known of these fields, occupying part of a caldera whose walls show acidic tuffs, basalt flows, and lacustrine sediments. Fumarole temperatures reach 95°. Extensive geological, geophysical, and geochemical work suggests an extensive high-temperature reservoir. Ixtlán, only 27 km to the northeast, occupies an east-west structural valley, and exhibits mud volcanoes and fumaroles to 100°. Drilling of deep exploration holes was in progress in 1970. Preliminary temperature-gradient drilling had encountered temperatures to 150°. Los Azufres also occupies part of a caldera formed in basaltic and rhyolitic flows and pyroclastics. Here, superheated steam at temperatures to 110° is emitted from powerful fumaroles. Geological and geophysical work is in progress. Los Humeros, the easternmost of these fields, exhibits fumaroles to 90°, in a Quaternary caldera containing rhyolite and basalt flows and pyroclastics. Here, too, geophysical and geochemical work is continuing.

El Salvador

In 1971 the national electricity agency (CEL) of El Salvador chose LC-Electroconsult of Milan, Italy, to design a 30-Mw geothermal plant for construction at Ahuachapán. This marked the successful completion of a geothermal exploration program jointly sponsored by the government and the United Nations. Even before this program began in 1965, the National Geologic Service of El Salvador had explored the Ahuachapán fumarole area and drilled two shallow wells. Over $4 million has been spent to date on this project. In 1972 CEL is expected to call for bids for construction of the plant, and it is anticipated that electricity will flow in 1975. Production capacity will be increased through additional drilling and plant construction by 1980.

The Ahuachapán region occupies some 30 km^2 on the northern slopes of a range of Quaternary andesitic volcanoes, and contains over a dozen areas of fumaroles, warm ground, and hot springs. The easternmost peak of the range, Izalco, is active. Geologic mapping and subsequent drilling suggest the presence of a caldera, largely buried by late- and post-volcanic debris. Fumaroles and other heat emissions at Ahuachapán are controlled by annular fractures forming the caldera wall and by northerly trending transverse fractures. The highest fumarole temperatures occur at the southern (uphill) end of the structure, and reach 125°C. No wells have been drilled in that area. Average well temperatures are about 225° to 230°, the maximum 235°, at depths of 600 to 900 m. However, geochemical indexes suggest a reservoir-equilibrium temperature approximately 10° to 20° higher. Several workers have thus speculated that the higher-temperature fumaroles to the south may more closely correspond to the reservoir source.

Reservoir fluid is a sodium-chloride brine of about 10,000 ppm concentration. Some 15 percent flashes to steam in the wellbore. Because average mass flow is about 320,000 kg/hr, it is expected that seven production wells will be needed for a 30-Mw plant. No pressure declines have been noted in production tests. The disposal of saline reservoir fluid presents a problem, since boron, chloride, and arsenic are present in amounts potentially harmful to agriculture. In reinjection tests, a production well has been used for disposal into a deep, hot aquifer. These tests have continued for a year and a half without definite effects upon the adjacent production wells. The question of possible silica deposition in reinjection wells remains unanswered. The alternative to reinjection is disposal to the sea via holding ponds and the natural river-drainage system.

A conservative estimate of developable potential at Ahuachapán is 100 Mw for a 50-year period. Drilling to date has explored only 2 km^3 of an estimated 40 km^3 field volume. Exploratory drilling has also been carried out at Berlín, in the eastern part of the country. A well drilled in 1968 to nearly 1,500 m encountered temperatures of at least 225°. Flow of hot water was not sustainable, indicating restricted permeability. Another hole drilled to 600 m, 10 km to the northwest, flowed strongly and continuously at about 100°. Both wells exhibited moderate chloride salinity. Further exploration is warranted.

Kenya

Kenya, Uganda, and Tanzania share a common electric-power system, operated in Kenya by the nationalized East African Power and Lighting Company (EAPL). In 1957 and 1958, EAPL drilled two test holes southwest of Lake Naivasha, in the Kenyan rift. These holes could not be brought into sustained flow, despite temperatures exceeding 200°C.

In 1970, exploration began again, jointly sponsored by the United Nations and the Government of Kenya, in the rift zone. Two broad target areas were chosen, Lake Hannington, where fumaroles and boiling springs issue at the edge of the tilted fault blocks, and near Lake Naivasha. Acid igneous domes and ejecta form an arc around the lake from northwest to south, and a few miles south of the lake an east-west fracture transects the rift. There are two fumarole complexes within the volcanic field around Naivasha. The northern complex, on the slopes of Mount Eburru, has been drilled to about 150 m. Low-pressure steam is produced and condensed for use in watering livestock, and pyrethrum leaves are dried in live steam for use as insect-repellant. The southerly field, Olkaria, was the site of the earlier drill holes by EAPL. One of these holes, about 900 m deep with temperatures to 230°, has been induced to flow, and tests are continuing. Extensive geological, geochemical, and geophysical surveys have been made, and three deep exploration holes are scheduled for the summer of 1973.

Ethiopia

A survey in 1970 and 1971 sponsored by the United Nations and the Government of Ethiopia catalogued the surface thermal phenomena of the Ethiopian rift zone. Over 500 separate hot spring and fumarole localities were reported. A second-stage exploration is planned to begin sometime in 1973. Three areas of greatest interest were noted: Dallol, in the Danakil depression of the northern rift; the Tendaho graben of the Afar region; and the Aluto caldera in the lakes district of the southern rift.

The Danakil depression is a north-south trench, mostly below sea level. At the north are the Dallol salt flats, an evaporite sequence over 1,000 m thick, producing salt-saturated springs at temperatures above 110°C. Immediately south is a chain of active volcanoes, with fumaroles to 220° on the flanks. At Tendaho, there are extensive boiling springs, fumaroles, Holocene cinder cones, and lava flows, in a situation of tilted and rotated fault blocks. In the lakes district several volcanic centers and calderas are characterized by hot springs and fumaroles.

Even at Addis Ababa, a city of 600,000 persons at 2,500 m elevation on the Ethiopian plateau, there is an extensive hot-water aquifer that might prove useful in space heating. Government-sponsored geothermal exploration is investigating the extensive potash reserves at Dallol and the need for electricity for pumped irrigation in the Tendaho area. Fresh water might be a by-product of either scheme, but currently hot water and steam condensate are used only locally for bathing and stock-watering.

The Philippines

At Tiwi, on Luzon in the Philippines, a 10-kw geothermal generator has operated since 1969 as a demonstration of the economic potential of the region. The area had been explored by the Philippine National Science Board and the Commission of Volcanology since 1966, when seven shallow, exploratory wells were drilled. The well that later drove the generator produced about 10,000 kg/hr of steam at 154° from 220 m depth.

In 1971 Philippine Geothermal Inc., a subsidiary of the Union Oil Company, contracted with the National Power Corporation of the Philippines to explore the 150 km^2 of geothermal resources in southeastern Luzon. Tests on a 1,500-m well drilled in 1972 indicated potential as a commercial power source by the mid-1970s. Two additional wells are scheduled to be drilled. Geologically, the region consists of Quaternary andesites and subsidiary dacites.

Indonesia

The island nation of Indonesia is among the most volcanically active regions of the world. On Java, where over 40 million people live, active volcanoes dot the island; and there are others on Sumatra, Bali, Flores, and the Celebes. As early as 1918, reports suggested that the volcanic heat of these islands could be harnessed. In 1928, the first geothermal test holes were drilled at Kawah Kamodjang on western Java. The deepest of these three holes was 128 m. The highest temperature was 140°C, and abundant low-pressure steam was encountered. Data from these wells were recorded continuously until World War II.

After the visit of a scientific team from UNESCO, in 1966, Indonesian geologists began compiling data from Java and the southern Celebes. In 1969, the U.S. Geological Survey began a program of assistance, which is still under way, concentrating on the Dieng area of central Java. The Dieng region comprises a series of constructional volcanic forms and their explosive and extrusive products, aligned roughly northwest and extending about 20 km. Surface temperatures to 95° are recorded in solfataras. Pyroclastic debris, lahars, and lacustrine sediments are believed to overlie Miocene sediments. Sandstones in the Miocene sequence are considered a possible reservoir.

In 1972, six slim exploratory holes were to be drilled to 200 m at Dieng. After tests, two are to be deepened to 600 m. An agreement also is anticipated between the governments of New Zealand and Indonesia, under which a New Zealand-owned company would conduct geothermal exploration of unspecified areas in Java and perhaps on other islands.

Chile

A United Nations-supported exploration program began in the summer of 1967 across an area of 100,000 km^2 in the volcanic interior of northern Chile. The elevation of this desert plateau is over 4,000 m, with Quaternary volcanic peaks reaching to almost 6,000 m. The region is extremely arid. On the basis of preliminary reconnaissance, three areas were chosen; of these, El Tatio has become the prime target. After detailed geochemical, geological, and geophysical surveys, several small-diameter holes were drilled. The deepest for which data are available went to about 650 m, and produced hot water from rocks at temperatures to 250°C. On the basis of these holes, areas were to be selected for drilling production wells.

Because of the altitude and remoteness of the field, exploration has progressed slowly—relative inaccessibility has made slim-hole drilling a practical necessity. Because of the regional aridity, it is considered likely that the production of fresh water will be a part of any development program. Initial estimates are for 20 to 30 Mw of electric power, perhaps for use in mining and smelting operations at copper deposits some 100 km distant, but also for use in the recovery of chemicals from the adjacent fumarole field.

Guadeloupe

The Caribbean islands of Guadeloupe and Martinique, and Cayenne on the north coast of South America, are classed as an Overseas Department of metropolitan France. Beginning in 1963, the French Bureau de Recherches Géologiques et Minières (B.R.G.M.) and Eurofrep, a French oil-exploration company, conducted geological, geochemical, and geophysical exploration for steam on both islands. On Martinique in 1969 an unsuccessful hole was drilled to 771 m, about 20 km southeast of the active volcano Mont Pelée.

Exploration then shifted to La Bouillante, near the fumarolic volcano Soufriere, on Guadeloupe. Three holes have been drilled, the deepest to 800 m. The first produced an intermittent flow of low-pressure steam. The second, at 338 m, yielded over 300,000 kg/hr of superheated water from rocks at 240°. Nearly one-third of the water by weight flashed to steam. The third well is reported to yield limited amounts of dry steam. The region consists of andesitic and dacitic flows, tuffs, and lahars, apparently cut by numerous fractures. Because the fractures supply the hot fluids—no reservoir formation is recognized—low permeability may prove to be a problem. Consideration is being given to further drilling and to attempts to utilize wet steam in a low-pressure turbine. A 30,000-kw plant, which is believed to be the goal of this project, would require, by analogy, seven or eight production wells.

Taiwan

In Taiwan, exploration has centered at the Tatun volcanic complex at the northernmost end of the island, where exploration began in 1965 and is still in progress. Thirteen fumarole and hot-springs areas occupy an area of over 50 km^2. Fumarole temperatures reach 120°C at one locality. Geologically, Pliocene and Pleistocene andesite and basalt lavas, tuffs, and volcaniclastic rocks overlie Miocene continental and marine sediments. Coarse sandstones make up the reservoir rock. These are exposed locally by folding and faulting.

Exploration has included detailed geologic mapping, geochemistry, and extensive geophysical surveying. Fifty-eight temperature-gradient holes have been drilled, to a maximum of 160 m. Temperatures to 174° have been encountered, and several holes produce very acid steam. At least eleven holes have been drilled to depths of 300 to 1,500 m, with a maximum reported temperature of 293°. Flow has varied up to 35,000 kg/hr of fluid, with steam flash-off ranging widely, from 10 to 83 percent. Fluid pH has usually been less than 5.0, and as low as 0.5, causing problems of corrosion and disposal of residual fluid. Of the eighteen areas at Tatun, Matsao is considered to be the most promising, on the strength of highest percentage of flash steam, most-moderate pH values, and temperatures above 200° at depths beyond 500 m. However, the relatively low mass-flows at Matsao may indicate lower permeability within the Miocene sandstone. Electric-power potential at Tatun has been estimated at 80 to 200 Mw, and construction of a 10-Mw pilot plant at Matsao is being considered.

Turkey

In 1961 the Turkish government began detailed studies of a series of geothermal manifestations in western Anatolia. The United Nations entered into a cooperative exploration venture in 1967, and the following year, after geological, geophysical, and geochemical surveys, the first well was drilled at Kizildere, in the Menderes (Meander) River valley. Hot springs are widespread through this region, with temperatures to 100°C. Mercury mineralization is also present locally.

At least seven deep holes have been drilled at Kizildere, another at Tekke Hamman, 10 km to the southwest. In the process, a hot-water reservoir has been discovered, with two horizons, at depths of about 350 to 400 m and below 600 m. Exploration has revealed a highly fragmented horst and graben terrain, primarily with east-west trend, step-faulted down to the south. A crystalline basement of schist, gneiss, marble, and quartzite is overlaid by Miocene and Pliocene fluviatile and lacustrine sediments. A highly fractured Miocene limestone in this sequence constitutes the upper reservoir; fracture permeability in the crystalline rocks provides the lower reservoir. Quaternary alluvium caps the Tertiary sequence. No youthful volcanic rocks are present. Rather, the heat source is postulated to be a cooling granitic mass at depths of several kilometers. Its upward movement is believed to have generated the horst and graben structure.

Average well depth is about 450 m, and reservoir temperatures are between 180° and 200°. Yield per well is between 25,000 and 300,000 kg/hr, averaging about 150,000 kg/hr, with about 10 percent flashing to steam. It is reported that the deeper wells, into the crystalline basement, have higher flow rates. The reservoir fluid is highly carbonated, and a major problem of calcite deposition must be solved before development can proceed. The well at Tekke Hamman, in fact, is inoperable because of calcite incrustation. Installation of a 10- to 20-Mw generator, utilizing a closed-system heat-exchanger, is being considered—reservoir fluid would be produced, utilized, and reinjected into the reservoir under pressure, thus avoiding the separation of calcium carbonate from the fluid.

Exploration in Other Areas

A pioneering geothermal development was begun in the 1950s at Kibukwe, in Katanga province of Zaire (Congo), where a 220-kw geothermal generating plant was installed at a metal mine. This plant, which used wet steam at about 95°, was costly and inefficient, but served satisfactorily for many years in a remote area far from cheap alternative power supplies. The plant and mine are now shut down.

In February 1972 it was reported that a small, experimental power station had operated successfully at Tengwu, in Kwangtung Province of the People's Republic of China. The station is reported to use steam flashed from hot water to turn turbines, but no details are available on plant size, water temperature, or operating characteristics.

In 1967 the United States undertook a project of geothermal exploration for the Government of Nicaragua, utilizing private U.S. concerns as contractors. Detailed geophysical surveys and geochemical and geologic evaluations were made, and a series of temperature-gradient holes was drilled in fumarolic terrain near Momotambo volcano in west-central Nicaragua. A deeper drill-hole encountered temperatures to 230°, but the project was allowed to lapse. In 1972 the United Nations entered into an agreement with the Government of Nicaragua for a second phase of exploration, which is now under way.

At Melun, France, about 50 km southeast of Paris, two holes drilled to depths of 1,800 m intersect an artesian aquifer with water at approximately 70°. This corresponds to a gradient of about 3°/100 m, and is not remarkable (see Table 2). Nonetheless, a district-heating scheme is being evaluated in which one well would be used for production and the other for reinjection of heat-depleted water. Heat energy produced from this scheme would be offered at prices competitive with other fuels. The artesian basin is believed to be quite vast.

Many other countries of southern and eastern Europe, Latin America, north Africa, and Asia have begun the collection of data for geothermal exploration. Prominent among them are Algeria, Colombia, Greece, Guatemala, India, Israel, Spain (Canary Islands), Venezuela, and Yugoslavia. A listing of some 50 countries active or interested in geothermal exploration is given in Table 3.

Temperature-gradient drilling in northeastern Algeria has outlined a region of anomalous gradient near the Tunisian border. Detailed geological and geochemical studies suggest a buried batholithic mass as the source of heat for the numerous hot springs of the region.

The Government of Guatemala has chosen a geothermal area near Moyuta volcano, near the border with El Salvador, for detailed exploration. Consideration is being given to contracting in 1972 for a deep exploration hole, and several engineering firms have been asked to submit proposals for a drilling and development program.

The Spanish government is supporting geothermal exploration in the Canary Islands, a Spanish territory off the coast of Morocco. Very-high-temperature fumaroles are reported at Lanzarote Island.

In Yugoslavia and Czechoslovakia, temperature-gradient and heat-flow studies are currently in progress, with an eye to locating sizable reserves of low-enthalpy fluids for use in space heating, agriculture, and industry.

In six underdeveloped nations (El Salvador, Chile, Turkey, Kenya, Ethiopia, and Nicaragua) the United Nations has sponsored geothermal exploration jointly with the national governments. United Nations missions have also been made to India, Greece, Peru, and Guatemala in the past year to evaluate potentials for exploration; cooperative exploration projects have been proposed in several cases. The United States has supported exploration in Nicaragua and Indonesia through the Agency for International Development. Colonial administrations have carried out exploration in such underdeveloped places as the Territory of Afars and Issas (French Somaliland), the Fiji Islands, and New Britain. In the United States and Japan, private and government organizations have engaged in exploration, either jointly or separately. Private concerns have assisted in geothermal exploration in the Philippines, Algeria, Guadeloupe, and elsewhere. Usually, the initial stages of data collection and the preparation of geologic-reconnaissance reports, power-demand projections, and transmission-grid maps, are undertaken by the local geological survey or electricity agency. Shallow borings are often made at this stage, too. Much of the exploration to date has been financed and carried out by agencies of the countries involved, for in many cases the geothermal resources are owned nationally and are thus not at the disposal of private landowners. But because many nations lack a sophisticated geological infrastructure, detailed exploration is likely to require cooperative ventures with the United Nations, with more developed nations, or with private concerns.

The Potential for Geothermal Energy Development to 1980

At Larderello and Monte Amiata (see Table 4), increases in power generation are likely to depend upon conversion from noncondensing to condensing turbines. This may increase capacity by 15 percent over the decade. But only if significant discoveries of steam are made at Travale, Roccastrada, or Radicofani is there likely to be new construction in the steam fields of Tuscany. If exploration elsewhere in Italy is successful, new generating facilities could be on line by 1980.

There are no plans to construct additional geothermal power plants in New Zealand in this decade. Industrial and municipal applications will be encouraged, however, and some generation of electricity may result incidentally to direct utilization of heat for industrial processes.

Plans have been made to increase generating capacity at The Geysers by 110,000 kw per year through 1975, at which time 630,000 kw will

TABLE 4
Expected Development of Electric-Power Capacity at Selected Geothermal Fields to 1980

Nation	Field	Installed capacity, late 1972, Mw	Expected development, Mw
El Salvador	Ahuachapán	—	30 by 1975; 60 by 1980
Iceland	Hengill (Hveragerdi)	—	Up to 32 by 1980
	Namafjall	3	None known
Italy	Larderello	365	15-percent increase possible
	Monte Amiata	25	
Japan	Hachimantai	—	Perhaps 10 by mid-1970s
	Matsukawa/Takinokami	20	Perhaps 60 by 1980
	Onikobe	—	Perhaps 10 by 1980
	Shikabe	—	7; salt-recovery works planned for 1970s
	Otake/Hachobaru	13	Perhaps 60 by 1980
Mexico	Cerro Prieto	75	150 by 1980
New Zealand	Kawerau	10	None planned
	Wairakei	160	None planned
U.S.S.R.	Pauzhetsk	6	Up to 25 by 1980
	Kunashir'	—	Up to 13 by 1980
United States	The Geysers	302	110 per year through 1980, to 1,180
	Imperial Valley	—	Demonstration, for desalination; power station by 1980

have been installed, making this the largest developed field in the world. By 1980, it is thought, installed generating capacity might be about 1,180,000 kw, if development continues to be successful. In the Imperial Valley it is likely that pilot electric-generation and desalination plants will be operating by 1980. Capacity is likely to be 10 to 20 Mw for demonstration plants. By then also, one or more small-scale, closed-system, heat-exchanging electric generators may be installed in that region. Conceivably, a few other pilot stations may be operating, or full-scale plants may be under construction, elsewhere in the western United States by 1980.

Several Japanese fields are likely to be developed in this decade. Matsukawa and Otake are scheduled for enlargement to perhaps 60 Mw each. This would include some development at Takinokami near Matsukawa and some at Hachobaru near Otake. Questions of scaling, well life, and water disposal may delay development somewhat. Onikobe, Hachimantai, and Nasu, on Honshu, and Shikabe, on Hokkaido, are also potential candidates for power-generation development; total installed capacity at these fields is not likely to exceed 30 Mw by 1980.

The plant at Pauzhetsk, in Kamchatka, may be expanded to as much as 25 Mw by 1980. Other fields in the Kurile Islands or on Kamchatka may be put into production, probably in the range of 10 to 20 Mw each. Space heating can be expected to increase greatly in the Caucasus and in regions of western and southern Siberia through the decade, so that by 1980 the consumption of hot water may surpass two million tons per year of fuel oil.

Similar increases in consumption of hot water can be expected in Iceland, with an equivalent of over one-third million tons of fuel oil used annually. Electric-power generation may increase slightly at Namafjall, and may begin at Hveragerdi in the Hengill area. Perhaps 15 to 35 Mw capacity will be installed by 1980.

In Hungary the consumption of hot water for space heating may more than double through the rest of this decade. Hot-water heating schemes may become operational in Yugoslavia, Czechoslovakia, France, and elsewhere in Europe by then.

A second 75-Mw plant may be installed at Cerro Prieto by 1980. And construction of generating facilities may be under way elsewhere in the central part of Mexico by then. Similarly, at Ahuachapán, El Salvador, a second 30-Mw facility may be operational at that time.

Small plants are likely to be operating or under construction in the Philippines, Kenya, Chile, Turkey, and Taiwan by 1980, and perhaps in Guadeloupe, Nicaragua, and elsewhere. Their aggregate output is likely to be between 70 and 150 Mw. But because of the 4- or 5-year minimum lead time required for exploration and construction, it is unlikely that extensive plant construction will have been undertaken elsewhere by 1980; conceivably, another 50 Mw of generating facilities will be erected in, for example, Indonesia, Ethiopia, or China.

Thus, a conservative projection of worldwide geothermal generating capacity by 1980 is on the order of 2,500 Mw, or three times present-day capacity. Because world consumption of electricity during this 8-year period is likely to double, the geothermal-power component of world output will remain at less than 1 percent of total generating capacity. Direct utilization of geothermal energy is likely to increase at a faster rate, especially in eastern Europe.

More rapid development is foreseen in the 1980s. It will depend in part upon improvements in geothermal drilling and utilization technology, increased knowledge of geothermal systems, and greater availability of funds for geothermal exploration and development.

SOURCE MATERIAL

Because of the great volume of source material from which this paper was drawn, citations are not given for individual references. Rather, the reader is directed to two massive proceedings that serve both as sources for data in this report and as bibliographies for continued research:

Proceedings of the United Nations Conference on New Sources of Energy, Solar Energy, Wind Power and Geothermal Energy, Rome, August 21–31, 1961; v. 2, 3, Geothermal energy. New York: United Nations, 1964.

Proceedings of the United Nations Symposium on the Development and Utilization of Geothermal Resources, Pisa, Italy, September 22–October 1, 1970; Geothermics, Special Issue 2, 2 volumes, 1971.

The former contains 77 individual papers and three summary articles by rapporteurs, and summarizes information to 1960. The latter contains 198 individual papers plus summary reports by rapporteurs of the eleven symposium sessions, and presents data for exploration and development through 1969.

Data more recent than these sources come largely from personal correspondence and discussion with informed professionals, especially L. J. P. Muffler, U.S. Geological Survey; Carel Otte, Union Oil Company of California; David N. Anderson, California Division of Oil and Gas; Gunnar Bodvarsson, Oregon State University; David Kear, New Zealand Geological Survey; G. V. Subba Rao, United Nations; and Giancarlo Facca, Lafayette, California.

3. Assessment of U.S. Geothermal Resources

ROBERT W. REX AND DAVID J. HOWELL

The heat of the Earth's interior is one of the most immense energy resources available to man. It is far larger than the energy available if all of the uranium and thorium in the Earth's crust were employed in breeder reactors. But more important than the sheer *size* of the resource are the extent of the resource that can be developed with present technology and at an acceptable cost; and the impact of practical constraints, especially those imposed by institutional and environmental problems. It is evident from currently operating geothermal plants in California that certain types of geothermal energy are not only technically feasible but also practical and economical. Present plants utilize natural underground steam. Hot-water geothermal fields are now under test in the Imperial Valley of California and will be on line in a similar geological setting in northwestern Mexico in 1973. Thus in the United States geothermal energy can be considered a competitively proved energy source.

Occurrence of Geothermal Energy in the United States

The western third of the United States, including Alaska and Hawaii, seems to be the most richly endowed in accessible geothermal-energy resources. There is little or no deep-well information on the hot-spring areas of Virginia, West Virginia, Georgia, Arkansas, or other parts of the eastern and central United States. And sources such as the hot, dry

At the time this paper was written, Robert W. Rex was Director and David J. Howell was Controller, Geothermal Resources Program, Institute of Geophysics and Planetary Physics, University of California, Riverside. They are both currently with Pacific Energy Corporation, Marina del Rey, California. This paper is Contribution No. 72-39, Institute of Geophysics and Planetary Physics.

rock reservoirs of the Appalachians may be quite large, but so little is known about them that it is necessary to project the resource in a very approximate manner. However, present analyses based on fairly conservative projections *assuming hot, dry rock systems to be technically exploitable* indicate that geothermal energy can be a major source of electrical energy for much of the United States.

Current geothermal developments center on dry-steam and hot-water systems. But almost all of the enormous heat content of the Earth's interior that is within reach of the drill is present in dry rock. As a basis for carrying out an order-of-magnitude appraisal of the geothermal-resource potential, let us take as a case study the volcanic area surrounding the Valles Caldera in the Jemez Mountains in northern New Mexico. Working from the basis that the energy is only in the hot rock of the system, we have utilized currently available temperature-gradient data, known lithology, and geophysical data to construct a simplified model. The measured temperature gradient to 700 m is 180°C/km (573°F/mi). Projecting this gradient and using the known topography and geology to construct a temperature model, we assumed a small increase in conductivity with depth, a heat capacity of 0.20 cal/°C/g, an average rock density of 2.75 g/cm^3, and a lower working temperature of 100°C. The seasonal air temperature averages about 0°–10°, making air cooling attractive for disposal of waste heat. Table 1 shows the results of this model for the Jemez Mountains case using the measured temperature gradient of 180°C/km.

If we further assume in this model that the usable thermal energy in the rock system is completely recovered over a time period of 100 years, then the recoverable thermal energy in this volcano is 8.48×10^{20} g-cal.

TABLE 1
Heat Characteristics of the Jemez Volcano Model
(Shallow $\nabla T = 180°C/km$)

Depth (km)	Volume available (km^3)	Mass (10^{18} g)	Average T (°C)	Useful ΔT (°C)	Useful ΔH (cal/g)	Total available enthalpy (10^{20} g-cal)
0–1	1,600	4.40	90	0	0	0
1–2	1,200	3.30	290	190	38	1.25
2–3	1,000	2.75	460	360	72	1.98
3–4	800	2.20	610	510	102	2.24
4–5	600	1.15	740	640	128	1.47
5–6	400	1.1	800	700	140	1.54
TOTAL						8.48

The conversion efficiency for producing electrical energy from geothermal energy is about 14 percent. Therefore, 1.19×10^{20} g-cal could be converted to electricity. Accordingly,

$$1 \text{ g-cal} = 1.163 \times 10^{-9} \text{ Mwhr}$$
$$1.19 \times 10^{20} \text{ g-cal} \times 1.163 \times 10^{-9} \text{ Mwhr/g-cal}$$
$$= 1.38 \times 10^{11} \text{ Mwhr} = 158{,}000 \text{ megawatt-centuries (Mwcen)}$$

There appear to be *at least* ten volcanic areas of this magnitude in Alaska, at least another ten in the contiguous 48 states, and at least five in Hawaii. Consequently, volcanic energy alone constitutes a reserve of energy of about 4×10^6 Mwcen, probably adequate to meet U.S. electrical energy needs for several centuries.

If the volcanic rock is in a fractured state and naturally water-filled, its water is carrying about 1.8 times as much energy as hot rock will. Dry steam in the fractures would be harder to appraise with respect to its energy content because of the uncertain density of the fluid. Although the net energy difference per unit volume between hot rock, hot water, and dry steam is not large enough to affect an order-of-magnitude resource-size calculation, steam and water are significantly less costly than hot, dry rock as sources of geothermal heat. The intense shattering of The Geysers geothermal field over a broad areal extent suggests that natural thermal stresses may play a major role in producing steam- or water-filled porosity in areas of major intrusives.

Von Herzen (1967) described two major heat-flow provinces in the western United States with an aggregate area of 2.7×10^6 km^2. Our work in the Imperial Valley of California and elsewhere and a general appraisal of the literature, as well as a review of unpublished industry data, lead us to suggest that about 5 percent of this area, excluding the volcanic areas discussed previously, is underlaid by hot rock or producible geothermal fluids. We have calculated the available resource using an average thickness of producible rock/sediment section of 3 km with an average temperature of 300°C. Using a minimum working temperature of 100°, we obtain an average ΔT of 200°. By conversion, $2.7 \times 10^6 \text{ km}^2 \times 0.05 \times 3 \text{ km} = 0.405 \times 10^{21} \text{ cm}^3$ of hot rock/water. The heat capacity of hot rock is about 0.20 cal/°C/g and ΔH is 40 cal/g. Assuming an average rock density of 2.65, the mass of rock = $2.65 \text{ g/cm}^3 \times 0.405 \times 10^{21} \text{ cm}^3 = 1.07 \times 10^{21}$ g. The total usable heat is therefore about 4.28×10^{22} g-cal. With a 14-percent efficiency for conversion of geothermal heat to electricity, 6.0×10^{21} g-cal converted to electricity would be about 7.0×10^{12} Mwhr or 8×10^{8} Mwyr or 8×10^{6} Mwcen.

Combined with volcanic energy, U.S. geothermal resources may amount to 12×10^{6} Mwcen. This is probably sufficient to supply U.S. energy needs for a millennium. If water replaces hot rock, producible energy would increase by about 1.8, not a significant factor. A need for additional energy could be met by deeper drilling or by expanding beyond the 5-percent figure the area tapped for geothermal energy. And development of other parts of the eastern and mid-continent United States would substantially expand these figures.

Price–Size Relationship

The key to understanding the extent and the importance of the U.S. geothermal potential is to relate the resource size of the various geothermal field types to the anticipated thermal-energy prices for plants exploiting these types. Geothermal plants such as those currently in use at The Geysers can be constructed for about $120 per kw of generating capacity, and it may be expected that this cost will rise to about $140–160 per kw in the next 5 years. Hot-water geothermal-generating systems using heat-exchangers will probably cost $200–220 per kw for generation facilities over the next 5 years. In contrast, nuclear units cost about $400–600 per kw installed, and coal-fired fossil-fuel plants with pollution-control equipment will run to $300–400 per kw over the same time period. It is evident, therefore, that solving electricity shortages by geothermal-energy development would place the lightest capital requirements on the U.S. economy if the energy price is competitive.

Fixed capital costs and annual operating expenses exclusive of fuel costs for a geothermal plant amount to approximately 2.2–3.3 mill/kwhr, depending on the type of geothermal resource being produced and on the usage or "load-factor" for each geothermal plant. The fixed costs include construction costs expressed on an annual basis as return of capital, depreciation, taxes on income, property taxes, and insurance. Other costs included in the 2.2–3.3 mill/kwhr range are operating costs, maintenance costs, and general expenses. The remainder of the cost of geothermally generated electricity is the "fuel" or energy price, which represents the fair market price for a risk-taking energy company in the private sector that explores for, drills, tests, and produces the steam or hot water and delivers it through a system of gathering lines to the utility.

We have made an estimate of the U.S. geothermal potential as a function of energy price and graded the estimates with respect to *known reserves*, *probable reserves*, and *undiscovered reserves* (see Table 2). Our estimate for probable and undiscovered reserves is based on our geothermal-exploration experience in various areas of the United States. The development of exploration technology over the past several years will undoubtedly lead to expanded exploration efforts. The earlier technology was developed in an environment of only limited interest in this vast energy resource. There have been substantial differences of opinion on how present technical knowledge should be extrapolated into unknown areas. For this reason and others a major effort is needed to estimate the U.S. geothermal potential that is available with present technology and to develop the technology for currently marginal geothermal resources. Our studies are based on present technology and research, current costs and projections, and historical rates of inflation.

The unit used in our appraisal is the megawatt-century—a megawatt of electrical energy produced for a century. For comparison purposes consider that current U.S. capacity is about 340,000 Mw. Adding the costs associated with generation (approximately 2.2 to 3.3 mill/kwhr) to the energy price given in Table 2 we can see that power at a cost of 5.10 to 6.30 mill/kwhr at the generating station is presently available from geothermal plants at The Geysers area of California. *We estimate that 400,000 Mw capacity with a projected life of a century could be*

TABLE 2

Amount of Producible Geothermal Energy in the United States

(Mwcen of electricity)

Energy price (mill/kwhr)[a]	Known reserves		Probable reserves		Undiscovered	
	Amount	Areas	Amount	Areas	Amount	Areas
2.90– 3.00	1,000	1	5,000	1	10,000	1
3.00– 4.00	30,000	1–2	400,000	1–4	2,000,000	1–5
4.00– 5.00	—	—	600,000	1–6	12,000,000	1–7
5.00– 8.00	—	—	—	—	>20,000,000[b]	[d]
8.00–12.00	—	—	—	—	>40,000,000[c]	[d]

AREAS: 1, Clear Lake–The Geysers; 2, Imperial Valley; 3, Jemez area, N.M.; 4, Long Valley, Calif.; 5, remainder of Basin and Range area of western U.S.; 6, Hawaii; 7, Alaska.

[a] In 1972 dollars.

[b] Hot, dry rock at less than 6.1 km (20,000 ft) depth.

[c] Hot, dry rock at less than 10.7 km (35,000 ft) depth.

[d] Development of hot, dry rock energy is assumed over 5 percent of the area of the western third of the U.S. Hot, dry rock systems development is based on hydraulic fracturing or cost-equivalent technology. Present drilling technology is assumed; new low-cost deep drilling could substantially improve these economics.

discovered and developed in the western United States in 20 years by the resource industry. This assumes that individual geothermal reservoirs are managed to be depleted in 100 years and that reservoir-pressure maintenance and water-reinjection programs are used whenever needed. And we must also respect the possibility that the utility industry may prove unable to keep pace with the resource industry.

The various economic assumptions used in calculating the steam price are given in the discussion of Table 2. In summary, these calculations involve continuance of present tax treatment, including maintenance of the present percentage-depletion allowance of 22 percent, present corporate and state tax structures and rates, a 5-percent annual rate of cost and price inflation, a reasonable and steady pace of generation-facility construction to match the drilling activity, and a reasonable return on investment to the venture operator. It is clearly inferable from Table 2 that geothermal energy should be capable of supplying a major portion of future U.S. electric-energy requirements. It should be noted that its development would involve costs lower than those for competing energy systems.

The environmental problems attendant upon the use of geothermal energy are primarily *housekeeping* problems for which technology either is now available or can be developed over the next decade if needed. All these problems appear to be technically feasible of solution, but there will be additional development costs associated with the individual solutions. A projection of these costs is included in the financial analysis described in the discussion of Table 2. Housekeeping problems include odor, noise, salt-spray emission, residual-waste recycling, and subsidence. It is unlikely that geothermal plants will be developed in high-population-density areas or National Parks, or in areas of particular environmental concern. But the size of the resource is so large that if Federal lands are released for geothermal development there will be many alternative sites for development.

Discussion of the Estimates

Several computer models were utilized to determine how sensitive the profitability—and, therefore, the economic feasibility—of developing the geothermal resource would be to variations in each cost element. Approximately 1,200 test cases were run for this analysis. The analysis covered the three types of geothermal reservoirs: vapor-dominated, hot-water, and hot, dry rock systems. However, the evaluation of hot, dry rock reservoir systems was based solely on the process of hydraulic fracturing; we did not attempt a financial analysis of steam production from nuclear-explosion-fractured hot, dry rock systems.

It is assumed that each development venture is undertaken by a risk-oriented energy company with private capital provided by the money markets. The venture operator is assumed to be a tax-paying entity with alternative sources of taxable income that will benefit from tax credits resulting from the large outlay for drilling and exploration during the 5 to 6 years that precede initial revenues from the sale of steam. The venture operator is rewarded with a 20-percent return on invested capital in order to compensate for the initial delay of revenue and for his accepting the investment risks associated with geothermal exploration. Return on investment is computed as that rate at which net cash flow after tax, less working capital, is discounted to the present (1972) to offset the discounted investment stream. Once reservoir performance has been proved by exploratory drilling and initial development wells, additional development drilling is funded, we assume, with capital from the private money markets in 10-year notes bearing 8.5 percent interest per annum. Each venture is evaluated over a 30-year period, and it is assumed that there is no salvage value to wellhead equipment or casing at the end of the thirtieth year. Steam-gathering lines are assumed to run $12 per kw-rating for each development well and each successful exploratory well. The landowner's royalty is 10 percent of gross revenue from the sale of steam. The load factor is, for vapor-dominated systems, 90 percent; for hot-water systems, 85 percent; and for hot, dry rock systems, 80–85 percent. Drilling costs are given in Table 3; and drilling depths for hot, dry rock systems in Table 4. Drill-

TABLE 3
Drilling Costs for Wells of Different Depths

Depth range (ft)	Cost range ($)
6,000–8,000	$300,000–520,000
8,000–10,000	425,000–770,000
10,000–15,000	635,000–1,055,000
15,000–20,000	940,000–2,750,000

SOURCE: Data compiled by K. E. Brunot, now with Phillips Petroleum Co.

NOTE: Costs are tangible and intangible, for exploratory, development, and reinjection wells. Add approximately $100,000 for each exploratory well. In hot-water systems, one reinjection well is drilled for each development well and each successful exploratory well.

TABLE 4
Costs at Different Drilling Depths for Hot, Dry Rock Systems

Energy price range (mill/kwhr)	Average well depths (ft)		
	Western U.S.	Mid-continent	Eastern U.S.
3.00– 4.00	14,000–16,000	—	—
4.00– 5.00	16,000–17,000	—	—
5.00– 8.00	17,000–20,000	15,000–17,000	—
8.00–12.00	20,000–30,000	17,000–22,000	15,000–18,000

NOTE: Calculations for mid-continent and eastern U.S. are not included in the resource estimate in Table 2.

ing-success ratios are, for exploratory wells, 40 percent; and, for development wells, 85 percent. All expenses increase at the rate of 5 percent per annum for the next 30 years; the sales price of the steam (energy price) increases at the same rate. Federal income-tax rate does not increase; state income taxes are at a level no higher than the California corporate rate (7.6 percent); depletion allowance is applicable to all three types of reservoir systems and remains at the present rate (22 percent of gross income, but not greater than 50 percent of net income); intangible drilling costs are expensed rather than capitalized; there are no severance taxes; and depreciation is straight-line over 14 years.

Some additional factors included in the analyses are geology, geophysics, engineering, maintenance, and operating expenses; overhead (fixed and variable); general and administrative expenses; and research, environmental-monitoring, and well-testing expenses. Construction of the generation facilities is completed and the field is in production within 3 years from completion of development drilling.

The reader should keep in mind that the energy prices given in Table 2 are listed in 1972 dollars and are assumed to escalate at 5 percent per annum. This facilitates comparing geothermal resources with the existing alternative "fuels" that are being utilized today in the United States.

Technology Needed

The key to the development of hot, dry rock systems is the technology of rock fracturing. A major program of fracturing research is needed to test present hydrofracturing technology and to develop additional, more effective methods.

Also needed is a national program of deep geothermal drilling in the range of 20,000 to 30,000 ft depth, with holes fairly evenly distributed across the United States, to appraise the total U.S. deep potential. This deep geothermal well-drilling program would provide data not available from oil wells and would require a substantial amount of new high-temperature instrumentation and technology.

A third area in need of research and development is the power-conversion technology for low-temperature hot-water geothermal systems. This would be needed to recover the energy from the hot, dry rock systems as well as from the numerous natural low-temperature hot-water geothermal systems found in much of the United States.

Recommendations

Four recommendations will be made:

1. The private sector should finance and develop dry-steam and high-temperature hot-water geothermal systems. (High temperatures are defined as those sufficient to yield 20-percent steam on flashing down to 100 psi pressure, or temperatures above 260°C/500°F.) Federal efforts should be focused on basic, environmental, institutional, desalination, and deep-basin research.
2. The Federal Government should fund a program on hot, dry rock geothermal systems with emphasis on engineering and field tests.
3. The Federal Government should fund a national deep geothermal drilling program to appraise the total U.S. geothermal energy potential for national energy planning.
4. The Federal Government should fund a program of low-temperature, hot-water, energy-recovery research, with research both on limiting technological factors and on demonstration plants to determine system economics and environmental influence. This program should also encompass bottoming-cycle research for fossil-fuel plants, which offers the possibility of eliminating thermal pollution by fossil-fuel-powered generating plants.

The Federal money invested would be returned rapidly and many times over by increases in the value of the massive Federal geothermal reserves on public lands.

REFERENCE

Von Herzen, R. P. 1967. Surface heat flow and some implications for the mantle. *In* T. E. Gaskell, ed., The Earth's mantle. New York: Academic Press, pp. 197–230.

4. Characteristics of Geothermal Resources

DONALD E. WHITE

Geothermal resources derive from the distribution of temperatures and thermal energy beneath the Earth's surface. Present-day technology emphasizes the production of electricity from geothermal steam; requirements include reservoirs of high temperature (at least 180°C, preferably above 200°C), depths of less than 3 km, natural fluids for transferring the heat to the surface and power plant, adequate reservoir volume (>5 km^3), sufficient reservoir permeability to ensure sustained delivery of fluids to wells at adequate rates, and no major unsolved problems. Because this configuration of characteristics occurs only rarely in the Earth's crust, other means of exploiting geothermal heat must be developed if the resource is to become much more than a curiosity.

The major known geothermal systems of the world are shown in Fig. 1. Table 1 summarizes the existing capacity of the major producing fields. These hydrothermal-convection systems, whether dominated by vapor or by hot water, are generally associated with tectonic-plate boundaries and volcanic activity. Fields along the belt of mountains extending from the Mediterranean eastward through Turkey to the Caucasus are related to the complex zone of collision between the Eurasian and African continents. Detailed plate-boundary relations of most geothermal areas of the western United States and central Mexico are not yet clearly established, but these areas seem to be associated with recent volcanism, high regional conductive heat flow, and relatively shallow depths to the mantle.

Donald E. White is Research Geologist with the U.S. Geological Survey, Menlo Park, California.

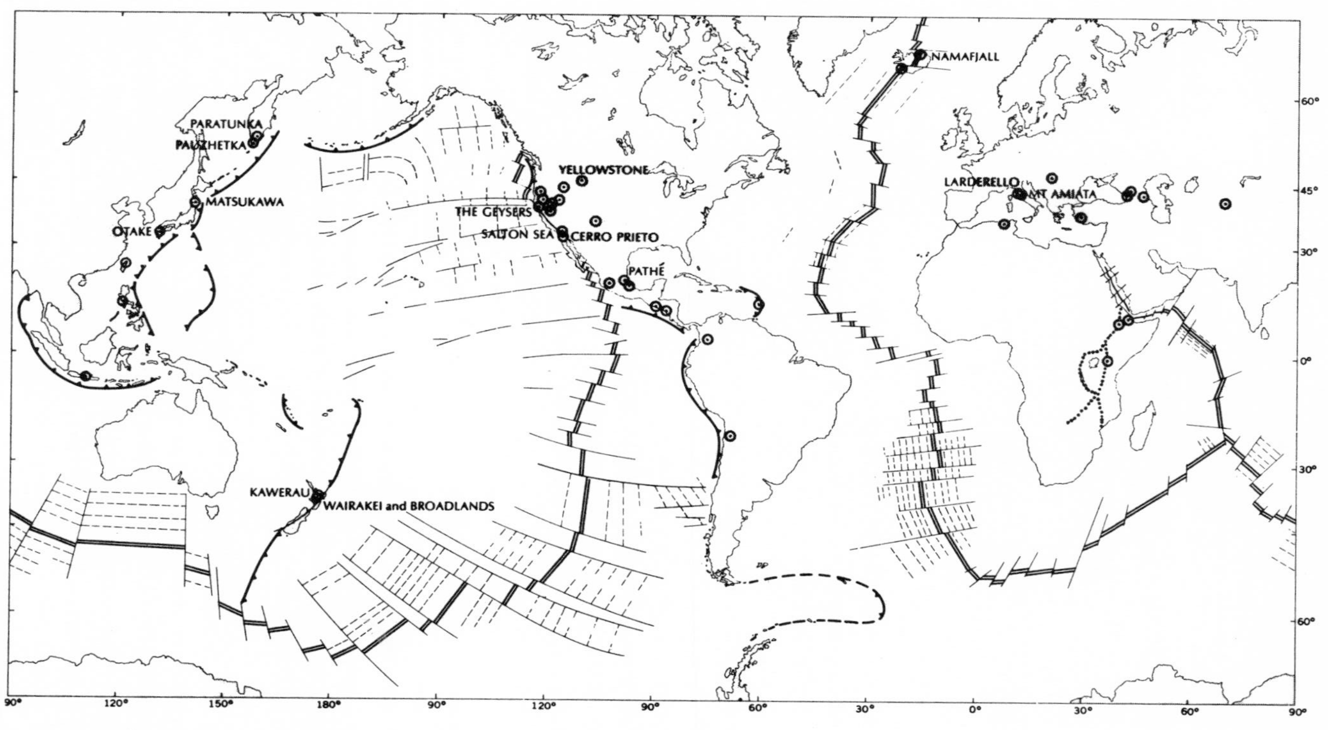

Fig. 1. The major known geothermal systems of the world. Named systems (other than Yellowstone) are those that are presently generating electricity or have power plants under construction (Muffler and White, 1972). Most fields are on spreading ridges (double lines) or rift valleys (heavy dotted lines) or are above subduction zones at plate boundaries (heavy barbed lines).

TABLE 1
World Geothermal Power-Generating Capacity, 1972

Country	Field	Electrical capacity, Mw			
		Operating	Under construction	Vapor-dominated systems	Hot-water systems
Italy	Larderello	358.6		358.6	
	Monte Amiata	25.5		25.5	
United States	The Geysers[a]	302	110	412	
New Zealand	Wairakei	160			160
	Kawerau	10			10
Japan	Matsukawa	20		20	
	Otake	13			13
Mexico	Pathé	3.5			3.5
	Cerro Prieto		75		75
Soviet Union	Pauzhetsk	5			5
	Paratunka	0.7			0.7
Iceland	Namafjall	2.5			2.5
TOTAL		900.8	185	816.1	269.7

SOURCE: Muffler and White, 1972. Modified for present purposes.
[a] As of November 1972, additional capacity of 110 Mw is scheduled to be completed in 1973 and another 110 Mw (not listed) in 1974.

Near many hot-spring areas of the world, geothermal waters are used for space heating, horticulture, industrial processes, and spas (Muffler, in press). Such uses generally are individually small in scale, with the important exception of sites in Iceland (Pálmason and Zoëga, 1970), Hungary (Boldizsár, 1970), and Japan, but they demonstrate the potential of such sites for more extensive utilization along these varied lines.

Future technological developments may greatly change the requirements and uses summarized above. The most useful of the possible "breakthroughs" are itemized toward the conclusion of this paper and elsewhere in this volume.

Conductive Thermal Gradients

Temperatures below the Earth's surface are controlled principally by conductive flow of heat through solid rocks, by convective flow in circulating fluids, or by mass transfer in magma. Other modes of heat transfer are minor and are hereafter ignored. Moreover, transfer in magma is considered only through its effects on conduction and hydrothermal convection.

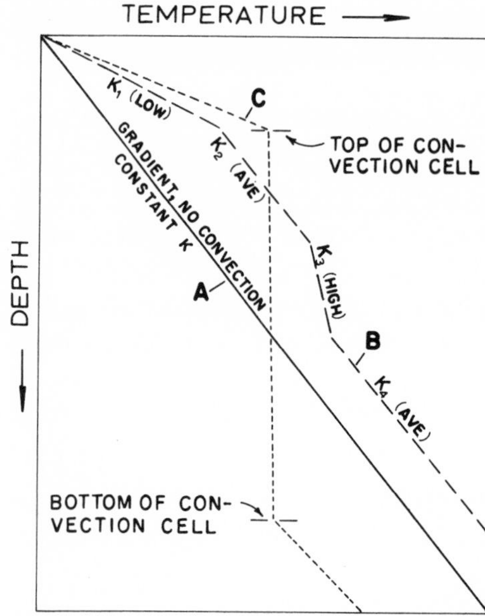

Fig. 2. Temperature/depth relations, where heat flow is controlled by thermal conduction in rocks of constant conductivity (A), or rocks of variable conductivity (B), or by major convective disturbance (C).

Conduction is the dominant mode of heat flow in most of the outer crust of the Earth. Where conduction is dominant, temperatures increase continuously with depth, but not at constant gradient. The important interrelations are those between thermal gradient, heat flow, and thermal conductivity of rocks, and the appropriate expression for the relationship is Fourier's law, $r = q/K$, where thermal gradient (r) is expressed in °C/km, heat flow (q) is in μcal/cm^2 sec, and thermal conductivity (K) is in mcal/cm sec °C. Thus, a measured thermal gradient is directly proportional to heat flow but inversely proportional to conductivity. Heat flow is the most fundamental parameter but ordinarily must be calculated from gradient and conductivity because, at low levels, heat flow cannot be accurately measured by any direct method.

In an area dominated by conduction and free from significant convective disturbances, heat flow is relatively constant in time and space, but conductivity of rocks differs greatly with depth as functions of mineralogy, porosity, and fluid content of pores. Therefore, temperature gradients may change greatly with depth, as in curve B of Fig. 2, which shows the effects of different thermal conductivities on thermal gradients, as compared with rocks of constant conductivity (curve A). A near-surface gradient cannot be reliably projected downward below explored depths because of likely changes in porosity and thermal conductivity with rock type and, especially, until possible convective influences can be evaluated.

Areas of near-"normal" conductive thermal gradient. The worldwide average heat flow is about 1.5 μcal/cm² sec (Lee and Uyeda, 1965; Sass, 1971), or 1.5 hfu (geothermal heat-flow units). This is about 1/2,000 of average solar energy at the Earth's surface, a very small quantity but an important one. For present purposes, we shall consider "normal" heat flow as ranging from 0.8 to 2.0 hfu. The thermal conductivities of most rocks range from 4 to 10 mcal/cm sec °C. Within these limits, temperatures can increase from 8° to 50°C/km (lines A and B, Fig. 3), averaging about 25°C/km (line C) or a bit more. At 3-km

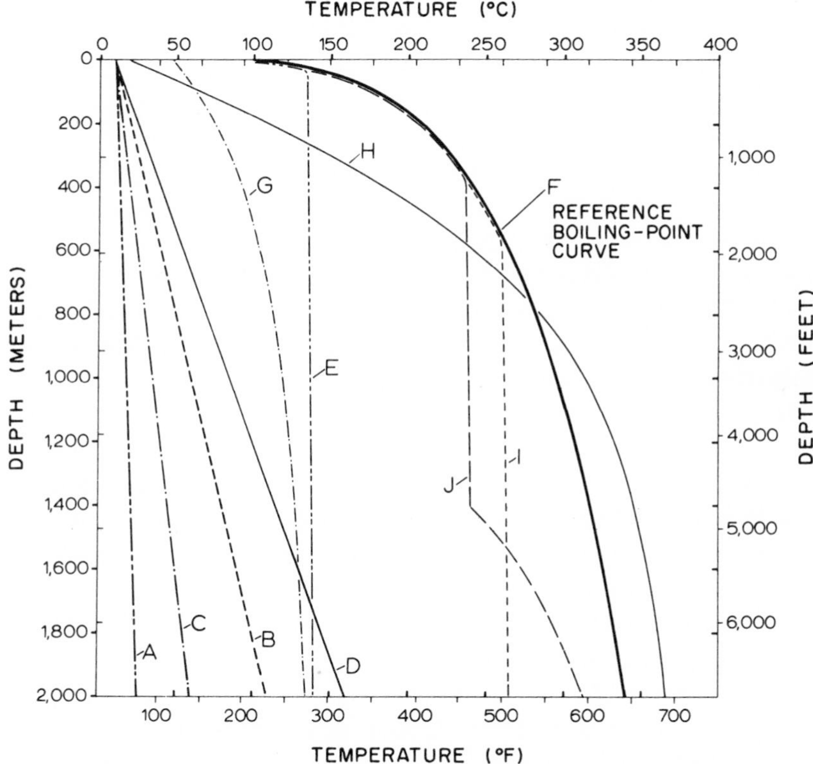

Fig. 3. Temperature profiles controlled by conductive gradients within the "normal" range (A, B, C) and above "normal" (D) and by convective transfer of heat in hot-water systems of different subtypes, compared with reference boiling-point curve (see text).

depth, with such gradients, temperatures range from 24° to 150°C above surface temperatures and average about 75°C. Most "normal" areas are not attractive for commercial geothermal exploitation, either now or in the immediate future, and their stored heat should not be considered as usable resources, much less as reserves recoverable under present conditions. Most, if not all, of this "low-grade" heat is as far removed from likely recovery as the trace contents of gold, copper, or uranium in average crustal rocks. Some areas may have a combination of heat flow in the higher part of the "normal" range and conductivities in the lower part of "normal"; such an area is the Gulf Coast of the U.S. (Jones, 1970), where gradients range up to 45°C/km or a bit higher (close to line B of Fig. 3). Such areas may prove to be exceptions warranting further study and evaluation, especially where existing oil and gas wells are already available for research utilization.

Areas of abnormally high conductive thermal gradient. Abnormally high thermal gradients result from unusually high heat flow, unusually low thermal conductivity, or favorable combinations of the two factors. In some large favorable areas, such as the Hungarian Basin (Boldizsár, 1970), conductive gradients range from 40° to 75°C/km (line D, Fig. 3) and perhaps locally even higher. Several rather large areas of high heat flow that seem unrelated to convection systems are now known in the United States. The "Battle Mountain High" of Sass et al. (1971) has an indicated average heat flow of about 3 hfu. But its thermal gradients, which range from 30° to 60°C/km, are not as high as would be expected, because of its relatively high thermal conductivities (~9 mcal/cm sec °C). An area near Marysville, Montana (Blackwell, 1969) has even higher heat flow (~7 hfu); rock conductivities are also high, and measured temperature gradients average about 75°C/km (line D, Fig. 3). Both areas may be related to large igneous intrusions. Another large area that is likely to have high conductive heat flow combined with somewhat lower, favorable conductivities is the region surrounding Clear Lake, California, where conductive gradients as high as 100°C/km would not be surprising.

Hydrothermal-Convection Systems

In hydrothermal-convection systems, most heat is transferred in circulating fluids rather than by conduction. Convection occurs because of the heating and consequent thermal expansion of fluids in a gravity field; heat, which is supplied at the base of the circulation system, is the energy that drives the system. Heated fluid of low density tends to rise and to be replaced by cooler fluid of higher density, which is supplied from the margins of the system. Convection, by its nature, tends to increase temperatures in the upper part of a system as temperatures in the lower part decrease. Moreover, convection (curve C, Fig. 2) obviously disturbs the pure conductive gradients (such as curves A and B, Fig. 2) that would otherwise obtain. Thus no single temperature gradient or heat flow can characterize a convection system. Gradients are commonly very high near the surface, and locally exceed 3°C/meter of depth; such a gradient, projected, exceeds 3,000°C at 1 km and is impossibly high, greatly exceeding the melting temperatures of all normal rocks (700° to 1,200°C). Where tested by drilling, temperature gradients in convection systems have been shown to decrease greatly with depth until the characteristic base temperature of the circulation system is attained. Locally, temperature reversals may occur.

Two major types of hydrothermal convection systems are recognized, differing in the physical state of the dominant pressure-controlling phase.

Hot-Water Systems

Hot-water systems are characterized by liquid water as the continuous, pressure-controlling fluid phase (White, Muffler, and Truesdell, 1971; White, 1970). Some vapor may be present, generally as discrete bubbles in the shallow, low-pressure zones. Continuity of liquid can be inferred with confidence from the distribution of pressures and from the abundance of constituents that are soluble in liquid water but have low vapor pressures and lack significant solubility in low-pressure steam. These include most of the constituents of ordinary water analyses, such as SiO_2, Na,* K, Ca, Mg, Cl, SO_4, HCO_3, and CO_3 (though B, CO_2, H_2S, and NH_3 are both volatile and soluble in water, and thus are not diagnostic).

Water in a major water-convection system (Fig. 4) serves as the medium by which heat is transferred from deep sources to a geothermal reservoir at shallower depths—shallow enough, perhaps, to be tapped by drill holes. Cool rainwater percolates underground from sur-

* The electrical charge of ionized constituents is not specified unless important to the discussion.

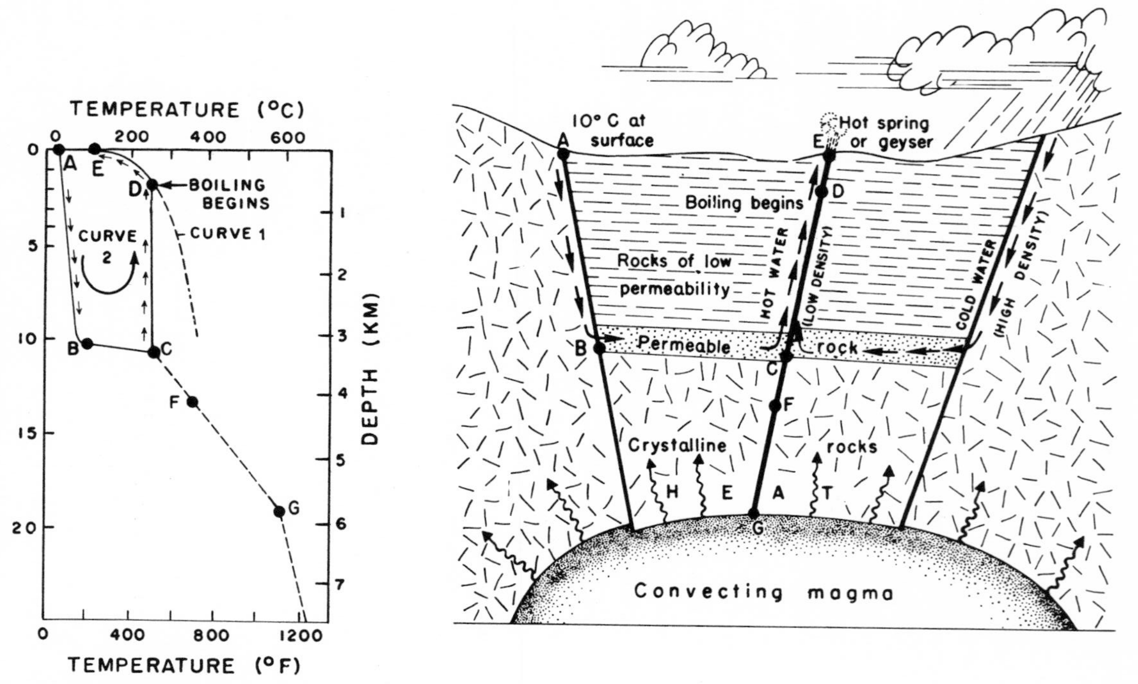

Fig. 4. Model of a high-temperature hot-water geothermal system. Curve 1 is the reference curve for the boiling point of pure water. Curve 2 shows the temperature profile along a typical circulation route from recharge at point A to discharge at point E.

face areas ranging across tens to possibly thousands of square kilometers, and then circulates downward. At depths of 2 to 6 km, the water is heated by conduction from hot rocks that, in turn, are probably heated by molten rock. The water expands upon heating and then moves buoyantly upward in a column of relatively restricted cross-sectional area (1 to 50 km^2). If the rocks have many interconnected pores or fractures of high permeability, the heated water rises rapidly to the surface and is dissipated rather than stored. However, if the upward movement of heated water is impeded by rocks with few interconnected pores or fractures, geothermal energy may be stored in reservoir rocks below the impeding layers. Heat, of course, accounts for the density difference between cold, downward-moving recharge water and hot, upwelling geothermal water.

Some subtypes. Hot-water systems actually include many subtypes that are not yet universally accepted or precisely defined. Different classifications can be based on total salinity, dominant chemical characteristics, temperature range, structural and stratigraphic environments, presence or absence of permeable reservoirs, and insulating cap-rocks. I shall not attempt systematic classification by any single system, but some subtypes of particular interest to geothermal exploration to date include:

1. Systems characterized by low to moderate temperatures, generally ranging from about 50° to 125°C in most cases but reaching as high as 150°C in Iceland (Bodvarsson, 1964); and by chemical similarity to surface and shallow ground waters of the region. Some systems in the higher part of this temperature range may be characterized by impressive boiling springs of high discharge. A drill-hole into such a system is likely to show a temperature profile similar to that of curve E of Fig. 3. This curve assumes rapid upflow of liquid water under pressures that are close to hydrostatic. Pressures are too high for boiling to occur at existing temperatures except in the upper 20 or 30 m, where curve E impinges upon curve F. The latter curve, a significant and useful one for many geothermal considerations, describes the calculated temperatures for boiling of pure water at pressures controlled by liquid densities up to the ground surface; the water densities are corrected for the assumed temperatures. Actual temperature profiles in such systems depend to a large extent on their deep temperatures and their rates of upflow. With a high upflow rate (as assumed for Curve E)

little heat is lost by conduction from the margins of a system, but with a low upflow rate (as for curve G), a higher proportion of the heat contained in the water is lost by conduction; with decreasing rates of upflow, such systems grade into purely conductive environments.

2. Systems in deep sedimentary basins, commonly bearing saline waters of moderate temperature similar to oilfield waters. These waters are, at least in part, nonmeteoric in origin (White, 1967b; White, Barnes, and O'Neil, 1973). Some brine systems, such as Wilbur Springs in California, are characterized by springs that discharge thermal water, but convective circulation in most deep basins is so inhibited (by low permeability, salinity gradients, or low temperature contrasts) that conductive thermal gradients are only slightly disturbed, and thermal discharge is lacking. Hydraulic gradients, however, commonly provide some evidence for subsurface circulation.

3. Hot-water systems known to contain brines of very high salinity. The Salton Sea geothermal system and the Red Sea brine pools (Craig, 1966b; White, 1968a; Helgeson, 1968; Muffler and White, 1969; Ross, 1972) both contain brines of about 26 percent salinity. But the two systems differ markedly: temperature relations and the bulk chemistry of the associated rocks and sediments differ greatly between these two systems, probably accounting for major differences in composition of the brines. Because of the effect of salinity on boiling (Haas, 1971), the temperatures deep in the Salton Sea system (curve H of Fig. 3, from Helgeson, 1968) are considerably above the reference curve for pure water.

4. Systems with natural cap-rocks that tend to inhibit discharge and also to insulate their reservoirs, thus conserving heat. The Salton Sea and Cerro Prieto systems of California and Mexico have cap-rocks of low-permeability, fine-grained sediments. This configuration largely explains the conductive-gradient dominance of the low near-surface temperatures of curve H (Fig. 3) from the Salton Sea system.

5. High-temperature hot-water convection systems that tend to create their own insulating cap-rocks by "self-sealing"; hydrothermal minerals are deposited in pore spaces, especially in near-surface parts where temperatures decrease abruptly upward because of the influence of the boiling-point curve. Figure 4 and curve I of Fig. 3 are idealized from Wairakei, New Zealand, and the geyser basins of Yellowstone Park. Because of thermo-artesian pressure and channels restricted by

self-sealing, actual near-surface pressure gradients may exceed hydrostatic, as in Yellowstone Park (White et al., 1968), and temperatures may plot somewhat above reference curve F.

General characteristics. The principal characteristics of hot-water systems have been discussed by White, Muffler, and Truesdell (1971) and White (1970) and may be summarized as follows:

1. Hot springs are a common but not universal indication of hot-water convection systems. Where the water table is at or very near the ground surface, all or most discharge from a system is visible as hot springs, but where near-surface rocks are permeable and the water table is low, as much as 100 percent of discharge may be subsurface, and therefore dispersed into surrounding ground water and not directly evident at the surface.

2. Springs that are highest in temperature and discharge are generally also highest in SiO_2, Cl, B, Na, K, Li, Rb, Cs, and As, relative to surrounding ground waters. An analysis of the SiO_2 content and the Na-K-Ca ratios of such springs (Fournier and Truesdell, 1970 [1972]; 1973) is generally the best means of evaluating subsurface temperatures.

3. A "base" temperature (Bodvarsson, 1964, 1970) characterizes the deeper parts of many hot-water convection systems. The zone of little temperature change between points C and D of Fig. 4 represents the base temperature of the Wairakei system, which was about 260°C prior to exploitation (Banwell et al., 1957). The base temperatures of some other systems with moderate to low salinity (<5,000 ppm total dissolved solids) are as high as 300°C, but no such systems with substantially higher temperatures are yet known.

4. The insulated Salton Sea brine system is as hot as 360°C (curve H of Fig. 3, from Helgeson, 1968), and the Cerro Prieto system of Baja California (with about two-thirds the salinity of sea water) may be as hot as 388°C (Mercado, 1970); such temperatures, which are near or above the critical temperature of pure water (373°C), can occur in brines because their physical properties differ from pure water (Sourirajan and Kennedy, 1962; Haas, 1971).

5. SiO_2 is the most important self-sealant of hot-water systems. Quartz and chalcedony are generally dominant at temperatures above about 140°C, but opal and β-cristobalite are characteristic of low-temperature margins and self-sealed cap-rocks (White, Brannock, and

Murata, 1956; Honda and Muffler, 1970); zeolites, clay minerals, and calcite may also be important.

6. Natural geysers and amorphous or recrystallized SiO_2 deposited at the surface by flowing hot water imply upflow of subsurface waters with base temperatures of 180°C or higher (White, 1970; Fournier, 1972). Travertine ($CaCO_3$ from hot-spring waters), by contrast, implies low subsurface temperatures (or, more rarely, solution of limestone after temperatures have decreased nearly to surface temperatures).

7. Low-temperature convection systems have little potential for self-sealing because their waters are not sufficiently high in SiO_2. In fact, systems with maximum temperatures below about 150°C may in general become *more* permeable with time because as much as 140 ppm of SiO_2 is dissolved during the heating of cold meteoric water of low SiO_2 content. A temperature of 150°C is high enough to increase the porosity of quartz-bearing reservoir rocks but may not be high enough to be offset by hydrated alteration minerals, which tend to decrease porosity. This may explain why some systems of moderate temperature tend to discharge from large, impressive single springs, with no evidence of self-sealing with time (as for curve E, Fig. 3).

8. In contrast, high-temperature systems with maximum temperatures above about 180°C tend to decrease in permeability in their upper parts with time because of self-sealing. Perfect sealing may seldom occur; leakage takes place through any available permeable channels. New channels may form and old channels may be reopened by tectonic forces or by thermo-artesian buildup of vertical pressure gradients. Local pressure gradients may greatly exceed hydrostatic. Water-pressure gradients in the upper few hundred meters of research drillings in Yellowstone Park averaged about 20 percent above hydrostatic, and the excess in one drill hole was nearly 40 percent (White et al., 1968). Such high gradients are probably rare, and may tend to be localized in the shallow parts of convection systems.

9. Wells in permeable reservoirs generally produce 70 to 90 percent of total mass flow as water; the proportion of steam that forms when pressure is reduced is related to initial fluid temperature and to final separating pressure (Muffler, in press, Fig. 1). For example, water flashed to separator pressure of 50 psig from an initial temperature of

300°C yields 33 percent steam; 200°C yields 11 percent steam, but 150°C yields none!

10. Wells in ground of low permeability, however, may first erupt water and steam, which may then change to wet steam and finally to dry steam, in the manner of the eruption stages of some geysers (White, 1967a). The increasing steam content may seem to be a favorable characteristic, but it is not; the stored heat of the reservoir rocks is not only evaporating all local water, thereby temporarily producing steam, but also precipitating dissolved matter, thereby decreasing permeability.

11. In most hot-water fields, the only fluid that enters a producing well is liquid water. This remains entirely liquid as it flows up the well, until pressure decreases enough for steam bubbles to start forming. With continued flow upward, more water "flash boils" to steam (as pressure and mixture temperature decrease). The buoyancy of the expanding steam displaces the remaining water upward, thereby increasing the velocity of the mixture and ejecting the residual water above the ground surface (unless diverted horizontally in pipes or separator), just as in natural geysers. The depth of first boiling in the well depends mainly on the initial temperature of the water, but also on formation-fluid pressure and separator pressure. Crude depths of first boiling can be calculated, but the principles are rather complex. A better method, where adequate equipment is available, is to measure in-hole pressure gradients under producing conditions; a break in slope of the pressure gradient indicates the depth of first boiling (where water flashes to steam).

12. The chloride content of water that has been above about 150°C is nearly always higher than 150 ppm. However, a very few hot-water systems with Cl content as low as 40 ppm have temperatures above 200°C.

13. Chloride is the most critical single constituent in distinguishing hot-water systems from vapor-dominated systems. Most metal chlorides are highly soluble in liquid water, and the chloride of most rocks is easily leached by water at high temperature (Ellis and Mahon, 1967). The common metal chlorides, however, have negligible volatility at temperatures as high as 400°C (Krauskopf, 1964) and are not appreciably soluble in low-pressure steam. Thus a chloride-bearing water

body definitely indicates a hot-water system (Cl > 50 ppm). Some hot-water systems, however, do produce surficial acid-sulfate springs that are low in chloride; such springs are sustained by steam boiling from an underlying chloride-bearing water body, but are otherwise chemically similar to springs associated with vapor-dominated systems.

Vapor-Dominated Systems

A few geothermal systems, including the important Larderello fields of Italy and The Geysers of California, produce dry or superheated steam with no associated liquid. For this reason they are commonly known as "dry-steam" systems, but White, Muffler, and Truesdell (1971) conclude that liquid water and vapor normally coexist in the reservoirs, with vapor as the continuous, pressure-controlling phase. Thus "vapor-dominated systems" seems to be a more appropriate term.

Opinions on the physical nature of the initial fluid(s) in the Larderello fields include initial saturation with liquid (Facca and Tonani, 1964; Marinelli, 1969) and superheated steam in a reservoir that, with production, is replenished by boiling from a deep water body (Elder, 1965; James, 1968) that may be a brine (Craig, 1966a). Sestini (1970) favors a reservoir filled largely with vapor but locally containing "disturbance water" (p. 637) that evidently has no significant influence on subsurface pressures. According to White, Muffler, and Truesdell (1971), vapor-dominated systems of the Larderello type develop initially from hot-water systems characterized by very high heat supply and very low rates of recharge; if and when the heat supply of a developing system becomes great enough to boil off more water than is being replaced by recharge, a vapor-dominated reservoir begins to form (in known systems, probably thousands to tens of thousands of years ago). The fraction of discharged fluids that exceeds recharge is supplied by water previously stored in the larger fractures and pores. Some liquid water is held in the smaller pores and on fracture surfaces by surface tension, and some additional water is no doubt locally retained in closed pores and various open spaces that terminate downward. Conductive heat losses from the margins of a reservoir, moreover, result in condensation of steam. And, finally, liquid not retained in all these ways drains downward under gravity to deeper water-saturated rocks. Vapor, however, is the continuous, pressure-controlling phase in large pores and open channels.

According to this model, a deep water table, perhaps crudely horizontal, separates an underlying zone, which is dominated by liquid, from the overlying zone, where pressures are controlled by the vapor phase. This deep water table continues to decline as long as the fluid discharged (mainly as steam) exceeds the recharge; and at the same time the reservoir of steam continues to develop. Only a part of the steam from the deep water table is discharged at the surface as vapor; probably most of the total upflow recondenses to liquid on the margins of the reservoir, where heat can be lost by conduction, as mentioned above. A conductive heat flow of 20 hfu from the margin of the reservoir to the surface, for example, requires the heat from 29 kg condensing steam per km^2 of surface area per second (White, 1970). Conductive heat flow through the reservoir cannot supply this much heat because of the reservoir's nearly constant temperature; the heat of vaporization of steam provides the only reasonable explanation for the phenomenon. An important implication of this conclusion is the downward draining of condensate through the reservoir, ensuring the coexistence of liquid and vapor in the natural systems prior to exploitation.

Two subtypes of the vapor-dominated system, the Larderello and the Monte Amiata, appear to be distinguishable.

Larderello subtype. The physical, chemical, and geologic characteristics of The Geysers, Larderello, and Matsukawa vapor-dominated systems, as known to date (White, Muffler, and Truesdell, 1971; White, 1970; Nakamura et al., 1970), are consistent with the model just described and may be summarized as follows:

1. Reservoirs occurring at or below about 350 m in depth tend to have *initial* temperatures near 240°C (curve J, Fig. 3) and pressures near 35 kg/cm^2. The published data for Larderello and Matsukawa are convincing (Penta, 1959; Burgassi, 1964; Sestini, 1970; Nakamura et al., 1970), but the fragmentary data from the deep reservoir of The Geysers system are less conclusive (Otte and Dondanville, 1968). Little initial physical difference is yet recorded for holes drilled at greater depths (changes in Larderello with exploitation are considered in item 9).

2. The relatively uniform initial temperatures and pressures are evidently strongly influenced by the maximum enthalpy of saturated steam (669.7 cal/g at 236°C and 31.8 kg/cm^2: James, 1968; White, Muffler, and Truesdell, 1971; Sestini, 1970, p. 625). As the gas content

of the vapor increases above a few percent, these physical characteristics change greatly (White, Muffler, and Truesdell, 1971). For example, at a constant temperature of 236°C for coexisting liquid and vapor, 1 percent of other gases in the vapor increases the total pressure only to 32.1 kg/cm^2; but corresponding pressure for 5 percent of other gases is 33.5 kg/cm^2, and that for 10 percent is 35.3 kg/cm^2.

3. Pressures in these vapor-dominated reservoirs are well below hydrostatic and, with few exceptions, the difference increases with depth (Truesdell, White, and Muffler, in preparation). In-hole pressures, of which few details are yet published, increase only slightly with depth because of the low density of the pressure-controlling vapor. Expected changes related to depth alone are given by White, Muffler, and Truesdell (1971, Table 3). Obviously, such a system could not form or persist if the water-saturated rocks that surround the reservoir could supply a high rate of recharge. The water thus supplied would flow into the reservoir under hydrostatic drive at a rate exceeding discharge, and the underpressured reservoir would "collapse."

4. Fumaroles, mud pots, mud volcanoes, turbid pools, and acid-leached ground characterize the discharge areas where surface activity is most intense. Springs in such areas are generally acid because of the H_2SO_4 produced by oxidation of H_2S in the escaping gas; pH's are as low as 2 to 3 except where NH_3 is abundant enough to neutralize the acid. Sulfate contents tend to be high, but Cl contents are uniformly low ($<$15 ppm). Likewise, the springs, streams, and ground water of the immediately surrounding area are low in chloride. Areas lacking intense surface activity are characterized by slightly acid to slightly alkaline bicarbonate-sulfate spring waters that may be high in total CO_2, B, or NH_4, but are low in Cl; some spring waters of such areas are also anomalously high in SiO_2.

5. Where the natural total discharge of fluids from vapor-dominated systems has been observed closely prior to exploitation, the discharge is consistently low, ranging from a few tens to several hundred liters per minute. Detailed descriptions of natural discharge at Larderello prior to initial subsurface exploitation in 1904 apparently do not exist. Sestini (1970), in attempting to evaluate the early records, concludes that "from documentation and other evidence, there is reason to believe that the total flow of water, steam, and gas was in the order of some hundreds of tons per hour." This flow is equivalent to a few thou-

sand liters per minute, expressing all H_2O as liquid, and is probably much too high for discharge solely from the reservoirs, judging from my observations of The Geysers and Matsukawa prior to exploitation, and of virgin systems in Yellowstone Park (Wyoming), Lassen Park (California), Steamboat Springs (Nevada), and Valles Caldera (New Mexico, perhaps a vapor-dominated system). This Larderello flow is *not*, however, unreasonably high if much of the credited discharge consisted of near-surface runoff heated by rising steam. And, in fact, most of the natural thermal discharge of liquid from vapor-dominated systems, especially at The Geysers and Lassen Park, consists of steam-heated surface waters.

6. Production wells normally produce dry to superheated steam (from a few degrees to more than 50°C of superheat); however, liquid water evidently occurs in some noncommercial wells on the borders of reservoirs and in the fluid initially produced from some wells that change from wet steam (i.e. steam containing a little water) to dry steam. Many shut-in wells (i.e. wells that have been capped off to recover earlier output pressure) contain vapor as the only fluid, but Sestini (1970, p. 636) notes that "fairly high-temperature water" flows into some Larderello wells that have been shut in for a while. Sestini calls this "disturbance water," without discussing its origin or characteristics in detail; it presumably is water locally perched on impermeable rocks or contained in downward-terminating fractures or caverns, and is surely *not* a part of a large liquid-dominated water body, since other, deeper drill holes are characterized by initial pressures of 31 to 34 kg/cm^2 (Sestini, 1970, p. 640) that are typical of vapor-dominated systems.

7. Most of the heat content of the reservoir is stored in solid phases (James, 1968; White, Muffler, and Truesdell, 1971; Truesdell, White, and Muffler, in preparation), which generally carry 80 to 90 percent of total heat.

8. Superheated steam forms from saturated steam by flow and decompression through hot rocks already dried by transfer of heat from solid phases to evaporating pore liquid. Critical aspects of these relations are the stored heat of solid phases and the decrease in boiling temperature with decrease in pressure (Truesdell, White, and Muffler, in preparation).

9. With long-term production, most Larderello wells show a rather

steady increase in well-head temperature (as high as 260°C in 1966, with most wells then starting to decline: Sestini, 1970). Enthalpy had increased to as much as 710 cal/g, along with increasing superheat. These gradual changes had been implied in earlier publications but are now clearly substantiated by Sestini, who proposes an increasing dependence on supercritical H_2O (>374°C) from a deep magmatic environment. Probably a more satisfactory explanation consists of increasing dependence on boiling from a deep, declining, saline-water body that becomes increasingly more saline as water is vaporized (Truesdell, White, and Muffler, in preparation). Curve J of Fig. 3 shows a possible distribution of temperatures within a deep brine-water body (assumed 25-percent salinity, starting at a depth of 1,400 m). The maximum enthalpy of steam coexisting with 25 weight percent NaCl brine, according to Haas (1971; written communication, 1972), is 681.5 cal/g at 275°C. Much higher steam temperatures and enthalpies are obtainable as residual salts become more highly concentrated and boiling occurs in environments of much greater initial temperature; Haas's data are computed only to 35 percent NaCl, where coexisting steam has its maximum enthalpy of 688.7 cal/g at about 295°C.

The previous discussion is concerned chiefly with systems such as Larderello, The Geysers, and Matsukawa (White, 1970) that are characterized by initial reservoir temperatures near 240°C, shut-in pressures near 35 kg/cm^2, and contents of gases (other than steam) of about 5 percent or less. My associates and I conclude that discharge areas seem essential for such systems, thus permitting the net loss of much initial pore water to establish domination by the vapor phase and the flushing out of gases other than steam. A large vigorous system is likely to have at least one prominent vent area that cannot be accommodated by discharge into ground water. Under some circumstances (see point 4 in the characteristics just concluded), less vigorous discharge of steam and gases can be accommodated. If CO_2, H_2S, and other gases are not permitted to escape, they are selectively concentrated on the cooler borders and tops of reservoirs as heat is lost by conduction; the decreasing of temperatures upward and outward requires some condensation of steam (see White, Muffler, and Truesdell, 1971, Table 5). The flow of fluid in systems of this subtype is limited by the low permeability of the *recharge* channels; these channels constitute the limiting impedance on fluid flow throughout the system.

Monte Amiata subtype. A second variety of vapor-dominated system, here called the Monte Amiata (Italy) subtype, is not yet well understood but is evidently similar in many respects to hot natural-gas fields (White, Muffler, and Truesdell, 1971). Temperatures in the Monte Amiata fields tend to be much lower (~150°C) and initial gas contents much higher (>90 percent: Burgassi et al., 1965; Cataldi, 1967) than for comparable initial pressures (20 to 40 kg/cm^2) in the Larderello fields. Thus, steam is a relatively minor initial constituent, presumably because of condensation of water vapor near the relatively cool borders of the reservoirs, as previously mentioned. With production and decompression, the initial vapor of high gas content is flushed out of the reservoir and is replaced by the relatively low-pressure steam of lower gas content that results from the boiling of water at only modest temperature. Another characteristic of the fluids produced from the Bagnore field of the Monte Amiata district is a trend from dry vapor to vapor plus liquid H_2O (Cataldi, 1967, Table 2).

The latter relations are interpreted to indicate water-flooding, or a rise in a deep water table that responds to the surrounding hydrostatic environment as reservoir pressures decline with production. The most restrictive impedance to fluid flow for the Monte Amiata subtype evidently occurs in the *discharge* part of the system, where low-permeability cap-rocks limit the discharge of gases to rates that are equal to or less than rates of generation or supply of gases. In contrast to the Larderello subtype, discharge of gases and steam is not required to form or maintain vapor-dominated reservoirs of the Monte Amiata subtype, although some leakage of gases is no doubt characteristic of most reservoirs.

Problems of Utilization

The problems attendant upon large-scale utilization of the various geothermal-field types and subtypes are manifold. A basic listing might proceed as follows:

1. Because of its special geologic and physical requirements, the commercially attractive Larderello subtype of vapor-dominated system is rare, accounting perhaps for only 5 percent of all geothermal systems with temperatures above 200°C; the advantages of this subtype for utilization are demonstrated by its dominance of present geothermal power-generating capacity (an estimated 73 percent of the world total, operating or under construction through 1973).

2. A discharge area is probably essential for the Larderello subtype,

with characteristic, recognizable manifestations of activity, chemistry, and ground bleaching. If this is so, then completely concealed deep systems are not available for future discovery.

3. The Monte Amiata subtype of vapor-dominated system, characterized by relatively high content of noncondensable gases and moderately low temperatures, may be more common than the Larderello subtype but will be more difficult to discover because of the absence of conspicuous surface characteristics. In any case, because of its physical and production characteristics this subtype is not as attractive for exploitation (2.3 percent of Table 1).

4. The high-temperature hot-water fields (25 percent of Table 1) are attractive for near-future increases in power production, but present utilization technology is not efficient, converting only about 1 percent of stored reservoir energy into equivalent electrical energy (Bodvarsson, 1970; Muffler, in press).

5. The water of many hot-water systems, when flash-erupted and cooled, deposits SiO_2 or $CaCO_3$ scale in wells and surface pipes; if similar flashing and mineral deposition occur in the reservoir immediately adjacent to wells, permeability and production rates decrease dramatically.

6. Some hot waters are corrosive because of high salinity, high CO_2 or O_2 content, or high acidity from H_2SO_4 or, rarely, HCl.

7. Some hot-water systems do not have adequate volume, temperature, or permeability to maintain commercial production. White (1968b) has suggested, from general experience at Broadlands and Waiotapu, New Zealand, and Beowawe and Steamboat Springs, Nevada, that inadequate permeability and reservoir characteristics may be as common as inadequate temperature.

8. Most hot-water effluents involve some environmental hazard, since they are generally higher in dissolved salts, B, NH_3, As, and heavy metals than most surface and ground waters. Such effluents will require disposal by some satisfactory means, with reinjection generally favored.

9. Some hot-water effluents may not be compatible with reservoir or other formation fluids, even though the fluids were initially identical. Compatibility and reliability of reinjection must be tested, and better principles for early recognition of the attendant problems must be developed. Reinjection has been tested for one year in the Salton Sea

system and for short intervals in the Long Valley system of California and the Ahuachapán field of El Salvador, but the only long-sustained test (about 3 years through 1972) has been at The Geysers in California. In the latter field, cool condensate is being reinjected successfully into an underpressured vapor-dominated reservoir; presumably much of the liquid is vaporized by transfer of heat from the still-hot rocks. These results are commonly interpreted as proof that reinjection elsewhere will be equally successful; but because individual hot-water systems vary greatly in fluid chemistry and precipitation potential, such a conclusion is hazardous without adequate testing. Production engineers estimate that if reinjection is successful for 1 year, it can be continued for several more years, indeed for the mechanical life of the injection equipment (Otte, private communication).

10. Most chemical problems are not serious for the low-temperature hot-water systems, but self-eruption will be unreliable or lacking for water that is too low in temperature or that must be "steam-lifted" from depths too far below the ground surface (Bodvarsson, 1970). The percentage of water that flashes to steam in a producing geothermal well depends mainly on the initial temperature and the pressure of steam separation from residual water, with liquid constituting 70 to 90 percent of most commercial production. The steam-lifting of moderate-temperature waters (150° to 200°C) becomes increasingly less effective as reservoir pressures and temperatures decline with production. As water levels (fluid potentials) decline below the ground surface, the energy required to lift liquid water increases. Thus, if produced at all, such water probably must be pumped.

11. Desalination of low-temperature waters involves more chemical and effluent-disposal problems, since the dissolved solids are concentrated into a small proportion of residual water. Soluble constituents, such as NaCl, normally will not precipitate, but constituents of low to very low solubility, such as SiO_2, $CaCO_3$, and $CaSO_4$, are potentially troublesome.

12. Thermal, noise, and air pollution (principally H_2S) may constitute environmental hazards requiring some control, depending on their severity.

13. Seismic hazards from reinjection must also be evaluated, especially for hot-water systems. But reinjection into underpressured vapor-dominated systems should involve little or no seismic hazard.

14. Subsidence will occur over hot-water reservoirs consisting in part of clay, silt, or shale where produced water is not locally replaced by reinjection. Sand and sandstone are less subject to compaction, unless pore fluids are overpressured. But subsidence over vapor-dominated reservoirs (initially already underpressured relative to hydrostatic) is likely to be slight.

Recoverability and Reserves of Geothermal Energy

This paper focuses attention on the physical and chemical nature of potentially useful concentrations of geothermal heat near the surface of the Earth. For the most part, the hydrothermal (vapor-dominated or hot-water) convection systems have been emphasized, since these include the principal geothermal resources that are usable and recoverable under present economic and technological conditions.

One or more of several potentially important breakthroughs in utilization technology may greatly expand the development of geothermal systems, hopefully in the immediate future. The most significant of the possible breakthroughs are:

1. Heat-exchange technology that would permit utilizing the heat from fluids down to 100°C or less (Jonsson, Taylor, and Charmichael, 1969), since total heat contained in easily recoverable natural fluids at temperatures of 100° to 180°C is far greater, perhaps by a multiple of 100, than total easily available heat above 180°C (see Anderson's discussion of the vapor-turbine cycle, this volume).
2. Multipurpose developments, including desalination and/or chemical recovery, that would yield significant sharing of total costs.
3. Low-cost mechanical, chemical, or nuclear fracturing of hot, dry rocks to increase permeability, thus permitting introduction of fluids and recovery of stored energy (these are described in later papers in this volume).
4. New methods for drilling low-cost holes to great depth.
5. New technology or other developments that favor wide applications to space heating, horticulture, and product processing.
6. Solution or control of all geothermal-resource problems at no greater cost than for corresponding environmental and other problems of competing sources of power.

Some of these breakthroughs could have profound effects on the recovering of geothermal energy from very large gradient-dominated volumes of rock, such as the deep sedimentary basins and hot, dry crystalline rocks, which are unlikely to be utilized within present prices and technology.

Strikingly disparate estimates have been made in recent years for the power potential, desalination potential, relative costs, and environmental-pollution aspects of geothermal energy. Depending on the source, the expressed view ranges from conservative (locally important, but with a relatively small potential for supplying national needs) to highly optimistic (very promising, with implied reliable potential for supplying a major part of all future needs for both power and desalinated water). The discrepancy is related largely to (1) a lack of agreement on the various categories of resources, with respect to certainty of existence and feasibility and cost of recovery (McKelvey, 1972; geothermal factors treated in detail by Muffler, in press); (2) differing assumptions on future technology and on whether hoped-for breakthroughs are likely to be realized with reliably predicted costs; and (3) a lack of agreement on the characteristics and nature of different types and subtypes of geothermal deposits, with respect to individual problems of discovery and energy recovery.

In my opinion, world geothermal power production is unlikely to exceed 30,000 Mw with present prices and technology. My estimate of proved, probable, and possible reserves recoverable in the United States under present conditions is approximately 600 Mw-centuries. Paramarginal reserves (recoverable with present technology but with as much as one-third increase in price) may be from 2,000 to 4,000 Mw-centuries. I am reluctant to offer estimates of geothermal resources that are now submarginal but that may be utilized with appropriate technological breakthrough; adequate cost data are completely lacking. However, major geothermal contributions (>10 percent of our energy needs) could result from such breakthroughs.

In contrast, the geothermal resource base (total stored heat, without regard to cost of recovery) can be estimated with some reliability, depending only on the assumed depth (3 km, 10 km, etc.) and thermal gradient, by utilizing the concept of volumetric specific heat (White, 1965). In that study I assumed an average gradient of 20°C/km and calculated 3×10^{26} cal of stored heat (i.e. heat above surface temperatures) under the surface of the Earth to a depth of 10 km, with 6×10^{24} cal of that worldwide total under the United States. The assumed average gradient may be too low, and 25°C/km may be a more likely average; but even if an improbable 30°C/km is assumed, the resource-base calculations are raised only to 4.5×10^{26} cal and 9×10^{24} cal, respectively. These calculations define upper limits for recoverable resources and are probably too high by at least two orders of magnitude.

ACKNOWLEDGMENTS

In large part, this paper summarizes the results of years of investigations made with associates of the U.S. Geological Survey in Yellowstone Park, Imperial Valley, and elsewhere. I am deeply indebted to L. J. P. Muffler, R. O. Fournier, A. H. Truesdell, and I. Barnes. Truesdell, especially, has contributed to our understanding of vapor-dominated systems. In addition, our knowledge of geothermal systems in other countries has progressed dramatically in recent years because of the collection and publication of data from New Zealand, Italy, and Iceland, and because of efforts elsewhere sponsored largely by the United Nations.

REFERENCES

Banwell, C. J., E. R. Cooper, G. E. K. Thompson, and K. J. McCree. 1957. Physics of the New Zealand thermal area. New Zealand Dept. Sci. and Industr. Res. Bull., v. 123, 109 pp.

Blackwell, D. D. 1969. Heat-flow determinations in the northwestern United States. Jour. Geophys. Res., v. 74, no. 4, pp. 992–1007.

Bodvarsson, G. 1964. Physical characteristics of natural heat resources in Iceland. *In* Geothermal energy I: Proc. United Nations Conf. on New Sources of Energy, Rome, 1961, v. 2, pp. 82–90.

——— 1966. Energy and power of geothermal resources. Ore Bin, v. 28, no. 7, pp. 117–24.

——— 1970. Evaluation of geothermal prospects and the objectives of geothermal exploration. Geoexploration, v. 8, no. 1, pp. 7–17.

Boldizsár, T. 1970 [1972]. Geothermal energy production from porous sediments in Hungary. Geothermics, special issue 2, v. 2, pt. 1, pp. 99–109.

Burgassi, R. 1964. Prospecting of geothermal fields and exploration necessary for their adequate exploitation performed in the various regions of Italy. *In* Geothermal energy I: Proc. United Nations Conf. on New Sources of Energy, Rome, 1961, v. 2, pp. 117–33.

Burgassi, R., R. Cataldi, J. Moutin, and F. Scandellari. 1965. Prospezione delle anomalie geotermiche e sua appricazione alla regione Amiatina. l'Industria Mineraria, v. 16, pp. 1–15.

Cataldi, R. 1967. Remarks on the geothermal research in the region of Monte Amiata (Tuscany-Italy). Bull. Volcanol., v. 30, pp. 243–70.

Craig, H. 1966a. Superheated steam and mineral-water interactions in geothermal areas. Trans. Am. Geophys. Union, v. 47, pp. 204–5 (abstract).

——— 1966b. Isotopic composition and origin of the Red Sea and Salton Sea geothermal brines. Science, v. 154, no. 3756, pp. 1544–48.

Elder, J. W. 1965. Physical processes in geothermal areas. *In* W. H. K. Lee, ed., Terrestrial heat flow, Am. Geophys. Union Mon., ser. 8, pp. 211–39.

Ellis, A. J., and W. A. J. Mahon. 1967. Natural hydrothermal systems and experimental hot water/rock interaction (pt. II). Geochim. et Cosmochim. Acta, v. 31, no. 4, pp. 519–38.

Facca, G., and F. Tonani. 1964. Theory and technology of a geothermal field. Bull. Volcanol., v. 27, pp. 143–89.

Fournier, R. O. 1972. Silica in thermal waters: Laboratory and field investigations (New York, Pergamon Press).

——— 1973. An empirical geothermometer based on Na, K, and Ca in natural waters. Geochim. et Cosmochim. Acta, v. 36, in press.

Fournier, R. O., and A. H. Truesdell. 1970 [1972]. Chemical indicators of subsurface temperature applied to hot spring waters of Yellowstone National Park, Wyoming, U.S.A. Geothermics, special issue 2, v. 2, pt. 1, pp. 529–35.

Haas, J. L., Jr. 1971. The effect of salinity on the maximum thermal gradient of a hydrothermal system at hydrostatic pressure. Econ. Geol., v. 66, pp. 940–46.

Helgeson, H. C. 1968. Geologic and thermodynamic characteristics of the Salton Sea geothermal system. Am. Jour. Sci., v. 266, pp. 129–66.

Honda, S., and L. J. P. Muffler. 1970. Hydrothermal alteration in core from research drill hole Y-1, Upper Geyser Basin, Yellowstone National Park. Amer. Mineral., v. 55, pp. 1714–37.

James, R. 1968. Wairakei and Larderello. Geothermal power systems compared. New Zealand Jour. Sci. and Techn., v. 11, pp. 706–19.

Jones, P. H. 1970 [1972]. Geothermal resources of the Northern Gulf of Mexico Basin. Geothermics, special issue 2, v. 2, pt. 1, pp. 14–26.

Jonsson, V. K., A. J. Taylor, and A. D. Charmichael. 1969. Optimisation of geothermal power plant by use of freon vapour cycle. Timarit VFI, v. 54, pp. 2–17.

Krauskopf, K. B. 1964. The possible role of volatile metal compounds in ore genesis. Econ. Geol., v. 59, pp. 22–45.

Lee, W. H. K., and S. Uyeda. 1965. Review of heat flow data. *In* W. H. K. Lee, ed., Terrestrial heat flow. Am. Geophys. Union Mon., ser. 8, pp. 87–190.

Marinelli, G. 1969. Some geological data on the geothermal areas of Tuscany. Bull. Volcanol., v. 33, no. 1, pp. 319–34.

McKelvey, V. E. 1972. Mineral resource estimates and public policy. Amer. Scientist, v. 60, pp. 32–40.

Mercado, S. 1970 [1973]. High activity hydrothermal zones detected by Na/K, Cerro Prieto, Mexico. Geothermics, special issue 2, v. 2, pt. 2, in press.

Muffler, L. J. P. 1973. Geothermal resources. *In* D. A. Brobst and W. P. Pratt, eds., Potential mineral resources: A geologic perspective. U.S. Geol. Surv. Prop. Paper.

Muffler, L. J. P., and D. E. White. 1969. Active metamorphism of Upper Cenozoic sediments in the Salton Sea geothermal field and the Salton Trough, southeastern California. Geol. Soc. Am. Bull., v. 80, pp. 157–82.

——— 1972. Geothermal energy. The Science Teacher, v. 39, no. 3, pp. 1–4.

Nakamura, H., K. Sumi, K. Katagiri, and T. Iwata. 1970 [1972]. The geological environment of Matsukawa geothermal area, Japan. Geothermics, special issue 2, v. 2, pt. 1, pp. 221–31.

Otte, C., and R. F. Dondanville. 1968. Geothermal developments in The Geysers area, California. Am. Assoc. Petrol. Geologists Bull., v. 52, p. 575 (abstract).

Pálmason, G., and J. Zoëga. 1970 [1972]. Geothermal energy developments in Iceland 1960–69. Geothermics, special issue 2, v. 2, pt. 1, pp. 73–76.

Penta, F. 1959. Sulle origini del vapore acqueo naturale e sull'attaule stato delle relative ricerche (ricerche per "forze endogene"). La Ricerca Scientifica, v. 29, no. 12, pp. 2521–36.

Ross, D. A. 1972. Red Sea hot brine area: Revisited. Science, v. 175, pp. 1455–57.

Sass, J. H. 1971. The Earth's heat and internal temperature. *In* Understanding the Earth, T. G. Gass, P. J. Smith, and R. C. L. Wilson, eds. (Sussex: Artemis Press), pp. 81–87.

Sass, J. H., A. H. Lachenbruch, R. J. Monroe, G. W. Greene, and T. H. Moses, Jr. 1971. Heat flow in the western U.S. Jour. Geophys. Res., v. 76, no. 26, pp. 6376–6413.

Sestini, G. 1970 [1972]. Superheating of geothermal steam. Geothermics, special issue 2, v. 2, pt. 1, pp. 622–48 (not available as Pisa Symposium preprint, 1970).

Sourirajan, S., and G. C. Kennedy. 1962. The system H_2O-NaCl at elevated temperatures and pressures. Am. Jour. Sci., v. 260, pp. 115–41.

Truesdell, A. H., D. E. White, and L. J. P. Muffler [in preparation]. Production of superheated steam from vapor-dominated geothermal reservoirs.

White, D. E. 1967a. Some principles of geyser activity, mainly from Steamboat Springs, Nevada. Am. Jour. Sci., v. 265, pp. 641–48.

——— 1967b. Mercury and base-metal deposits with associated thermal and mineral waters. *In* H. L. Barnes, ed., Geochemistry of hydrothermal ore deposits (New York: Holt, Rinehart, and Winston), pp. 575–631.

——— 1968a. Environments of generation of base-metal ore deposits. Econ. Geology, v. 63, no. 4, pp. 301–35.

——— 1968b. Geothermal energy reservoirs. Am. Assoc. Petrol. Geol., v. 52, no. 3, p. 568 (abstract).

——— 1970 [1973]. Geochemistry applied to the discovery, evaluation, and exploitation of geothermal energy reservoirs. Geothermics, special issue 2, v. 1, in press.

White, D. E., I. Barnes, and J. R. O'Neil. 1973. Thermal and mineral waters of non-meteoric origin, California Coast Ranges. Geol. Soc. Am. Bull., in press.

White, D. E., W. W. Brannock, and K. J. Murata. 1956. Silica in hot-spring waters. Geochim. et Cosmochim. Acta, v. 10, pp. 27–59.

White, D. E., L. J. P. Muffler, R. O. Fournier, and A. H. Truesdell. 1968. Preliminary results of research drilling in Yellowstone thermal areas. Trans. Am. Geophys. Union, v. 49, p. 358 (abstract).

White, D. E., L. J. P. Muffler, and A. H. Truesdell. 1971. Vapor-dominated hydrothermal systems compared with hot-water systems. Econ. Geol., v. 66, pp. 75–97.

5. Exploration for Geothermal Resources

JIM COMBS AND L. J. P. MUFFLER

Broadly considered, geothermal resources are the natural heat of the Earth's crust. This natural energy is economically significant, however, only where it is concentrated into restricted volumes in a manner analogous to the concentration of valuable metals into ore deposits or of oil into commercial petroleum reservoirs. In this paper we discuss the various methods available for the exploration, discovery, and delineation of economic reservoirs of geothermal energy.

The basic principles of heat transfer, hydrothermal systems, hot springs, and geothermal energy have been covered in detail by other writers, in this volume and elsewhere, and will not be repeated here. For the reader who wishes to learn more about these principles, we suggest the following papers: Muffler and White (1972), Grose (1971), Grose (1972), Jaffé (1971), White (1965), McNitt (1965), Ellis (1969), Barnea (1972), Beck (1965), and Sass (1971).

Geothermal Exploration Philosophy

If an exploration manager, lacking in specific geothermal experience but trained in geology and geophysics and experienced in (for example) petroleum exploration, were asked to design and carry out a geothermal exploration program, where should he start? What assumptions should he make? What background knowledge would he need?

Certainly he should start by recognizing that the commodity being sought is *heat*. In contrast to oil, gas, coal, or uranium, geothermal en-

Jim Combs is with the Department of Geological Sciences and Institute of Geophysics and Planetary Physics, University of California, Riverside; and L. J. P. Muffler is with the U.S. Geological Survey, Menlo Park, California. Publication of this paper is authorized by the Director, U.S. Geological Survey.

ergy can be used directly; it does not require combustion or fission to produce usable thermal energy. Thus, techniques proved successful in exploration for minerals or fuels are not necessarily the best techniques for geothermal exploration. Conversely, techniques of little use in petroleum exploration may be ideal tools in the search for natural heat.

Our hypothetical exploration manager should then recognize that, although the Earth's crust is an immense source of heat (White, 1965, has calculated that the amount of geothermal heat above mean Earth-surface temperature in the outer 10 km is 3×10^{26} calories), most of this heat is far too diffuse ever to be recovered economically. Regions of the Earth's crust with only an average heat content grade into pockets where heat is more or less concentrated. In some of these pockets, heat energy can be extracted and used. The degree of concentration (i.e., the "grade" of the geothermal resource) used as an exploration cutoff point is defined in terms of use, technology, and economics, and will vary with the situation, location, and time.

For example, if the use of geothermal resources is restricted to the generation of electricity by conventional steam-fed turbines, the temperature of the geothermal reservoir must be greater than 180°C (preferably greater than 200°). Furthermore, with existing technology, such a reservoir must have adequate permeability and water to allow the heat to be extracted. Fluids at temperatures of 40° to 180° have been used for space heating, agricultural heating, product processing, and air conditioning, and are potentially useful in desalting, electricity generation via a heat exchanger and a cold-vapor (freon or iso-butane) cycle, and mineral recovery. The technological and economic feasibilities of these uses will dictate the minimum reservoir temperature to be sought.

Our exploration manager will undoubtedly perceive that, under present technology and economics, geothermal resources at great depths in the crust may be vast and may be at high temperatures, but nonetheless cannot be economically exploited, owing to the high costs of deep drilling. The deepest geothermal well to date (at The Geysers, California) is about 3 km. How deep one may drill economically in the future is not certain, but an outside limit is perhaps 10 km.

Finally, our exploration manager should recognize that although all geothermal anomalies involve the concentration of heat, naturally productive geothermal systems require the circulation of water. This water serves as the means by which heat is transferred from a deep igneous heat source to a geothermal reservoir shallow enough to be tapped by drill holes, and as the means by which heat is transferred from rock to a well and thence to the surface. Accordingly, naturally productive geothermal systems must have adequate porosity and permeability. If either of these factors is inadequate, the geothermal system will require artificial stimulation by hydrofracturing, explosions, thermal cracking, or other methods discussed in this volume. Reinjection of the produced fluids or of surface fluids may also be necessary. Furthermore, if "dry" geothermal reservoirs do exist at depths accessible to drilling, the techniques available to locate such reservoirs are considerably fewer than those available to locate naturally productive systems.

Geological Considerations

Geothermal reservoirs are not uniformly distributed in the Earth's crust. High-temperature shallow geothermal reservoirs (essentially all fields explored to date) are near the margins of crustal plates (Muffler and White, 1972), where scientists consider that crust is being either created or consumed (see, for example, Dewey and Bird, 1970). In both situations, molten rock is generated and moves buoyantly upward in the crust. The resulting intrusive rock provides the heat that is transferred by conduction to the convecting systems of meteoric water. Plate margins are tectonically active areas, commonly characterized by recent volcanism and regionally or locally high heat flow.

Some large sedimentary basins not at crustal-plate margins do contain large amounts of hot water, generally at depths greater than 3 km and commonly at pressures greater than hydrostatic. Such geothermal resources have been described from the Gulf Coast (Jones, 1970), Hungary (Boldizsár, 1970), and the U.S.S.R. (Makarenko et al., 1970). The relatively high temperature gradients in these areas are due to a combination of somewhat elevated heat flow and relatively low thermal conductivity (White, this volume).

The continental nuclei, the Precambrian shields, are particularly unfavorable for geothermal exploration. The shields have uniformly low heat flow, are not tectonically active, display no recent volcanism, and are composed of old metamorphic and igneous rocks of low permeability.

The general location of a geothermal system is determined by the location of a deep igneous mass (at perhaps greater than 5 km) that is the probable heat source driving the overlying meteoric convection system (White, 1968). Accordingly, the anomalous temperature regime of the upper few kilometers has been superimposed on whatever rock units happened to be there, regardless of their age or mode of formation (i.e., sedimentary, igneous, or metamorphic). An example of this circumstance is The Geysers, where a Quaternary geothermal system occurs in Franciscan metasedimentary and metavolcanic rocks of Jurassic and Cretaceous age. The nature of the resultant geothermal systems, however, is strongly dependent on the physical characteristics, e.g., porosity and permeability, of the reservoir rocks. Moreover, the geometry of the convection system can be greatly influenced by structural and stratigraphic details in the upper few kilometers. Hot water will rise buoyantly up the path of least impedance; this need not be vertical, and accordingly the plume of hot water can be displaced laterally from the heat source perhaps by several kilometers.

The presence of a high-temperature water-convection system can itself greatly change the mineralogy and physical characteristics of the rocks (Browne and Ellis, 1970; Steiner, 1968). Hot fluids commonly contain silica in solution derived from rocks at depth (Fournier and Rowe, 1966). As these waters move upward and cool, they precipitate silica as quartz, β-cristobalite, or opal, thus plugging up or "self-sealing" the cool margins of the hydrothermal plume (Bodvarsson, 1964; Facca and Tonani, 1967; Bodvarsson, 1970; White, 1970). In the Salton Sea geothermal system, sediments of the Colorado River delta have been hydrothermally metamorphosed to hard, dense rocks of the greenschist facies (Muffler and White, 1969).

Exploratory Techniques

The objectives of any geothermal exploration venture are to locate areas underlain by hot rock, to estimate the volume, temperature, and permeability at depth, to predict whether wells will produce dry steam or a mixture of water and steam, and to predict the chemical composition of the produced fluid. Because each prospect represents a unique combination of geological, hydrological, geochemical, geophysical, technical, and financial characteristics, no one exploration technique suffices for all situations. A given procedure may be informative in one area but not in another. Consequently, we recommend against the use of sophisticated exploratory techniques until one has worked through a logical progression of preliminary steps; we suggest the following steps, or phases, each successively more sophisticated and/or expensive:

Phase 1. Literature search
Phase 2. Airborne survey
Phase 3. Geologic and hydrologic survey
Phase 4. Geochemical survey
Phase 5. Geophysical survey
Phase 6. Drilling

We shall take up each in turn.

Literature Search

In most areas, some topographical, hydrological, meteorological, geological, geochemical, and geophysical information already exists and should be thoroughly evaluated before new investigations are undertaken. This information may eliminate the need to conduct certain surveys, and will assist in the interpretation of other data collected during the course of the exploration venture.

Airborne Survey

Interpretation of modern aerial photography is an important first step in geothermal exploration. Photographic surveys aid in structural analysis and geologic mapping, and in areas of poor topographic control are essential for geographic location during geophysical surveys.

If the regional structural and stratigraphic framework is poorly known, an aeromagnetic survey might be a relatively inexpensive means of providing additional data. Similarly, an airborne infrared survey is useful for mapping surface thermal manifestations. But if information of this sort is available, we see little need for aerial magnetic or infrared surveys, for they give almost no direct information about the subsurface distribution and nature of geothermal resources.

Geologic and Hydrologic Survey

Field work should begin with regional geologic and hydrologic surveys (Healy, 1970). The aim is to delimit the thermal area geographically. The geologic survey therefore should emphasize the tectonic and stratigraphic setting of the area, recent faulting, the distribution and age of young volcanic rocks, and the location and character of thermal manifestations, including hydrothermally altered rock. The hydrologic reconnaissance should include temperature and discharge measurements on hot and cold springs, chemical analysis of the springs (see below, on geochemical survey), determination of the water table in available wells, and evaluation of surface and subsurface movements of water. Basic meteorologic data (temperature, humidity, precipitation, etc.) should not be neglected, for these data are important in the planning of geophysical and drilling programs to follow.

As geothermal prospects are identified, these regional studies are gradually transformed into detailed studies around potential drilling targets, with the aim of predicting the geologic and hydrologic conditions to be encountered at depth. Aerial photography (particularly color and color infrared), thermal infrared imagery, and radar imagery can be of considerable use in defining active faults that may be potential drilling targets (e.g. Grindley, 1970). Vegetation type and character vary with soil temperature and can be used to delimit thermal areas.

From the geologic and hydrologic observations, one should develop a realistic structural, stratigraphic, petrologic, and hydrologic model to serve as a guide to further exploration and development.

Geochemical Survey

The use of geochemistry in geothermal exploration has been recently reviewed by White (1970). Geochemical reconnaissance involves sampling and analyzing waters and gases from hot springs and fumaroles in the area under investigation. The data obtained are then used to determine whether the geothermal system is hot-water or vapor-dominated, to estimate the minimum temperature expected at depth, to estimate the homogeneity of water supply, to infer the chemical character of the waters at depth, and to determine the source of recharge water.

Chloride analyses can be used to discriminate between hot-water and vapor-dominated geothermal systems. Chloride contents in excess of 50 ppm characterize most high-temperature hot-water systems, whereas springs associated with vapor-dominated systems consistently display chloride contents less than 20 ppm (White, 1970).

Several constituents or ratios of constituents can be used to estimate minimum reservoir temperatures of hot-water geothermal systems. These include SiO_2 content, Na-K-Ca relationships, $Cl/(HCO_3 + CO_3)$ ratio, Cl/F ratio, and Mg content (White, 1970). Of these, SiO_2 content and the relationship between Na, K, and Ca have been quantified to yield numerical estimates of reservoir temperature (Fournier and Rowe, 1966; Fournier and Truesdell, 1973).

Measurement of the isotopic composition of hydrogen and oxygen in waters serves to specify the origin of the water and to evaluate the hydrology of the region. For example, studies in the Imperial Valley have demonstrated that most of the geothermal water in the central and southern parts of the Salton Trough was derived from the Colorado River underflow (Coplen, 1971), whereas the geothermal fluids of the Salton Sea geothermal field were derived from runoff from the nearby Chocolate Mountains (Craig, 1966).

Geophysical Survey

Geophysics is drawn upon chiefly to define target areas for drilling. The existence of a geothermal reservoir can be inferred from the indirect measurement of various physical parameters at depth. These physical parameters include temperature, electrical conductivity, propagation velocity of elastic waves, density, and magnetic susceptibility. The most useful geophysical techniques for geothermal exploration are temperature or geothermal-gradient surveys, heat-flow determinations, electrical-conductivity surveys, and passive seismic methods such as microearthquake measurements or seismic-noise detection. These methods can delimit the geothermal reservoirs and furnish data on subsurface thermal processes. Active-seismic, gravimetric, and magnetic surveys may prove justifiable in refining a regional geologic model, but they generally provide little useful information for defining the geothermal reservoir.

Thermal techniques. The average worldwide conductive heat flow is about 1.5 $\mu cal/cm^2$ sec (Lee and Uyeda, 1965; Simmons and Horai, 1968). Heat flows slightly in excess of normal can be due to exothermic chemical reactions, high content of radioactive materials, friction along faults, or migration of waters of different origins in areas of nearly normal geothermal gradient. Elevated heat flow owing to these phenomena are usually of restricted extent and of limited duration. Geothermal areas that are economically attractive under present conditions, however, can have heat flows that are up to several thousand times normal (White, 1969) and can persist for many thousands of years (White, 1968). Heat flows of such magnitude and duration are possible only where rocks of near-magmatic temperature are nearby (White, 1968).

Thermal-exploration techniques provide a direct method for assessing the size and potential of a geothermal system; they can be ordered in terms of increasing cost as follows:

1. Surface and shallow temperature measurements, generally at depths of 6 m or less.
2. Geothermal-gradient surveys, generally at depths of 15 to 100 m.
3. Heat-flow determinations, generally requiring hole depths of at least 100 m.

Temperature and gradient measurements at depths on the order of 1 m (Thompson et al., 1964) are quick and inexpensive, and can be used to detect anomalously hot areas. These measurements are strongly influenced, however, by near-surface effects, including insolation, topography, precipitation, and movement of ground water. The last of these effects is particularly important, for a relatively slow movement of ground water across even a strong thermal anomaly can carry away the conductive heat flow, displacing surface-temperature patterns and grossly distorting gradient measurements.

Temperatures as a function of depth in boreholes 15 to 150 m deep have been used by a number of investigators in geothermal exploration (Lovering and Goode, 1963; Burgassi, Battini, and Mouton, 1964; Duprat, 1970; Burgassi et al., 1970; Combs and Rex, 1971; Combs, 1971). At these depths most near-surface thermal disturbances are avoided and gradient measurements can be made with considerable precision and reliability; nevertheless, the data must always be evaluated for the effects of lateral movement of ground water. Over most economically attractive geothermal areas, the gradients at these intermediate depths are greater than 7°C per 100 m, compared to a normal geothermal gradient of about 3° per 100 m.

Although gradient measurements do define the areal extent of thermal anomalies, the explorer must be very cautious in extrapolating these gradients to depth. Two factors combine to ensure that a linear extrapolation will be in error on the high side. The first is the variation in conductivity of the rock with depth. In a sedimentary section, such as in the Imperial Valley of southern California, the porosity decreases strikingly with depth (Bur. Rec., 1972). Inasmuch as the thermal conductivity of minerals is 3 to 10 times that of water, the bulk conductivity therefore increases greatly with depth and decreasing porosity.

Since heat flow is the product of gradient and thermal conductivity, the gradient must decrease with depth as the thermal conductivity increases. The second factor, convection, will have an even greater effect in reducing thermal gradients at depth. White (this volume, Fig. 2) shows that convection produces high conductive gradients in the rock above the convection cell, but very low thermal gradients within the convection cell. In an area where convection of water within the pore space of rocks is possible, the extrapolation of measured high, near-surface gradients is clearly unjustified and likely to suggest erroneously high temperatures at depth.

When the mean thermal conductivity of the subsurface is essentially constant throughout the complex in which boreholes are drilled, the thermal gradients measured are obviously proportional to the heat-flow values. The essential advantage of heat-flow measurements, as opposed to purely gradient measurements, is that heat flow is independent of the in-situ thermal conductivity of each rock type. Therefore, in nonhomogeneous terrains, only heat-flow measurements enable us to obtain accurate information on the potentially productive zone (see, for example, Sestini, 1970).

Exploration based only on gradient measurements is often considered sufficient to indicate the presence of a geothermal area as a whole. The lowest gradients measured over a geothermal anomaly are about 5 to 10 times higher than the average regional geothermal gradient. But in order to distinguish between the central productive zone and a less productive marginal zone, heat-flow measurements should be made; thus, heat-flow measurements will define the potential drilling targets. The best demonstration of these arguments is from the natural-steam field at Larderello, Italy, where Boldizsár (1963) has shown that the productive field is more accurately delineated by the heat-flow data than by the gradient data alone.

Broad regions of anomalously high heat flux can be defined by a few carefully located heat-flow determinations—for example, in bedrock at the margins of sedimentary basins. If the regional heat flow so determined is significantly higher than normal, but the heat-generation values measured on bedrock samples are normal, we can infer the presence of hydrothermal-convective systems and/or young, hot intrusive rocks in the region. Further detailed exploration would be indicated.

Over the past few years, attempts have been made to map thermal activity by means of the radiation emitted principally in the near- or intermediate-infrared region (Friedman et al., 1969; Gomez Valle et al., 1970; Hochstein and Dickinson, 1970; Hodder, 1970; Pálmason et al., 1970). The infrared scanners normally operate in the 3- to 5-micron or 8- to 14-micron transmission windows in the atmosphere. Thermal-infrared imagery at present is a noise-limited system. Most of the "thermal" anomalies on the imagery are the results of outcrop, slope direction, slope magnitude, soil moisture, fog or condensation, difference in rock properties, vegetation, etc. At present, heat-flux anomalies less than about 100 to 150 times normal cannot be accurately detected with thermal-infrared techniques (White and Miller, 1969; Pálmason et al., 1970; Hochstein and Dickinson, 1970).

Electrical and electromagnetic techniques. Electrical and electromagnetic techniques in geothermal exploration measure electrical conductivity at depth. Temperature, porosity, salinity of interstitial fluids, and/or content of clays and zeolites tend to be higher within geothermal reservoirs than in surrounding ground. Consequently, the electrical conductivity in geothermal reservoirs is relatively high. But only at the Wairakei field are there sufficient drill holes inside *and outside* the resistivity boundary to demonstrate a precise correlation between low apparent resistivity and high subsurface temperature.

There are a number of electrical and electromagnetic methods that measure electrical resistivity at depth. The telluric and magnetotelluric methods depend on measuring variation in natural electrical or electromagnetic fields. Several electrical techniques involve putting current into the ground at two electrodes and measuring the resultant potential at two other electrodes. The various electromagnetic methods involve the generation of a magnetic field that varies with time, and the detection of either the electrical or the magnetic field arising from currents induced in the earth (Keller, 1970).

Most of the published data on the use of electrical techniques in geothermal areas are from New Zealand, where experience over the past 10 years has indicated that in the volcanic environment of the Taupo graben the most useful technique is dc-resistivity profiling using linear arrays, either Wenner or Schlumberger (Banwell and Macdonald, 1965; Hatherton, Macdonald, and Thompson, 1966; Risk, Macdonald, and Dawson, 1970). Effective probing depth increases as electrode spacing increases; a Schlumberger-array spacing (AB/2) of 500 m, giving an effective probing depth of about 700 m, is perhaps the most satisfactory. A number of geothermal areas in the Taupo Volcanic Zone have been outlined using dc-resistivity profiling (Hatherton, Macdonald, and Thompson, 1966), and the Wairakei, Broadlands, and Kawerau fields have been delimited in detail (Banwell and Macdonald, 1965; Risk, Macdonald, and Dawson, 1970; Macdonald and Muffler, 1972). Dc-resistivity methods are preferred to ac methods, owing to the skin effects present at large spacings with the ac methods.

An electrical-prospecting technique that is being increasingly used in geothermal exploration is the dipole-dipole array. This technique has been used by Risk, Macdonald, and Dawson (1970) to outline the Broadlands field at depths of 1 to >3 km. Greater depths can be attained (Keller, 1970) using very powerful sources and exceptionally well-grounded current electrodes. Effective dipole-dipole investigations require complicated data analysis and careful interpretation, but the method is logistically simple and is insensitive to rugged topography (Harthill, 1971).

Electromagnetic methods have been used in geothermal exploration only during the past 5 years. Although instrumentation and interpretation are complex, electromagnetic (inductive) methods have two theoretical advantages over electrical methods (Keller, 1970): with an inductive method, signal size increases with decreasing resistivity, making measurements easier and more accurate in geothermal areas; and inductive methods are not adversely affected by near-surface high-resistivity zones. Three electromagnetic methods have been used to date in geothermal areas: (1) a two-loop method (the EM gun), used in New Zealand and in Chile to investigate depths of 15–50 m (Lumb and Macdonald, 1970); (2) an audio-frequency magnetotelluric method, used in New Zealand, Nicaragua, Indonesia, and Kenya, which appears to be an effective, rapid, and easy reconnaissance tool (Keller, 1970; Harthill, 1971); and (3) an inductive method using a long-wire source and a loop detector, used by Keller (1970) to investigate the Taupo Volcanic Zone to depths of 10 to 30 km.

Electrical and electromagnetic techniques are ideal for geothermal exploration in a region such as the Taupo Volcanic Zone, where resistivities within the geothermal areas are one-fifth to one-twentieth of those outside. The data are less easy to interpret in areas such as Kizildere, Turkey (Duprat, 1970) and the Imperial Valley, California (Meidav and Furgerson, 1971). In both areas the resistivity contrast is only one-fourth to one-half, owing to occurrence of the geothermal system in sediments with relatively saline waters and abundant shales.

Passive seismic techniques. Many geothermal reservoirs are characterized by abundant microearthquakes and a relatively high level of seismic (background) noise. Accordingly, the precise location of microearthquakes and seismic noise may aid in delimiting fractured and permeable zones in geothermal areas.

Passive seismics involves recording local, naturally generated microearthquakes and thus determining faults that may be associated with geothermal systems. Microearthquakes believed to arise from fault movements in geothermal areas have been described by Lange and Westphal (1969) and Hamilton and Muffler (1972) at The Geysers, California; by Ward, Pálmason, and Drake (1969) and Ward and Björnsson (1971) in Iceland; by Brune and Allen (1967) in the Imperial Valley; and by Ward and Jacob (1971) at Ahuachapán, El Salvador. The incidence of microearthquakes may be increased by the production of geothermal fluids or by injection of waste water, and monitoring of a field under development may be necessary.

Another passive seismic method, seismic-noise detection or geothermal-noise detection, records acoustic-noise patterns within certain frequency ranges. Seismic-noise measurements seem to provide a relatively simple method for detecting and mapping certain types of geothermal areas (Clacy, 1968; Whiteford, 1970; Goforth, Douze, and Sorrells, 1972). Initial studies suggest that individual geothermal systems have characteristic seismic signatures. The data also suggest that there is an empirical relationship between reservoir depth, high temperature gradients, and high seismic-noise level. If this relationship proves to be reliable, geothermal-noise detection could be useful in future geothermal exploration, owing to the relative speed, mobility, and economy of the seismic technique.

Other geophysical techniques. Active seismic methods, gravity surveys, and magnetic surveys all fall under the category of "structural" or "indirect" methods, as applied to geothermal exploration (Hatherton, Macdonald, and Thompson, 1966; Bodvarsson, 1970). In contrast to the thermal, electrical, and electromagnetic methods described above, these "structural" methods do not study the properties of the hot fluids sought, but instead investigate the attitude and nature of the host rocks.

Active seismic methods involve the use of explosions to generate elastic waves. The reflection method utilizes energy reflected from subsurface interfaces between rocks of different physical properties. The refraction method utilizes seismic waves refracted horizontally along an interface and thence back to the surface. Both methods are used to determine subsurface structure and the configuration and depth of basement rocks. Hayakawa (1970) has used common-depth-point techniques and analysis of attenuation in a reflection seismic survey at the Matsukawa geothermal area in Japan.

Gravity surveys have been used both to outline major structural features and to delineate local positive anomalies that may be related to the geothermal system. Such local positive anomalies can be caused by local structural highs, buried volcanic rocks, intrusive rocks, or hydrothermally metamorphosed rock (Hochstein and Hunt, 1970; Kresl and Novak, 1970; Biehler and Combs, 1972). But because local positive anomalies can be produced by factors other than an active geothermal system, gravity surveys in geothermal areas are open to gross misinterpretation unless used in conjunction with other exploration techniques.

In general, magnetic surveys are probably the geophysical tool least useful in defining geothermal drilling targets. In some areas negative magnetic anomalies appear to be caused by the hydrothermal conversion of magnetite to pyrite (Studt, 1964). In other areas, positive magnetic anomalies can be related to very young intrusive and volcanic rocks associated with a geothermal system (Griscom and Muffler, 1971). In most areas, however, so many factors influence the character of a magnetic map that it is difficult to interpret in terms of geothermal resources.

Drilling

Surface and near-surface geophysical techniques serve primarily to site exploratory drill holes to depths of up to 3 km. The final phase of any geothermal exploration is the drilling of these exploratory wells, which provide the only way to determine actual geothermal-reservoir characteristics and thus to evaluate the potential for heat, power, minerals, or fresh water. Data acquired from the drill holes ideally should include temperature-depth distribution, pressure-depth distribution, permeability, porosity, lithology and stratigraphy, and fluid composition. A full set of geophysical logs combined with production tests will provide the necessary data and allow evaluation of the site.

Information that can be obtained only from a borehole is essential in:

1. Estimating the ability of the geothermal reservoir to produce sufficient energy over a sufficiently long time to be economically attractive.
2. Distinguishing between different models of the geothermal system, with the aim of accurately predicting production characteristics under varying exploitation conditions.
3. Calibrating and refining geophysical and geochemical methods for recognizing and delimiting geothermal systems.

Two Case Histories

Broadlands geothermal field, New Zealand. Published case histories of geothermal fields are few and are generally incomplete. Perhaps the two best documented geothermal case histories come from New Zealand, where much scientific study of geothermal phenomena over the past 25 years has resulted in an impressive series of publications. The Wairakei geothermal field has been summarized by Grindley (1965). The Broadlands field has been under exploration since the early 1960s; exploration drilling began in 1965 and continued into 1971. Although no single monograph on Broadlands has yet been published by the New Zealand government, a great array of technical data is available in the papers presented at the United Nations Symposium on the Development and Utilization of Geothermal Resources held in Pisa, Italy, in September of 1970. These papers are gradually being published as a special issue of the new journal *Geothermics*. The following discussion of the Broadlands geothermal area is a summary of the New Zealand work, based mainly on the papers presented at Pisa.

The Broadlands geothermal field is located in the southern part of the Taupo Volcanic Zone on the North Island of New Zealand. This zone is a major volcano-tectonic depression that extends for 260 km from the active volcanoes of Tongariro National Park to White Island Volcano in the Bay of Plenty (Fig. 1). From vents in and near the zone nearly 11,000 km^3 of lava, ash-flow tuff, and air-fall tuff (all dominantly of rhyolitic composition) have been erupted in Pliocene to Holocene time (Healy, 1962). The Taupo Volcanic Zone contains all of New Zealand's visibly active volcanoes and all of its boiling springs and large thermal areas (Healy, 1964).

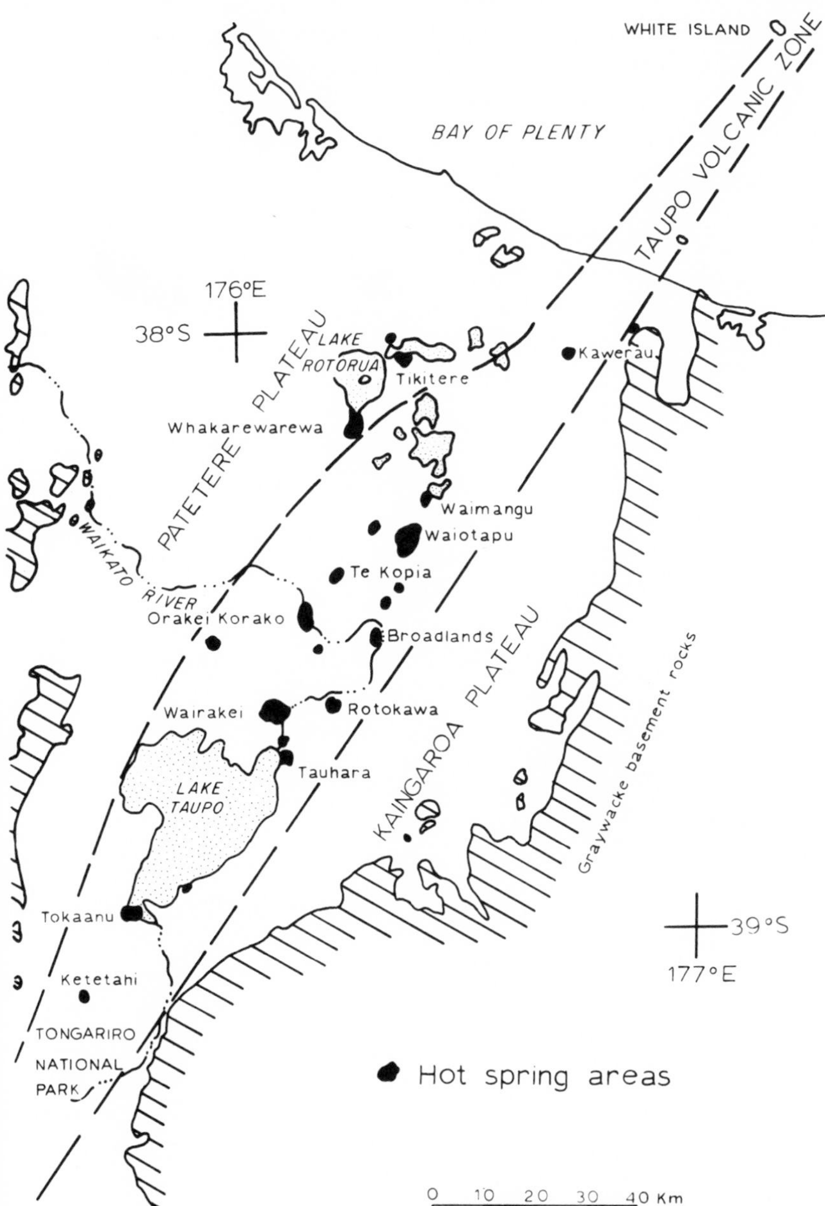

Fig. 1. Map of the Taupo Volcanic Zone (from Healy, 1964).

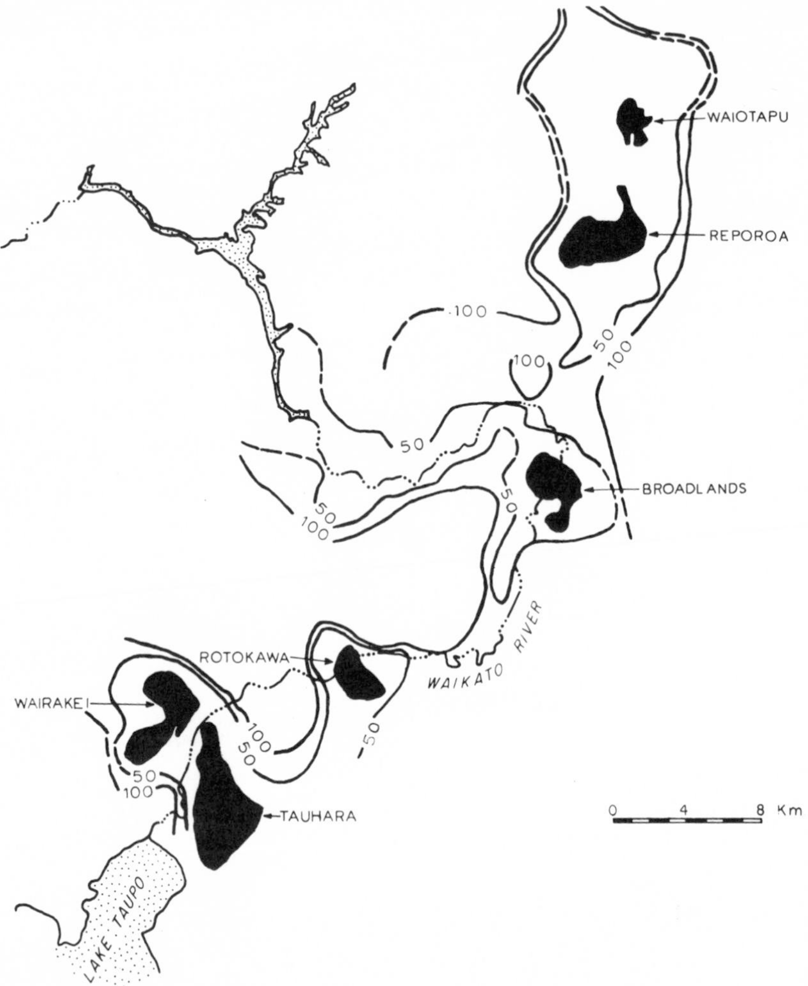

Fig. 2. Apparent-resistivity map (Wenner array; A = 550 m) of the southern part of the Taupo Volcanic Zone. Areas of apparent resistivity less than 5 ohm-meters are shaded. (From Hatherton, Macdonald, and Thompson, 1966, Fig. 6.)

Broadlands is just one of at least a dozen major geothermal areas in the Taupo Volcanic Zone. Subsequent to exploration and development of the Wairakei geothermal field in the 1950s and early 1960s (Grindley, 1965), the Broadlands area was chosen for exploration on the basis of a regional dc-resistivity survey (Smith, 1970, p. 236; Hatherton, Macdonald, and Thompson, 1966). This survey (Figs. 2 and 3) indi-

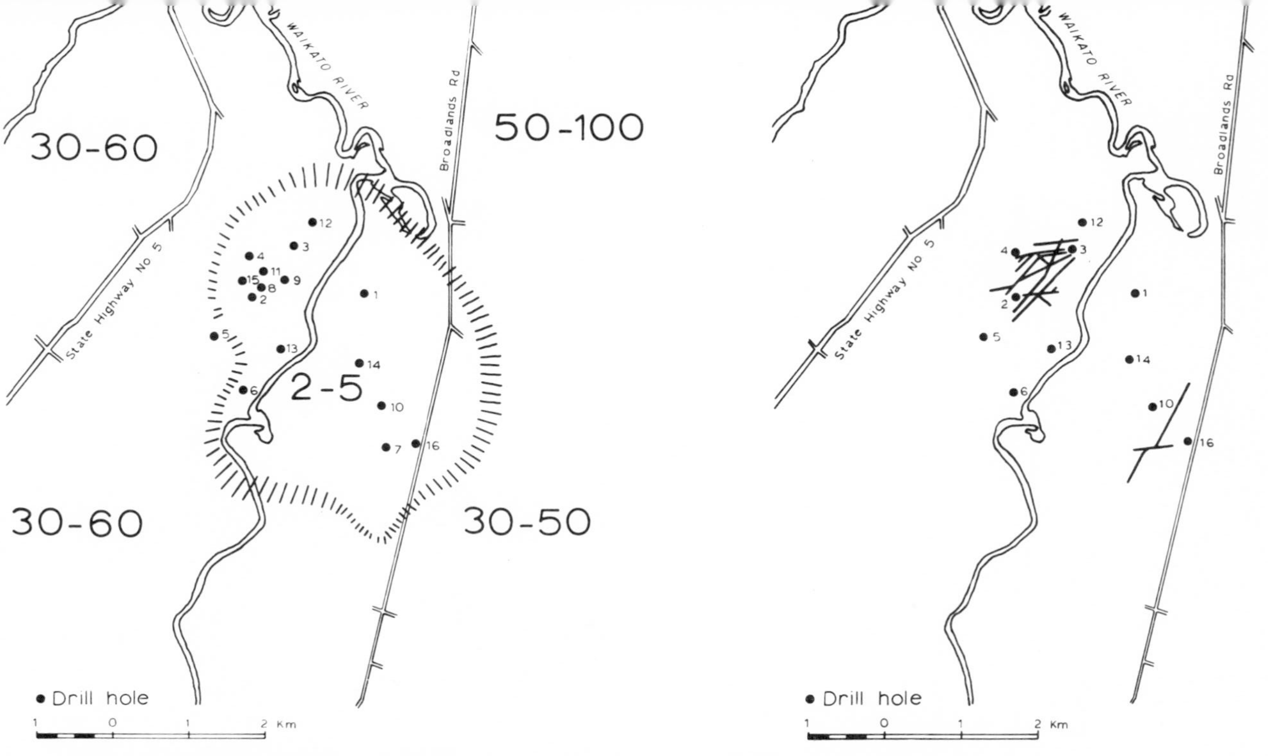

Fig. 3. Boundary of Broadlands geothermal field, as derived from dc-resistivity survey using Wenner configuration with electrode spacing of 550 m. Average apparent resistivities are in ohm-meters; effective probing depth is 0.75 km. (From Risk et al., 1970, Fig. 11.)

Fig. 4. Map of Broadlands geothermal area, showing faults inferred from photogeology. (From Grindley, 1970, Fig. 1.)

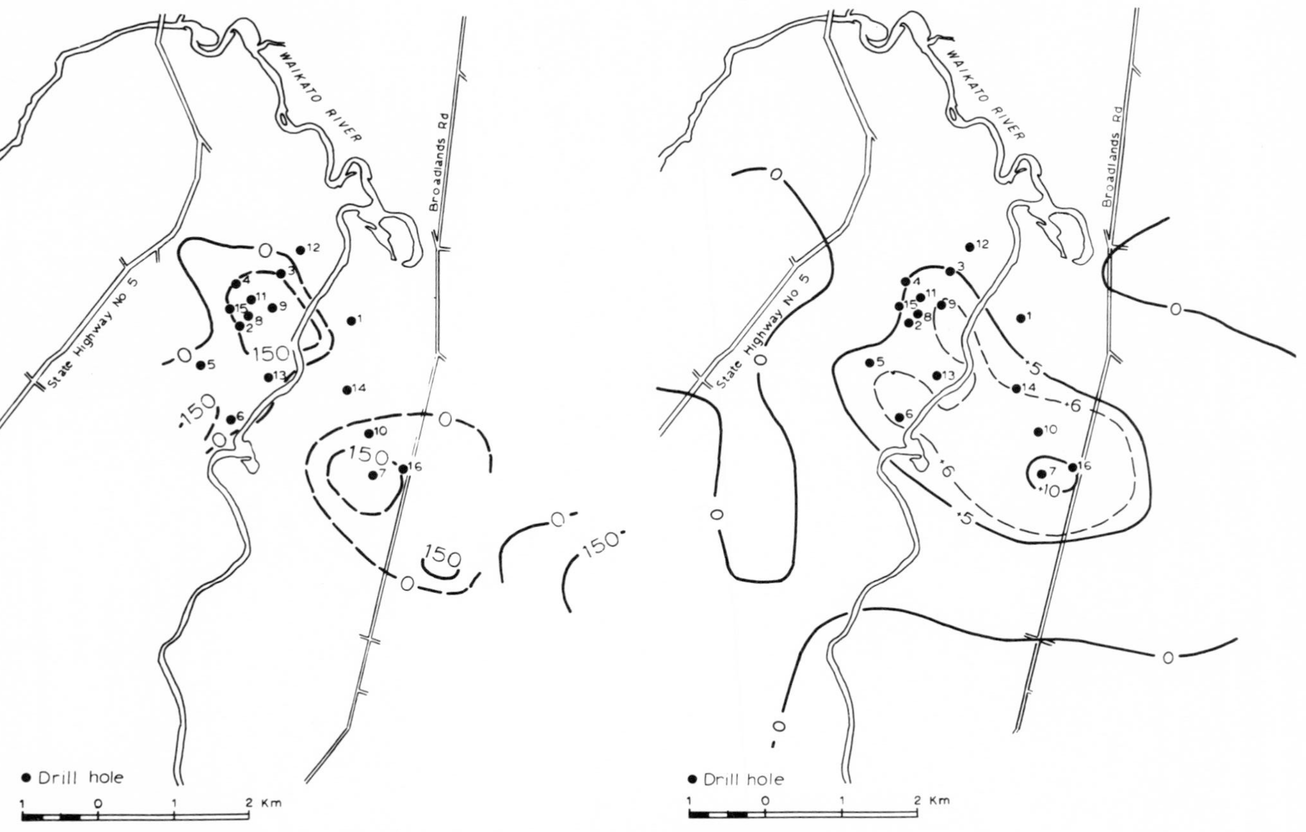

Fig. 5. Upper surface of buried rhyolite domes at the Broadlands geothermal field. Contours are in meters above sea level. (From Hochstein and Hunt, 1970, Fig. 1.)

Fig. 6. Second-order residual-gravity contour map of the Broadlands geothermal area.

cated that the Broadlands geothermal area was larger than the few surface thermal manifestations suggested, and, furthermore, that the lateral boundary of the field at depths near ¾ km was abrupt.

Chemical analyses of water discharged from nine hot springs in the area indicated that the geothermal fluid at depth was a neutral-pH chloride-bicarbonate water of low salinity. Silica concentrations and Na/K ratios suggested a minimum underground temperature of 200°C, with actual temperatures probably 50° to 60° higher (Mahon and Finlayson, 1972, p. 53).

Geological analysis aided by aerial photography suggested that the surface escape of hot water was governed by recently active faults (Fig. 4). Proximity to these fault traces, and in particular to intersections of faults, proved to be important in siting productive drill holes (Grindley, 1970).

A refraction seismic survey (Hochstein and Hunt, 1970) outlined a prominent interface at depths of 50 to 400 m. Petrologic work on drill cores and cuttings subsequently showed that this interface corresponds roughly to the upper surface of buried rhyolite domes (Fig. 5). These domes may be related to the conjugate pattern of recent faults at the surface (Grindley, 1970, p. 260). But the seismic survey provided little information about the depth to graywacke basement beneath Broadlands, mainly because of the high seismic attenuation in the hot subsurface (Hochstein and Hunt, 1970, p. 344).

The gravity survey also proved of little use in detecting basement relief beneath the field, mainly because of the masking effect of lateral density variations within the volcanic sequence (Hochstein and Hunt, 1970, p. 344). The Bouguer anomaly map was modified by removing both the regional anomalies (those owing to density changes at depths greater than 5 km) and the effect of a simply dipping basement surface. The resulting second-order gravity map (Fig. 6) shows three positive gravity anomalies, only a small part of which can be attributed to the rhyolite domes. The remaining part of these second-order anomalies is considered by Hochstein and Hunt (1970) to be due to local increase in density of the volcanic rocks resulting from hydrothermal alteration.

Various electrical and electromagnetic surveys have provided a good indication of the boundaries of the Broadlands geothermal system. Electromagnetic surveying using a horizontal coplanar-loop technique (Lumb and Macdonald, 1970) proved to be a quick, low-cost method

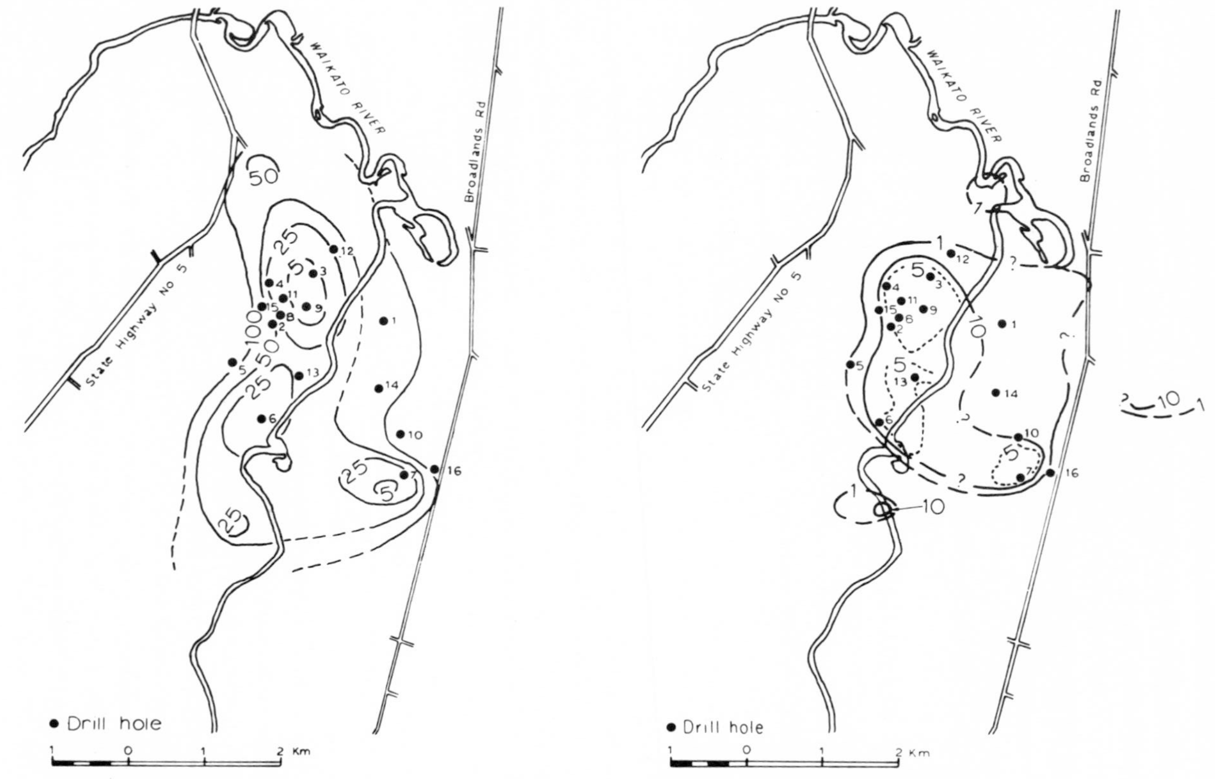

Fig. 7. Contour map showing apparent resistivities derived from the horizontal coplanar-loop electromagnetic (EM gun) survey of the Broadlands geothermal field. Contours are in ohm-meters. (From Lumb and Macdonald, 1970, Fig. 5.)

Fig. 8. Map showing temperature contours in °C at 1 m (dotted lines) and 15 m (solid

for determining resistivities to depths of about 30 m. The resulting map (Fig. 7) shows a close relationship of low-resistivity zones to areas where temperatures at 15 m depth are at least 5°C above ambient (Fig. 8).

Perhaps the most useful geophysical tool in outlining the area of the Broadlands geothermal reservoir at depth was the dc-resistivity survey using a Wenner array with spacing A = 550 m (Fig. 3). Effective probing depth at this spacing is about 760 m (Risk, Macdonald, and Dawson, 1970, p. 293). A similar Wenner-array dc-resistivity survey with A = 180 m (effective probing depth, 250 m) gave results more nearly like those of the electromagnetic map (Fig. 7).

Dipole surveys were used in delimiting the Broadlands geothermal field at effective probing depths of 1.5 and 3.0 km (Fig. 9). But the analysis of the raw dipole data is complex and subject to considerable misinterpretation because of the boundary effects and because of the variation in effective probing depth with electrode spacing and orientation (Risk, Macdonald, and Dawson, 1970, p. 293). Risk determined the field boundaries shown on Fig. 9 by noting where a sudden resistivity drop occurred between adjacent field stations along a line of current flow. Resistivity values determined from a dipole source located outside the field had to be reduced by 50 percent to agree with apparent resistivity measurements made with the source inside the geothermal field.

The magnetic survey failed in its objective of outlining productive areas within the resistivity low (Hochstein and Hunt, 1970). Although the magnetic low in the center of the map (Fig. 10) may be due in part to hydrothermal alteration of magnetic rock to nonmagnetic rock, the anomaly extends well beyond the geothermal-field boundary (as defined by the resistivity survey) and probably is a composite feature related to many factors. Hochstein and Hunt (1970) suggest that the magnetic high to the east may be due to unaltered rocks on the eastern edge of the buried rhyolite dome.

As of 1971, eighteen wells had been drilled within the Broadlands resistivity low. Of these wells, all but one had measured temperatures greater than 270°C (Mahon and Finlayson, 1972, Table 5), but some (wells 1, 10, 12, and 14) have low production rates, owing to low formation permeability (Smith, 1970). Only one well has been drilled in the Broadlands region outside the resistivity low. This well, no. 5, just outside the southwest resistivity boundary, had a maximum temperature of 244°, but inadequate permeability.

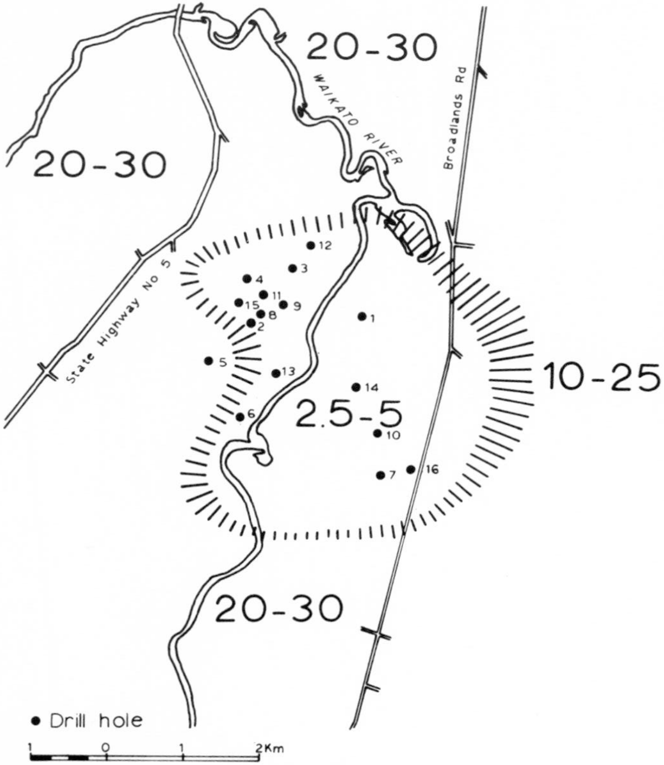

Fig. 9. Boundary of the Broadlands geothermal field, as derived from dipole-resistivity surveys with effective probing depth of 3 km. Corrected apparent resistivities are in ohm-meters. (From Risk, Macdonald, and Dawson, 1970, Fig. 13.)

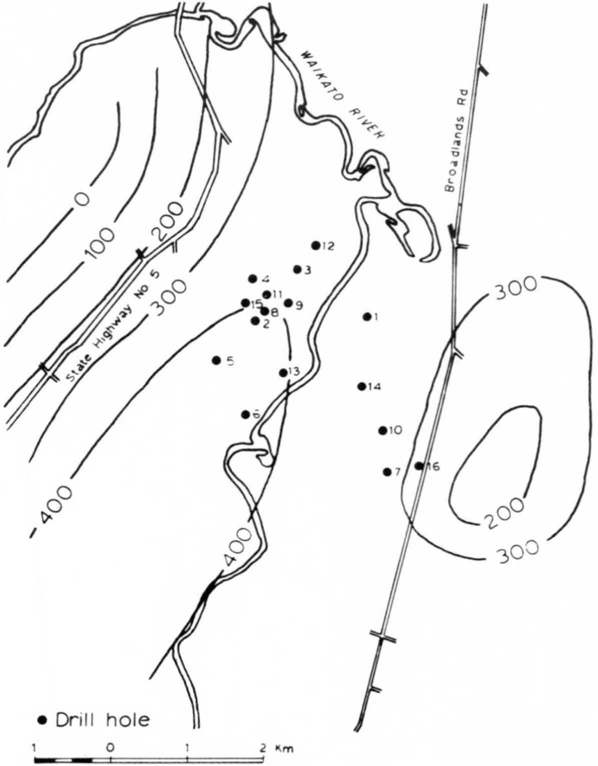

Fig. 10. First-order total-force magnetic-contour map of the Broadlands geothermal

Mesa geothermal anomaly, southeastern California. Until recently, geothermal exploration has been directed primarily toward areas of surface leakage of water or steam. One of the greatest challenges in geothermal exploration is the exploration for large exploitable geothermal reservoirs that have no hydrothermal manifestations at the surface.

The Mesa geothermal anomaly (Fig. 11) appears to be such a hidden reservoir. The anomaly is near the eastern edge of the Salton Trough, the sediment-filled northern extension of the Gulf of California. The trough is filled with up to 4 km of late Cenozoic sediments and sedimentary rocks overlying a basement of higher seismic velocity (Biehler, Kovach, and Allen, 1964). Most of the sedimentary rocks were derived from the Colorado River drainage basin (Merriam and Bandy, 1965; Muffler and Doe, 1968). There are no surface thermal manifestations or volcanic rocks at the Mesa anomaly, and the nearest hot springs and volcanic rocks are at the southeast end of the Salton Sea geothermal field (Helgeson, 1968; Muffler and White, 1969), where the late Cenozoic sedimentary section is intruded at depth by large quantities of igneous rock (Griscom and Muffler, 1971).

Refraction seismic profiling (Biehler, Kovach, and Allen, 1964)

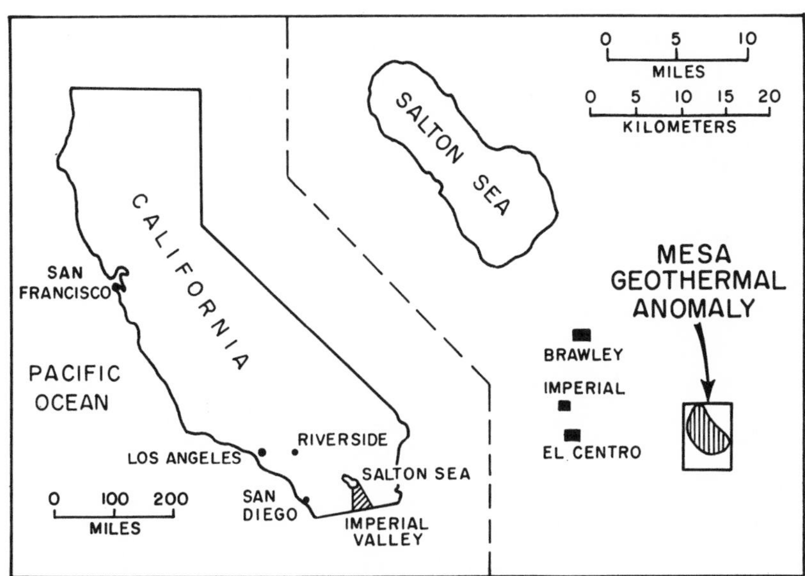

Fig. 11. Location map showing the Imperial Valley and the Mesa geothermal anomaly. (The frame defines the area of the next figures.)

indicates that the crystalline basement at the Mesa anomaly is at a depth of about 3.7 km. An oil-test well 6 km northeast of the Mesa anomaly was drilled to 2,444 m, and a second oil test 4.5 km southeast of the Mesa anomaly was drilled to 3,238 m. Electric logs of both wells show a sequence of sand and clay, probably deltaic; sand predominates throughout the section (Walter Randall, personal communication, 1971).

Geophysical data were obtained through the following field surveys during the period 1969–71: geothermal-gradient measurements in approximately 50 test holes ranging in depth from 30 to 450 m; gravity measurements with station-spacing ranging from several kilometers throughout much of the East Mesa area to about 0.5 km on the Mesa anomaly; seismic-noise survey of the Mesa anomaly with 1.5-km spacing between observations; and widely spaced dc-resistivity soundings.

Figure 12 is a geothermal-gradient map showing increases in temperature with depth. Thirty-seven boreholes ranging in depth from 30 to 450 m were used (Combs, 1971). These specially designed test holes were drilled by conventional rotary methods and completed with small-diameter (1.9 to 5.1 cm) pipe, with a seal at the base. The cavities were backfilled, and the pipes were cemented in place, filled with water, and capped. Temperature measurements made with a thermistor probe were repeated in each hole until thermal equilibrium was attained. The temperature measurements were then processed and evaluated to produce a map (Fig. 12) showing the general shape of the upper part of the thermal feature on the Mesa anomaly. Preliminary analysis indicates heat-flow values from 4 to 10 heat-flow units (Combs, 1972). The thermal high may be a single manifestation of a deeper, substantially larger, convective hydrothermal system.

Detailed gravity and geothermal-gradient measurements (Biehler and Combs, 1972) in the Imperial Valley indicate a very close areal relation between seven very prominent gravity maxima (closures of 2 to 22 mgal) and areas of high geothermal gradients (greater than 11°C/100 m). These geophysical anomalies have no geological expression at the surface.

Fig. 12. Geothermal-gradient map of the Mesa anomaly. Open circles indicate thermal test holes; stars are wildcat oil tests. Contours are in °C per 100 m. (Data compiled and interpreted by J. Combs.)

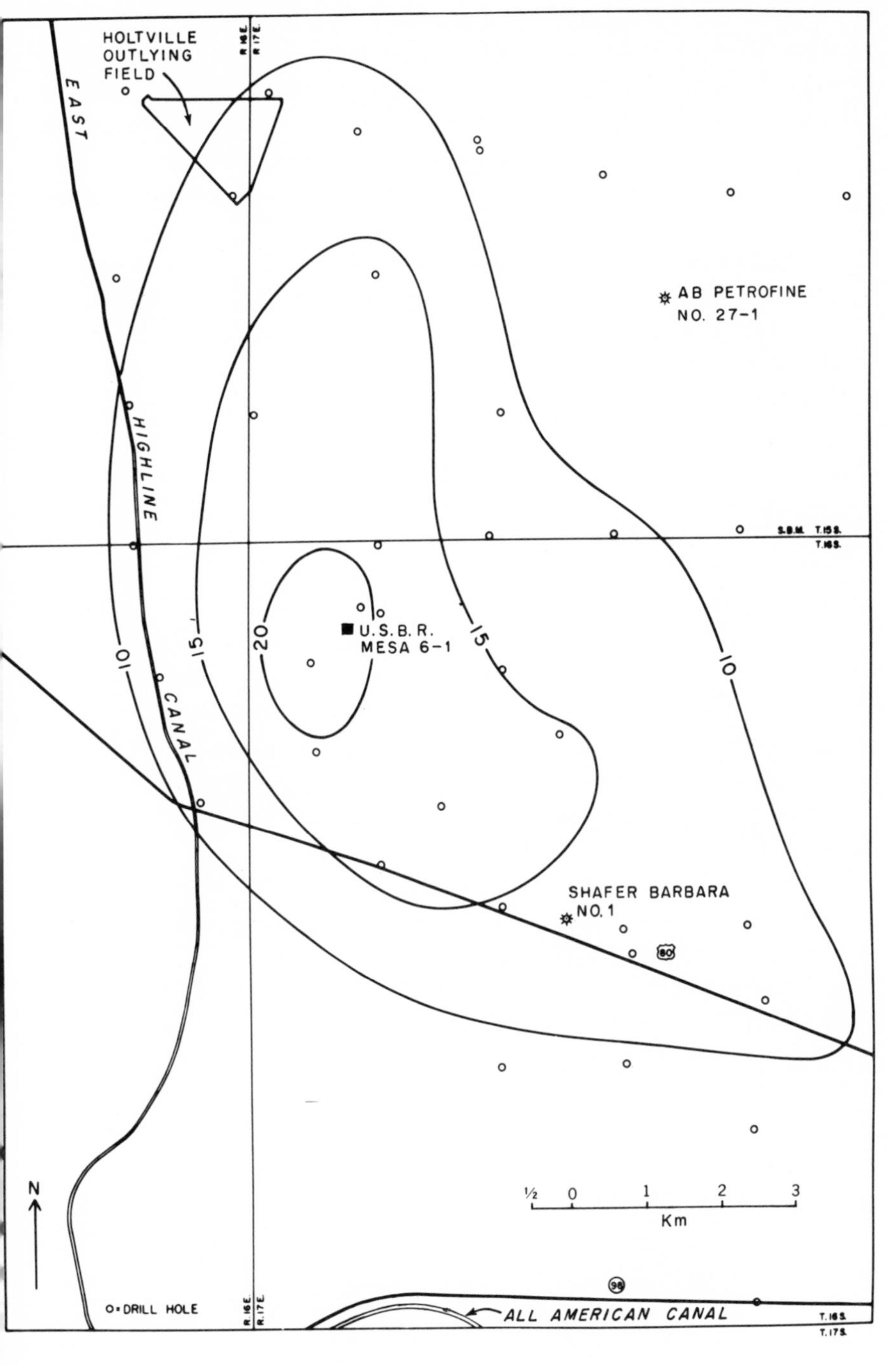
HOLTVILLE
OUTLYING
FIELD
EAST
HIGHLINE
CANAL
R.16E.
R.17E.
AB PETROFINE
NO. 27-1
S.B.M. T.15S.
T.16S.
U.S.B.R.
MESA 6-1
20
15
10
SHAFER BARBARA
NO.1
80
½ 0 1 2 3
Km
N
98
O=DRILL HOLE
ALL AMERICAN CANAL
T.16S.
T.17S.

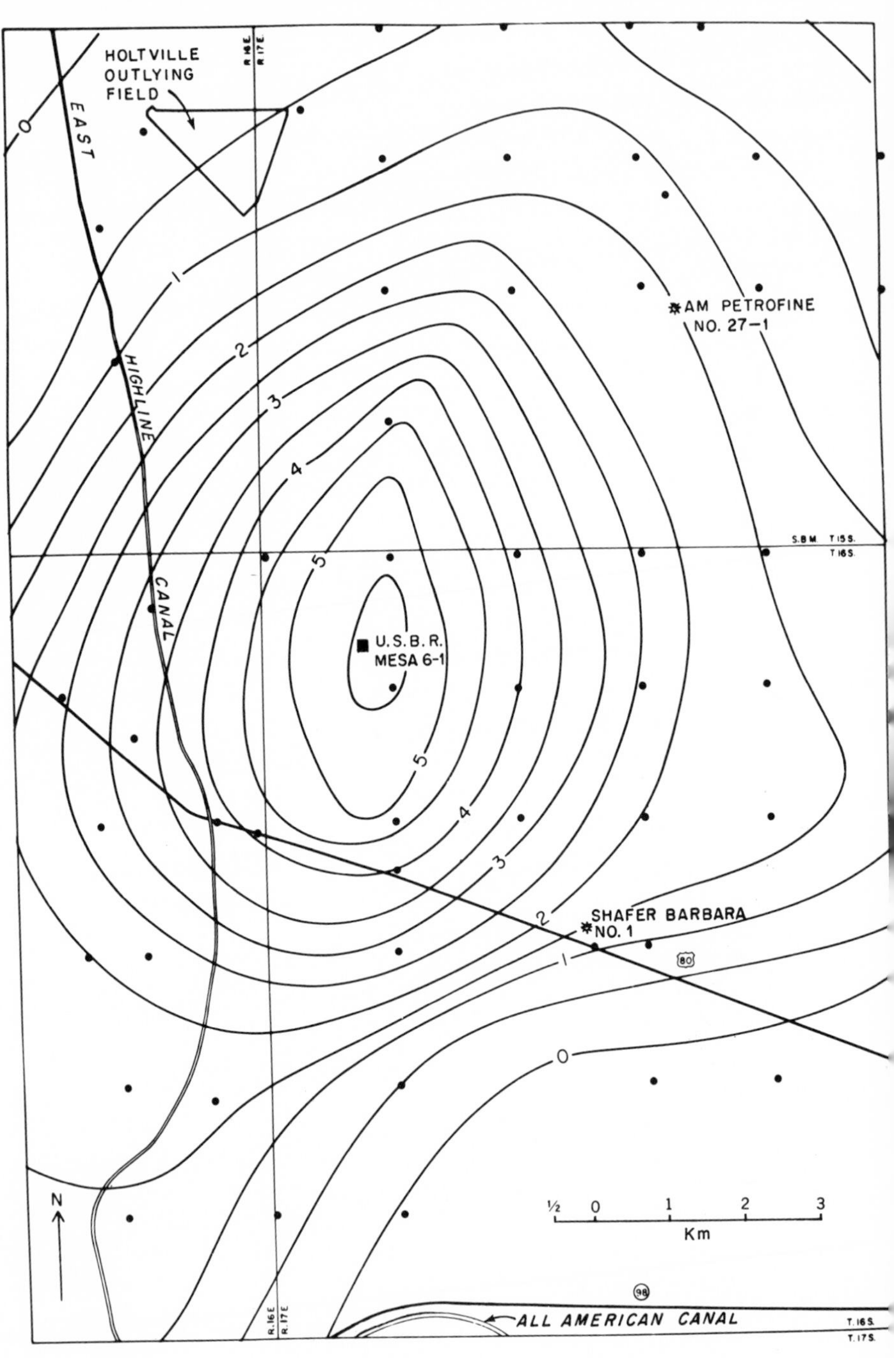

HOLTVILLE
OUTLYING
FIELD
EAST
HIGHLINE
CANAL
R.16E.
R.17E.
0
1
2
3
4
5
U.S.B.R.
MESA 6-1
AM PETROFINE
NO. 27-1
S.B.M.
T.15S.
T.16S.
SHAFER BARBARA
NO. 1
80
N
1/2
0
1
2
3
Km
98
ALL AMERICAN CANAL
T.17S.

After the removal of regional effects arising from basement and upper crustal rocks (Biehler, 1971), the residual-gravity map for the Mesa anomaly (Fig. 13) shows a 4-mgal closure over about 2.5 km^2, suggesting that a mass surplus exists under the area. Biehler and Combs (1972) postulated that the higher gravity near the thermal anomaly reflects one or both of the following: an increase in density of the sediments, owing to cementation and thermal metamorphism by circulating hot brines; or the emplacement of higher-density igneous rocks.

The results of a seismic ground-noise survey conducted by Teledyne-Geotech of Dallas, Texas, are shown in Fig. 14. This survey shows high ground noise in the southern and southeastern parts of the Mesa anomaly. The data suggest an empirical relationship between a possible geothermal reservoir, high thermal gradients, and high seismic-noise level.

A reconnaissance dc-resistivity survey (Meidav and Furgerson, 1971) was made of a large part of the Imperial Valley, including the Mesa anomaly (Fig. 15), using a Schlumberger array with AB/2 ranging from 300 to 2,500 m. The results show a 5-ohm-meter closure over about 50 km^2 around the Mesa anomaly, as defined by the other geophysical methods. Owing to the large distance between data points, this electrical survey does not allow the precise determination of a drilling target in the East Mesa area.

From these geological and geophysical data, an exploratory geothermal well, United States Bureau of Reclamation Mesa #6-1 (see Fig. 11), was sited and drilled in the SW 1/4, NE 1/4, SE 1/4 of Section 6, Township 16 South, Range 17 East. The drilling was completed to a depth of 2,445 m during August 1972. Bottom-hole nonequilibrium temperatures of approximately 200° were measured a few days after the circulation of drilling mud ceased.

ACKNOWLEDGMENTS

We thank B. Bell, L. H. Cohen, B. Greider, D. L. Peck, D. E. White, and two anonymous reviewers for their very helpful comments and criticisms. Financial support for most of the work pertaining to the Mesa geothermal anomaly was provided by the United States Department of

Fig. 13. Residual Bouguer gravity-anomaly map of the Mesa area, contour interval 0.5 mgals. Dots indicate gravity stations; stars indicate wildcat oil wells. (From Biehler, 1971, Fig. 5.)

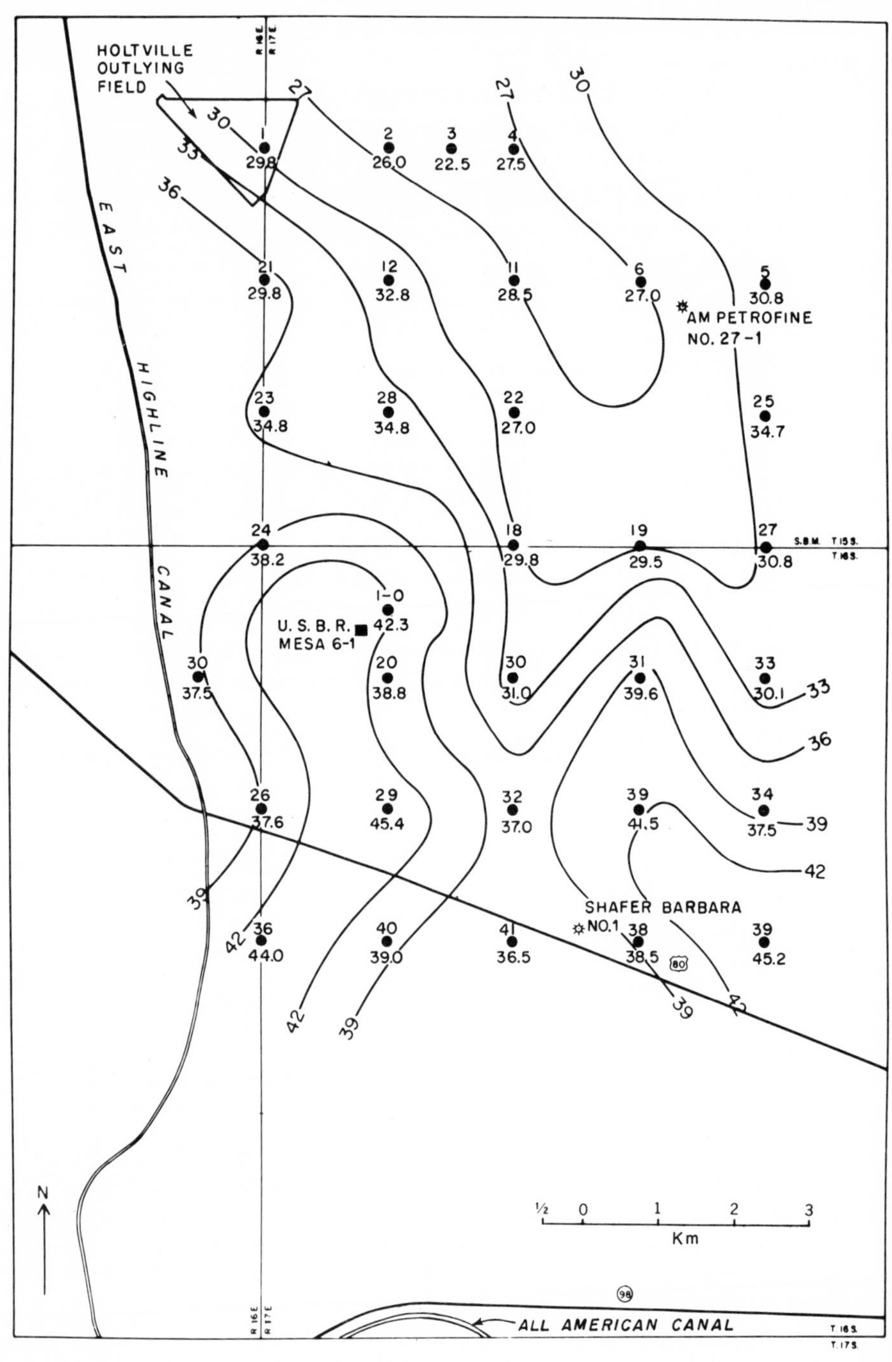
HOLTVILLE
OUTLYING
FIELD
EAST HIGHLINE CANAL
AM PETROFINE
NO. 27-1
U. S. B. R.
MESA 6-1
SHAFER BARBARA
NO.1
ALL AMERICAN CANAL
R 16 E
R 17 E
S.B.M.
T 15 S.
T 16 S.
T 17 S.
N
Km

Interior through Bureau of Reclamation contracts number 14-06-300-2194 and number 14-06-300-2258.

REFERENCES

Banwell, C. J., and W. J. P. Macdonald. 1965. Resistivity surveying in New Zealand thermal areas. Eighth Commonwealth Mining and Metall. Congr., Australia and New Zealand, New Zealand Section, paper no. 213, pp. 1–7.

Barnea, J. 1972. Geothermal power. Scientific American, v. 226, no. 1, pp. 70–77.

Beck, A. E. 1965. Techniques of measuring heat flow on land. *In* W. H. K. Lee, ed., Terrestrial heat flow. Am. Geophys. Union Mon., ser. 8, pp. 24–57.

Biehler, S. 1971. Gravity studies in the Imperial Valley. *In* Cooperative geological-geophysical-geochemical investigations of geothermal resources in the Imperial Valley area of California. Univ. Calif. Riverside, Education Research Service, pp. 29–41.

Biehler, S., and J. Combs. 1972. Correlation of gravity and geothermal anomalies in the Imperial Valley, Southern California (abs.). Geol. Soc. Amer. Abstracts with Programs, v. 4, no. 3, p. 128.

Biehler, S., R. L. Kovach, and C. R. Allen. 1964. Geophysical framework of northern end of Gulf of California structural province. *In* T. H. Van Andel and G. G. Shor, Jr., eds., Marine geology of the Gulf of California—a symposium. Amer. Assoc. Petrol. Geol. Mem. 3, pp. 126–43.

Bodvarsson, G. 1964. Utilization of geothermal energy for heating purposes and combined schemes involving power generation, heating, and/or by-products. Proc. U.N. Conf. New Sources of Energy, Rome, 1961, v. 3, Geothermal Energy, pp. 429–36.

——— 1970. Evaluation of geothermal prospects and the objectives of geothermal exploration. Geoexploration, v. 8, pp. 7–17.

Boldizsár, T. 1963. Terrestrial heat flow in the natural steam field at Larderello. Geofs. Pura Appl., v. 56, pp. 115–22.

——— 1970. Geothermal energy production from porous sediments in Hungary. Geothermics, special issue 2, v. 2, pt. 1, pp. 99–109.

Browne, P. R. L., and A. J. Ellis. 1970. The Ohaki-Broadlands hydrothermal area, New Zealand: Mineralogy and related geochemistry. Am. J. Sci., v. 269, pp. 97–131.

Brune, J. N., and C. R. Allen. 1967. A micro-earthquake survey of the San Andreas fault system in southern California. Seismol. Soc. Am. Bull., v. 57, pp. 277–96.

Bureau of Reclamation. 1972. Geothermal resource investigations, Imperial Valley, California, January 1972—developmental concepts. U.S. Bur. Reclamation, 58 pp.

Fig. 14. Seismic ground-noise map of the Mesa anomaly. Dots indicate stations; stars indicate wildcat oil tests. Contours are given in decibels relative to 1 (millimicron/sec)^2 per Hertz in the passband of 3.0 to 5.0 Hz. (Data compiled and interpreted under the technical direction of E. J. Douze, Teledyne-Geotech, Dallas, Texas, August 1971.)

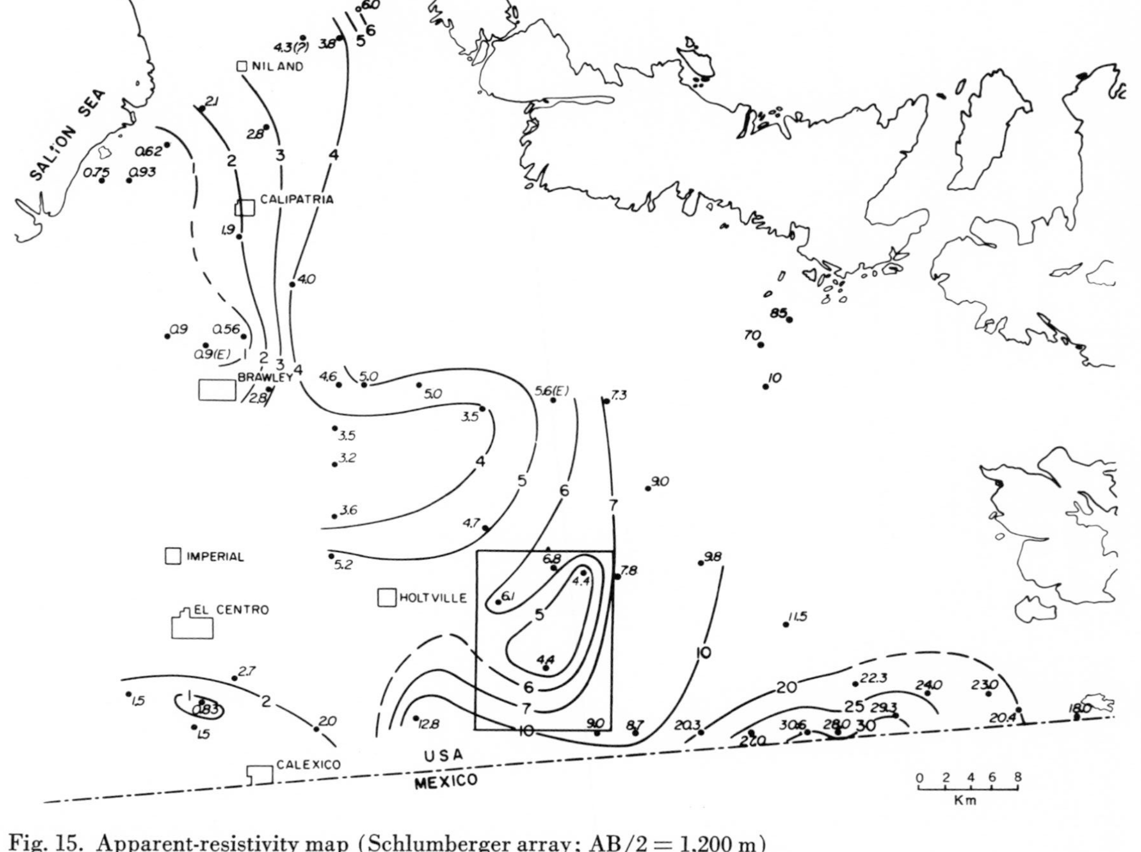

Fig. 15. Apparent-resistivity map (Schlumberger array; AB/2 = 1,200 m) for the Imperial Valley. Dots indicate resistivity soundings; numerical values are apparent resistivity in ohm-meters. Enclosed areas are basement outcrop. (After Meidav and Furgerson, 1971, Fig. 3, p. 47.)

Burgassi, R., F. Battini, and J. Mouton. 1964. Prospection géothermique pour la recherche des forces endogènes. Proc. U.N. Conf. New Sources of Energy, Rome, 1961, v. 2, Geothermal Energy, pp. 134–40.

Burgassi, P. D., P. Ceron, G. C. Ferrara, G. Sestini, and B. Toro. 1970. Geothermal gradient and heat flow in the Radicofani region (east of Monte Amiata, Italy). Geothermics, special issue 2, v. 2, pt. 1, pp. 443–49.

Clacy, G. R. T. 1968. Geothermal ground noise amplitude and frequency spectra in the New Zealand volcanic region. J. Geophys. Res., v. 73, pp. 5377–83.

Combs, J. 1971. Heat flow and geothermal resource estimates for the Imperial Valley. *In* Cooperative geological-geophysical-geochemical investigations of geothermal resources in the Imperial Valley area of California. University of California, Riverside, Education Research Service, pp. 5–27.

——— 1972. Preliminary heat flow values and temperature distributions associated with the Mesa and Dunes geothermal anomalies, Imperial Valley, southern California (abs.). EOS, v. 53, no. 4, pp. 515–16.

Combs, J., and R. W. Rex. 1971. Geothermal investigations in the Imperial Valley of California (abs.). Geol. Soc. Am. Abstracts with Programs, v. 3, no. 2, pp. 101–2.

Coplen, T. 1971. Isotopic geochemistry of water from the Imperial Valley. *In* Cooperative geological-geophysical-geochemical investigations of geothermal resources in the Imperial Valley area of California. University of California, Riverside, Education Research Service, pp. 113–18.

Craig, H. 1966. Isotopic composition and origin of the Red Sea and Salton Sea geothermal brines. Science, v. 154, pp. 1544–48.

Dewey, J. F., and J. M. Bird. 1970. Mountain belts and the new global tectonics. Jour. Geophys. Res., v. 75, pp. 2625–47.

Duprat, A. 1970. Contribution de la geophysique a l'étude de la région géothermique de Denizli-Sarayköy, Turquie. Geothermics, special issue 2, v. 2, pt. 1, pp. 275–86.

Ellis, A. J. 1969. Present-day hydrothermal systems and mineral deposition. Ninth Commonwealth Mining and Metall. Congr., Mining and Petroleum Geology Section, paper no. 7, pp. 1–30.

Facca, G., and F. Tonani. 1967. The self-sealing geothermal field. Bull. Volcanol., v. 30, pp. 271–73.

Fournier, R. O., and J. J. Rowe. 1966. Estimation of underground temperatures from the silica content of water from hot springs and wet-steam wells. Am. Jour. Sci., v. 264, pp. 685–97.

Fournier, R. O., and A. H. Truesdell. 1973. An empirical Na-K-Ca geothermometer for natural waters. Geochim. et Cosmochim. Acta, in press.

Friedman, J. D., R. S. Williams, Jr., G. Pálmason, and C. D. Miller. 1969. Infrared surveys in Iceland—preliminary report. U.S. Geol. Survey Prof. Paper 650-C, pp. C89–C105.

Goforth, T. T., E. J. Douze, and G. G. Sorrells. 1972. Seismic noise measurements in a geothermal area. Geophysical Prospecting, v. 20, pp. 76–82.

Gómez Valle, R. G., J. D. Friedman, S. J. Gawarecki, and C. J. Banwell. 1970. Photogeologic and thermal infrared reconnaissance surveys of the Los Negritos–Ixtlan de los Hervores geothermal area, Michoacan, Mexico. Geothermics, special issue 2, v. 2, pt. 1, pp. 381–98.

Grindley, G. W. 1965. The geology, structure, and exploitation of the Wairakei geothermal field, Taupo, New Zealand. New Zealand Geol. Survey Bull. n.s. 75, 131 pp.

——— 1970. Subsurface structures and relation to steam production in the Broadlands geothermal field, New Zealand. Geothermics, special issue 2, v. 2, pt. 1, pp. 248–61.

Griscom, A., and L. J. P. Muffler. 1971. Aeromagnetic map and interpretation of the Salton Sea geothermal area, California. U.S. Geol. Survey Geophys. Inv. Map GP-754.

Grose, L. T. 1971. Geothermal energy: Geology, exploration and developments. Part 1. Colorado School of Mines, Mineral Industries Bull., v. 14, no. 6, pp. 1–14.

——— 1972. Geothermal energy: Geology, exploration and developments. Part 2. Colorado School of Mines, Mineral Industries Bull., v. 15, no. 1, pp. 1–16.

Hamilton, R. M., and L. J. P. Muffler. 1972. Microearthquakes at The Geysers geothermal area, California. Jour. Geophys. Res., v. 77, pp. 2081–86.

Harthill, N. 1971. Geophysical prospecting for geothermal energy. The Mines Magazine, Colorado School of Mines, June 1971, pp. 13–18.

Hatherton, T., W. J. P. Macdonald, and G. E. K. Thompson. 1966. Geophysical methods in geothermal prospecting in New Zealand. Bull. Volcanol., v. 29, pp. 485–97.

Hayakawa, M. 1970. The study of underground structure and geophysical state in geothermal areas by seismic exploration. Geothermics, special issue 2, v. 2, pt. 1, pp. 347–57.

Healy, J. 1962. Structure and volcanism in the Taupo Volcanic Zone, New Zealand. *In* G. A. Macdonald and H. Kuno, eds., The crust of the Pacific Basin. Am. Geophys. Union Mon., ser. 6, pp. 151–57.

——— 1964. Volcanic mechanisms in the Taupo Volcanic Zone, New Zealand. N.Z. Jour. Geol. and Geophys., v. 7, no. 1, pp. 6–23.

——— 1970. Pre-investigation geological appraisal of geothermal fields. Geothermics, special issue 2, v. 2, pt. 1, pp. 571–77.

Helgeson, H. C. 1968. Geologic and thermodynamic characteristics of the Salton Sea geothermal system. Am. Jour. Sci., v. 266, pp. 129–66.

Hochstein, M. P., and D. J. Dickinson. 1970. Infrared remote sensing of thermal ground in the Taupo region, New Zealand. Geothermics, special issue 2, v. 2, pt. 1, pp. 420–23.

Hochstein, M. P., and T. M. Hunt. 1970. Seismic, gravity and magnetic studies, Broadlands geothermal field, New Zealand. Geothermics, special issue 2, v. 2, pt. 1, pp. 333–46.

Hodder, D. T. 1970. Application of remote sensing to geothermal prospecting. Geothermics, special issue 2, v. 2, pt. 1, pp. 368–80.

Jaffé, F. C. 1971. Geothermal energy, a review. Bull. Ver. Schweiz. Petrol.-Geol. u. -Ing., v. 38, no. 93, pp. 17–40.

Jones, P. H. 1970. Geothermal resources of the Northern Gulf of Mexico Basin. Geothermics, special issue 2, v. 2, pt. 1, pp. 14–26.

Keller, G. V. 1970. Induction methods in prospecting for hot water. Geothermics, special issue 2, v. 2, pt. 1, pp. 318–32.

Kresl, M., and V. Novak. 1970. Terrestrial heat flow in the territory of Czechoslovakia and the measurement of thermal conductivity with fully automatic apparatus. Geothermics, special issue 2, v. 2, pt. 2, in press.

Lange, A. L., and W. H. Westphal. 1969. Microearthquakes near The Geysers, Sonoma County, California. Jour. Geophys. Res., v. 74, pp. 4377–78.

Lee, W. H. K., and S. Uyeda. 1965. Review of heat flow data. *In* W. H. K. Lee, ed., Terrestrial heat flow. Am. Geophys. Union Mon., ser. 8, pp. 87–190.

Lovering, T. S., and H. D. Goode. 1963. Measuring geothermal gradients in drill holes less than 60 feet deep, East Tintic district, Utah. U.S. Geol. Survey Bull. 1172, 48 pp.

Lumb, J. T., and W. J. P. Macdonald. 1970. Near-surface resistivity surveys of geothermal areás using electromagnetic method. Geothermics, special issue 2, v. 2, pt. 1, pp. 311–17.

Macdonald, W. J. P., and L. J. P. Muffler. 1972. Recent geophysical exploration of the Kawerau geothermal field, North Island, New Zealand. N.Z. Jour. Geol. and Geophysics, v. 15, pp. 303–17.

McNitt, J. R. 1965. Review of geothermal resources. *In* W. H. K. Lee, ed., Terrestrial heat flow. Am. Geophys. Union Mon., ser. 8, pp. 240–66.

Mahon, W. A. J., and J. B. Finlayson. 1972. The chemistry of the Broadlands geothermal area, New Zealand. Am. Jour. Sci., v. 272, no. 1, pp. 48–68.

Makarenko, F. A., B. F. Mavritsky, B. A. Lokchine, and V. I. Kononov. 1970. Geothermal resources of the USSR and prospects for their practical use. Geothermics, special issue 2, v. 2, pt. 2, in press.

Meidav, T., and R. Furgerson. 1971. Electrical resistivity for geothermal exploration in the Imperial Valley. *In* Cooperative geological-geophysical-geochemical investigations of geothermal resources in the Imperial Valley area of California. University of California, Riverside, Education Research Service, pp. 43–88.

Merriam, R., and O. L. Bandy. 1965. Source of upper Cenozoic sediments in Colorado delta region. Jour. Sed. Pet., v. 35, pp. 911–16.

Muffler, L. J. P., and B. R. Doe. 1968. Composition and mean age of detritus of the Colorado River delta in the Salton Trough, southeastern California. Jour. Sed. Pet., v. 38, pp. 384–99.

Muffler, L. J. P., and D. E. White. 1969. Active metamorphism of upper Cenozoic sediments in the Salton Sea geothermal field and the Salton Trough, southeastern California. Geol. Soc. Am. Bull., v. 80, pp. 157–82.

——— 1972. Geothermal energy. The Science Teacher, v. 39, no. 3, pp. 40–43.

Pálmason, G., J. D. Friedman, R. S. Williams, Jr., J. Jónsson, and K. Saemundsson. 1970. Aerial infrared surveys of Reykjanes and Torfajökull thermal areas, Iceland, with a section on the cost of exploration surveys. Geothermics, special issue 2, v. 2, pt. 1, pp. 399–412.

Risk, G. F., W. J. P. Macdonald, and G. B. Dawson. 1970. D.C. resistivity surveys of the Broadlands geothermal region, New Zealand. Geothermics, special issue 2, v. 2, pt. 1, pp. 287–94.

Sass, J. H. 1971. The earth's heat and internal temperatures. *In* I. G. Gass, P. J. Smith, and R. C. L. Wilson, eds., Understanding the earth. Sedgwick Park, Eng.: Artemis Press, pp. 81–87.

Sestini, G. 1970. Heat-flow measurement in non-homogeneous terrains. Its application to geothermal areas. Geothermics, special issue 2, v. 2, pt. 1, pp. 424–36.

Simmons, G., and K. Horai. 1968. Heat flow data 2. Jour. Geophys. Res., v. 73, pp. 6608–29.

Smith, J. H. 1970. Geothermal development in New Zealand. Geothermics, special issue 2, v. 2, pt. 1, pp. 232–47.

Steiner, A. 1968. Clay minerals in hydrothermally altered rocks at Wairakei, New Zealand. Clays and Clay Minerals, v. 16, pp. 193–213.

Studt, F. E. 1964. Geophysical prospecting in New Zealand's hydrothermal fields. Proc. U.N. Conf. New Sources of Energy, Rome, 1961, v. 2, Geothermal Energy, pp. 380–85.

Thompson, G. E. K., C. J. Banwell, G. B. Dawson, and D. J. Dickinson. 1964. Prospecting of hydrothermal areas by surface thermal surveys. Proc. U.N. Conf. New Sources of Energy, Rome, 1961, v. 2, Geothermal Energy, pp. 386–401.

Ward, P. L., and S. Björnsson. 1971. Microearthquakes, swarms, and the geothermal areas of Iceland. Jour. Geophys. Res., v. 76, pp. 3953–82.

Ward, P. L., and K. H. Jacob. 1971. Microearthquakes in the Ahuachapan geothermal field, El Salvador, Central America. Science, v. 173, pp. 328–30.

Ward, P. L., G. Pálmason, and C. Drake. 1969. Microearthquake survey and the mid-Atlantic ridge in Iceland. Jour. Geophys. Res., v. 74, pp. 665–84.

White, D. E. 1965. Geothermal energy. U.S. Geol. Survey Circular 519, 17 pp.

——— 1968. Hydrology, activity, and heat flow of the Steamboat Springs thermal system, Washoe County, Nevada. U.S. Geol. Survey Prof. Paper 458-C, 109 pp.

——— 1969. Rapid heat-flow surveying of geothermal areas, utilizing individual snowfalls as calorimeters. Jour. Geophys. Res., v. 74, pp. 5191–201.

——— 1970. Geochemistry applied to the discovery, evaluation, and exploitation of geothermal energy resources. Geothermics, special issue 2, v. 1, in press.

White, D. E., and L. D. Miller. 1969. Calibration of geothermal infrared anomalies of low intensity in terms of heat flow, Yellowstone National Park (abs.). Am. Geophys. Union Trans., v. 50, no. 4, p. 348.

Whiteford, P. C. 1970. Ground movement in the Waiotapu geothermal region, New Zealand. Geothermics, special issue 2, v. 2, pt. 1, pp. 478–86.

6. Steam Production at The Geysers Geothermal Field

CHESTER F. BUDD, JR.

The production of geothermal fluids from a subterranean reservoir is comparable to the production of oil and gas. The extractive system is basically a well, drilled to a depth sufficient to encounter geothermal fluids, cased and completed to provide a stable conduit through which the fluids can flow to the surface, and furnished with the facilities necessary to control and transport the fluids to their point of utilization.

Geothermal fluids are found to exist either as liquid water or as gaseous steam. Most known geothermal systems are hot-water reservoirs at high pressure and temperature, and produce steam only as a result of flashing to a lower pressure in the wellbore (Koenig, 1972). A few systems that exist at pressures substantially below hydrostatic contain gaseous steam at saturation conditions, probably coexisting with water. The dominant phase, steam or water, governs the production mechanism operating within the reservoir and dictates reservoir-management principles and production practices. At The Geysers in northern Sonoma County, California, the dominant phase is gaseous steam, a condition existing in only a few major commercially producing fields in the world, e.g., the Larderello field, in Italy, and the Matsukawa field, in Japan.

Union Oil Company of California has been operating The Geysers geothermal field since entering into a joint venture with the original developers, the Magma Power Corporation and Thermal Power Company, in 1967. During this time, the installed generating capacity has risen from 54 Mw to 302 Mw and the proved capacity of the field has

Chester F. Budd, Jr., is with the Union Oil Company of California, Big Geysers Station, Cloverdale.

risen from 192 Mw to 750 Mw. Exploratory drilling is under way to extend the known productive limits of the field. Development drilling within the proved area is continuing to add generating capacity at a rate of approximately 100 Mw per year while maintaining steam-production rates for the existing plants.

Planning and Development

The steam produced at The Geysers is presently sold to the Pacific Gas and Electric Company for utilization in condensing turbines driving electric-power generators. The reservoir pressure of 450–500 psig reduces to about 125 psig at the wellhead, as the steam, in flowing from the reservoir to the surface, expands and cools; and the turbines are thus designed to operate at 80–100 psig intake pressure. Geothermal steam, because it rapidly loses its heat energy when transported through a pipeline, must be utilized at the steam field rather than at a distant market, as is the case with more conventional fuels. The utility, then, must build its generating facilities close to the wells. This requires that an early agreement be reached between operator and utility on the location of a site on which the generating plant is to be built.

The turbines require a constant throughput of steam to operate at optimum capacity. Each of the generating units now being installed at The Geysers requires a throughput of about 1 million lb/hr of superheated steam, from which it generates about 55 Mw of electricity. Since the average initial per-well production rate at The Geysers is approximately 150,000 lb/hr, seven wells are required for each unit. As production from these wells declines, new wells are drilled and added into the system, ensuring a steady throughput of steam to the generator. With this continued demand for new drilling to maintain production, the operator and the utility must agree on the minimum life to be expected of each generating unit, so that sufficient productive acreage on which to drill wells for that unit can be set aside. The acreage thus set aside must be adjacent to the area originally developed for that unit and held in reserve specifically for its use, so that the new wells can be easily fed into the existing flow-gathering system.

The total capacity of a geothermal field depends on how many such "blocks-of-power" can be proved up within the productive field limits, the size of each block being related directly to the output of the generating unit and the equipment life over which the utility and operator agree to amortize their investment. Exploration techniques used to locate geothermal prospects usually identify the limits of potentially productive areas only in very general terms (Combs and Muffler, this volume). For example, at The Geysers the known gravity anomaly, as described by Garrison (1972), covers 100 mi^2. Within this anomaly, there are four widely scattered wells drilled below 7,000 ft that nonetheless did not produce steam. Their failure to produce is attributed to a local lack of permeability within the steam reservoir. As in oil and gas development, the final and definitive exploration tool in geothermal development is the drill bit that must penetrate the formations to ascertain the presence of heat, fluid, and permeability, the necessary ingredients of a commercial geothermal resource.

Proving up and confirming the production capacity necessary for a block-of-power follows a logical sequence of steps. The plants now being built for The Geysers are of 110-Mw capacity, in pairs of 55-Mw units, one such installation to be completed each year. The general area in which to build each plant is located by exploratory drilling. The areal limits for a proposed plant site are confirmed by drilling a few wells to determine the depth and productivity of the steam reservoir. These early tests provide engineers with the information necessary to determine how much productive area should be dedicated to the proposed plant site, how many wells are initially required, and how many additional wells will be required to maintain a level rate of production for the plant life. This has been accomplished with as few as two widely spaced wells.

Under current California Public Utility Commission regulations, utility companies are required to submit applications for plant sites 1 year in advance of commencing construction, which in practice means at The Geysers about 4 years before anticipated operation. The required steam supply, moreover, must be available for about 1 year to allow for pipeline construction. Thus, allowing time for evaluation subsequent to discovery, a lead time of 5 years between confirmation drilling and commercial operation is required. All of these delays combine to lower the rate of return on the investment.

Drilling and Completion Practices

Drilling geothermal steam wells has presented problems not encountered in drilling oil and gas wells. The first is that of the formation itself. At The Geysers, the producing zone is the graywacke sandstone of the Franciscan formation. (In oilfield country the Franciscan formation is called "basement rock.") The graywacke sandstone is a tight metamorphosed sandstone that causes extremely slow penetration rates even with bits specifically designed for hard-rock drilling. Drilling with mud causes "lost circulation" to permeable zones with fluid pressures far below hydrostatic. Drilling with air has improved penetration rates over that with mud and eliminated the lost-circulation problems, but air drilling can be done only below segments where significant water-bearing strata have been cased off, since the water would enter the bore-hole, causing sloughing of the formations. The heat and abrasiveness of the formation take their toll of subsurface tools and equipment. And once steam is encountered, air-circulation pressures rise and annular velocities approach sonic velocity, causing tool joints to suffer the erosional wear of high-speed particle impact. Finally, if mechanical failures following completion of the well are to be avoided, the engineering design of the casing and the choice of cementing materials must be carefully matched to the circumstances existing in the hole.

The production rate that can be attained from a completed steam well is highly sensitive to hole and casing sizes. These wells are generally capable of very high mass-flow rates, and the frictional resistance of the casing and hole may limit the flow rates. Much attention has been given to the design of an optimum drilling and completion program at The Geysers in order to ensure a favorable balance between cost and flow rate. From back-pressure testing (Craft and Hawkins, 1959) and isochronal testing (Cullender, 1955), it has been found that a steam well at The Geysers conforms to the classic formula for the performance of a gas well. The equation may be expressed as

$$W = C(P_s^2 - P_f^2)^n$$

where

W = steam flow rate, in lb/hr
C = factor, a function of time, reservoir matrix, fluid properties, flow rate, wellbore condition, etc.
P_s = static reservoir pressure, in psia
P_f = wellbore pressure at steam entry, in psia
n = exponent, usually constant with W within the range of usual production rates, $0.5 \leq n \leq 1.0$

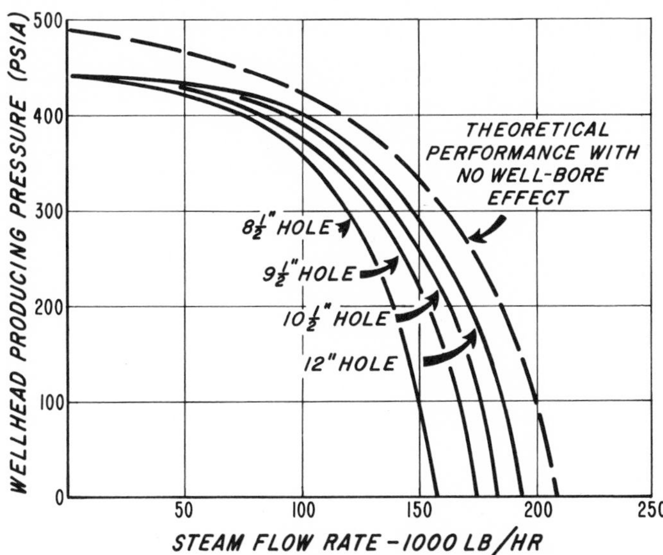

Fig. 1. Steam-flow rate vs. wellhead pressure for various equivalent pipe diameters for a typical 5,000-ft steam well at The Geysers.

Production, W, is therefore dependent upon the bottom hole pressure, P_f, which is the sum of the wellhead flowing pressure, the total frictional resistance pressure, and the column weight of the fluid. Fig. 1 shows the variation in theoretical performance of a typical geothermal steam well at The Geysers when completed with various casing and hole sizes. Smaller-diameter casings produce higher friction loss and lower total steam flow, other conditions being equal.

At The Geysers, the cost per foot of drilling increases rapidly with depth. Completed wells are as shallow as 600 ft and as deep as 9,000 ft. The probability of finding an economic steam-producing well—i.e. one yielding a reasonable return on investment—diminishes rapidly with increased depth simply because the length of the steam-flow conduit increases. This is illustrated in Fig. 2; with other factors held constant, a well completed at 10,000 ft delivers almost 20 percent less steam than the same well completed at 5,000 ft. The problem, then, is to determine the bore configuration that yields the highest flow rate per dollar of drilling investment. As it happens, a combination of wellbores and open hole, such as that shown in Fig. 3, has been found to be optimum at The Geysers. The wellbore consists of 1,900 ft of 13⅜-inch casing,

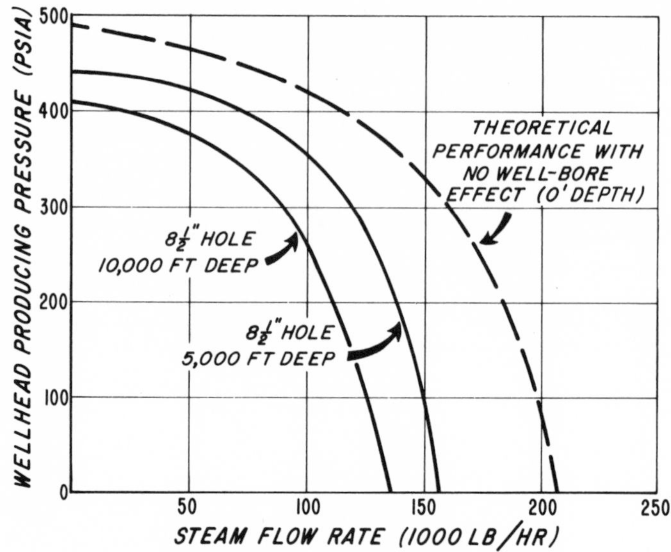

Fig. 2. Steam-flow rate vs. wellhead pressure for two well depths for a typical steam well at The Geysers.

2,000 ft of 9⅝-inch casing, and 2,000 ft of 8¾-inch open hole to the steam entry.

Obviously, an optimization analysis can be undertaken only after a well is drilled and tests are performed to determine values for the constants of the steam-flow equation. At The Geysers, as at Larderello, there appears to be no way to predict the productivity of a well prior to drilling, hence the values for C and n cannot be accurately predicted. Thus the optimization calculation must draw upon typical values from typical wells so that conditions close to average will be encountered.

Reservoir Performance

As is the case with the exploitation of an oil or gas field, the potential of a geothermal field can be predicted only after the producing characteristics of the reservoir have been determined. And as with oil and gas fields, each geothermal field can be expected to have its own reservoir matrix and fluid properties and thus its own producing characteristics. The vapor-dominated system at The Geysers has been the subject of a continuing reservoir study by Union Oil Company and others since late 1967 (Ramey, 1968). The performance of the field

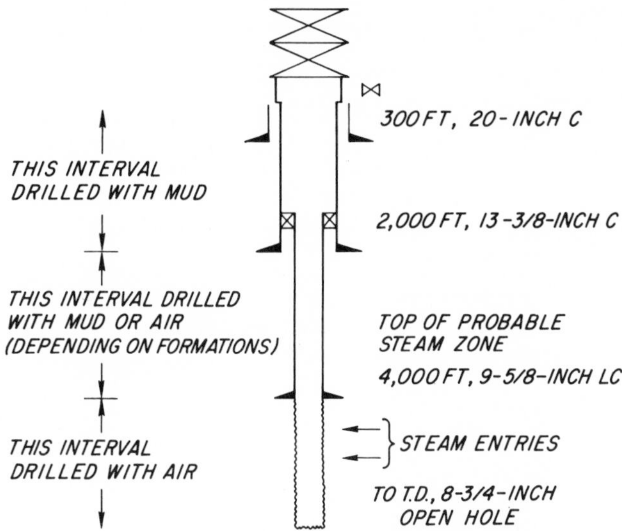

Fig. 3. Typical well-completion configuration at The Geysers.

has shown that, contrary to earlier speculation, a constant rate of production cannot be sustained for an essentially infinite time. Individual wells have declined in production, and additional wells have been required to maintain the supply of steam to the generating units. This has been the experience at the Larderello field as well, where continued drilling is required to maintain a constant generating capacity of 365 Mw.

In oil and gas work, new fields are often evaluated by observing the production performance of the new wells relative to the performance of similar wells in fields that have been producing over a greater time span, some of which are in their terminal stages of production. Such documented history of vapor-dominated geothermal systems does not exist. Even though the Larderello field, the other major producing vapor-dominated geothermal reservoir, has been producing since the turn of the century, its production rate has been quite low with respect to its capacity and reserves. Consequently, Larderello cannot be considered a complete model of performance history. Furthermore, records of well production, static reservoir pressure, and pressure buildup tests have been maintained there only recently.

At The Geysers field, as well, records of production and pressure were

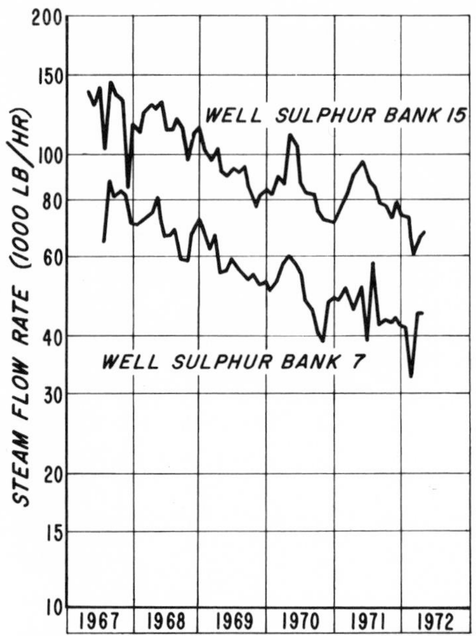

Fig. 4. Steam-production rate vs. time for geothermal steam wells at The Geysers.

not maintained until mid-1967. But examination of older data, such as diaries and correspondence, has enabled Union Oil engineers to piece together a credible production and pressure history of the wells that produced prior to that time. These data and the post-1967 records constitute a fairly comprehensive case history of a vapor-dominated system. The application of fundamental physical principles to this history has been the basis of a reservoir study directed toward predicting future performance.

Typical production performance of two wells at The Geysers is shown in Fig. 4. The two wells, with other producing wells, are part of a block-of-power that has been producing steam for generating unit 3 since 1967. The average well density is one well per 5 acres; and as the figure shows, productivity from both wells has declined with time.

Throughout their producing life, the wellhead flowing pressure at the two wells has been roughly constant. Thus the variation and decline in productivity is not due to changing surface conditions. Pressure buildup tests on these wells have been performed periodically since February 1967. These tests have shown that no near-wellbore permeability change has occurred, as would be the case if the drop in production were caused by incrustation of the wellbore or some other mechanical plugging phenomenon in the wellbore or in the formation. These data suggest, rather, that pressure depletion in the reservoir is a factor in the production decline, and field-wide static reservoir-pressure surveys have confirmed this suggestion.

Pressure interference occurring between wells is another phenomenon observed in the particular area of the field where the two wells are located. Owing to the matrix connectivity of the reservoir, the performance of one well is greatly influenced by the status of nearby wells. The static pressures of shut-in wells are reduced by production from nearby wells. And when producing wells are closed-in, the static pressures of the nearby shut-in wells usually recover to values near the static pressure prevailing in the reservoir. Conversely, producing wells manifest this interference in decreased production rates when nearby wells are placed on production, or increased rates when nearby producing wells are closed-in. This condition has led engineers to attempt a mathematical simulation of the flow conditions within the steam reservoir. The model is used to optimize well spacing in future expansion of the field and to set out a development program requiring a minimum of new wells during the life of the project while developing the field to its fullest potential.

Various spacing alternatives have been studied; typical results are shown in Fig. 5. As expected, the interference effect is greatest with the highest well density, one well per 5 acres. The exact shape of each production curve is dependent upon completion technique and depth, as discussed earlier. The curves show clearly that the production rate falls off more sharply with more closely spaced wells. Such data enable engineers to select an optimum spacing program and to hold sufficient proved acreage in reserve on which to drill make-up wells and thus maintain full production for the life of the plant.

Surface Facilities

The gathering system of a geothermal field collects the steam of several wells and delivers it to the generating plant free of moisture and particulate matter and with as little reduction in energy as is practical. It must be equipped with insulation sufficient to conserve heat, and of such a size as to offer minimum resistance to flow. It must be equipped

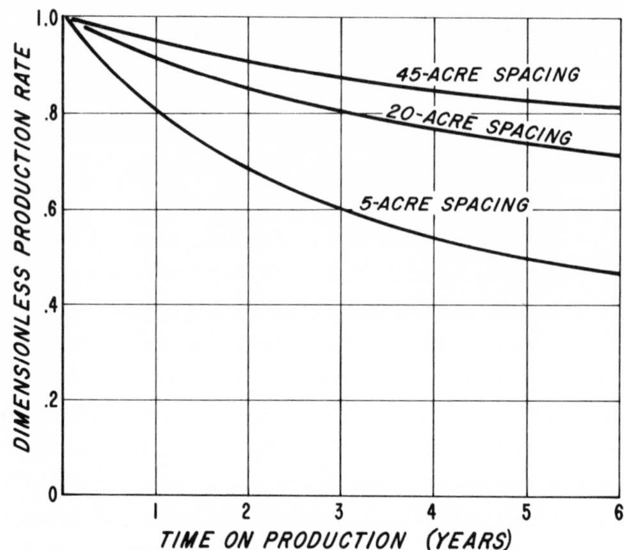

Fig. 5. Effect of well density on production rate as determined by a reservoir simulation model.

with separating devices and vessels for the removal of liquid and solid particulate matter, and safety features adequate to relieve line pressure in event of a plant shutdown. In addition, the facility must be designed and operated to minimize dead-end line segments in which steam could be trapped, cooled, and condensed; condensate cannot be allowed to drip back into the line and further deplete the steam's energy. The layout of the line should anticipate the continued growth of the system by providing for the subsequent accommodation of make-up wells.

The selection of line size must consider the influence of size and therefore surface area and diameter on both heat loss and pressure loss. The objectives of maximizing steam flow and minimizing cost and heat loss are to some extent conflicting. The use of large-diameter piping tends to maximize production; the frictional pressure loss is reduced; the wellhead pressure is lowered; and a greater formation draw-down is possible. But heat loss is also increased because of increased surface area; and large-diameter pipe increases the cost of the pipeline. Smaller pipe increases friction and pressure drop and reduces flow. Consequently, the design engineer, faced with the task of optimizing both production and cost, is confronted with a set of mutually antagonistic variables.

A system that to a considerable degree optimizes these variables was developed at The Geysers for the purpose of delivering 2,000,000 lb/hr of dry superheated steam through two separate systems to two units jointly generating 110 Mw of electricity. A schematic diagram of this system is shown in Fig. 6. The apparently circuitous routing of the lines derives from the irregular mountainous terrain of the area. The largest pipe size is 36 inches nominal diameter at the final runs into the two separate turbines of the installation, telescoping down to as little as 10 inches in some wells. The main trunk lines are oversized at their extremities to allow for the insertion of lines from additional wells when the original group of fourteen wells declines in productivity.

Near each wellhead is a vessel for the separation of liquid and solid particulate matter from the steam. Because the steam produced at The Geysers is dry and superheated, liquid separation is necessary only under such unusual conditions as mechanical failure of the well casing. There is, however, a continuous but minute scattering of particulate matter borne along with the steam. It consists largely of formation dust and corrosion particles. These materials are handled by centrifugal-type horizontal separators that are 99-percent effective in removing particles 10 microns and larger in diameter.

Also near the wellhead of each well is a meter supplying a continuous recording of the well's production rate and producing pressure. This information is used not only in calculating royalty disbursements but also in assembling accurate production records for long-term reservoir studies. Finally, enthalpy measurements and steam-quality measurements, as well as particulate-content tests, are routine scheduled operations carried out with specialized adaptations of conventional equipment and instrumentation.

Wellhead pressures reach 480 psig when wells are shut-in. If the generating plant were shut down, this shut-in pressure would be exerted on the line. To relieve the line of excessive pressure, air-actuated control valves are positioned in the line near the plant intake to discharge the steam to the atmosphere when the operating pressure increases from the normal 100–120 psig to 150 psig. This valve system is designed to pass the full line load, i.e. the full capacity of all wells fed into the system, at 170 psig. Backup safety is provided by rupture discs installed at selected points along the line; these protect the line against pressures above 180 psig.

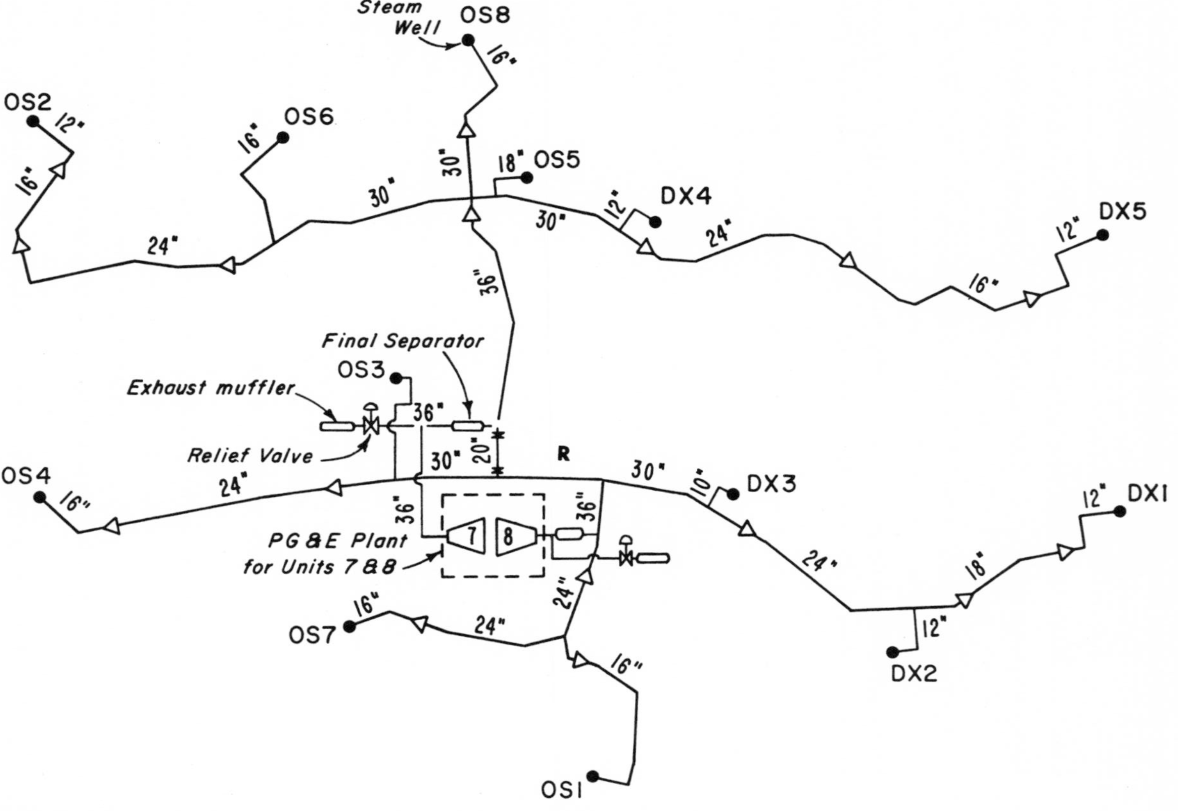

Fig. 6. Schematic diagram of a steam-gathering system at The Geysers

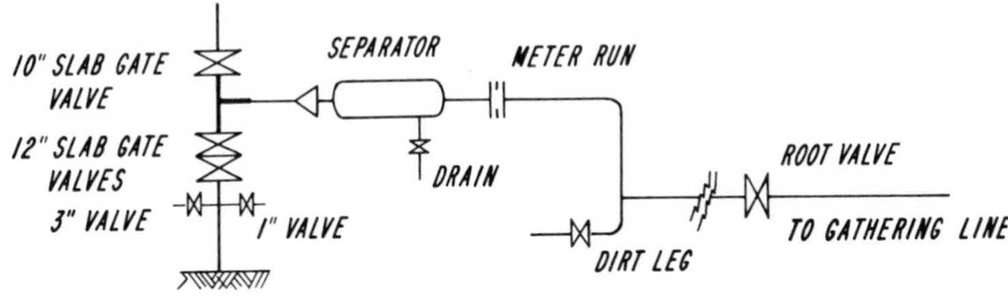

Fig. 7. Schematic diagram of typical wellhead configuration at The Geysers.

The control-valve exhausts are equipped with mufflers for the attenuation of a particularly difficult noise spectrum. The low pressure and high mass flow of the steam-relief system produce a noise spectrum with an unusually high intensity of low frequencies. Attenuation of this noise is difficult, and much costly experimentation has been done at The Geysers to reduce the nuisance effect of the noise.

Dirt-legs, which afford a place for larger particles to come to rest, are furnished at selected points in the line. A dirt-leg is a tee through which the steam flow is shifted 90 degrees. The remaining leg of the tee is a dead-end length of pipe forming a receptacle for the heavier particles. A valve permits blowing-down the dirt-leg to expel the accumulated debris.

In addition to the dirt-legs, wellheads are equipped with blow-down tees through which wells can be opened to the atmosphere and blown-down to line pressure before being produced into the gathering line. This provision prevents the larger particles (up to fist size) from entering the pipeline system during the initial blow-down period. The use of these blow-down tees and dirt-legs has greatly reduced the excessive erosion that earlier damaged some of the generating units at The Geysers.

A typical wellhead and pipeline configuration, showing the blow-down tee, centrifugal separator, meter run, and dirt-leg, is shown in Fig. 7. This type of installation is currently in use at The Geysers on new installations, and old installations are being converted to this arrangement.

Production Management

When a block-of-power has been developed with the required wells and gathering system, and the utility's plant has been constructed, the operations utilizing the geothermal resource require close communication and cooperation between the steam-field operator and the utility company.

The initial warm-up of the steam line is started some 24 to 36 hours before the steam is needed at the plant. All wells on the line are opened up gradually to the atmosphere through their blow-down tees to evacuate the formation dust, rocks, and condensed water that always accompany the first passage of steam. Thermal shock on the formation and casing is minimized by reducing pressure and temperature gradually. As wells are cleared of debris and moisture and their flowing pressures brought down to the line pressure of 100–120 psig, the blow-down valves are closed and the steam is directed into the line. With the main steam valve at the plant closed, the main control valve is automatically actuated to vent the steam to the atmosphere through the muffling devices. Moisture that has accumulated in the line is blown from the steam line vents and drain valves as the line is brought up to operating temperature.

When the wells are producing into the line and the line is "hot," the plant is ready to draw steam to purge its own lines and, finally, to roll the turbines. At this point, when the plant operator is in control of the steam wells, coming up to load too quickly could cause an excessive instantaneous draw-down on the line. This could be quickly reflected in pressure and temperature reductions at the wells, causing dust, dirt, and debris to be brought up from the wellbore. In seeking to minimize this hazard, we have found it effective to open all the wells connected to a pipeline system rather than, say, two or three, into the line at once, even if the plant is operating at reduced load, as when initially starting up a unit to check the turbine balance. The excess steam is vented in the meantime. Distributing the demand in this manner reduces the possibility that a sudden drain on the steam line brought on by human or mechanical error would be transmitted violently to one or a few wells.

Once the steam-gathering system is in operation and the plant is running, the operation settles down to periodic inspection of meters and scheduled blow-down of separators. Scheduled and unscheduled shutdowns, however, do occur. Ideally, when a shutdown occurs unexpectedly, as when a mechanical malfunction of the utility's plant causes an outage, the control valves begin to open on response to a 10- to 15-psig increase in pressure. The line pressure is thus raised only slightly and the wells do not experience great fluctuations in temperature or pressure. This has been found to be the best means of preserving the clean production necessary to the operation. When the shutdown is to be for an extended period, such as for major repair of the power-operating unit, it has been the practice at The Geysers to make use of the downtime by shutting-in the wells—no better opportunity is accorded the testing of pressure buildup and static reservoir pressure for use in reserve studies, or for performance of preventive line maintenance.

Reinjection

The utility's power-plant liquid effluent constitutes about 25 percent of the total condensed steam throughput, the remainder being evaporated in the cooling towers. This effluent contains most of the water-soluble constituents of the steam and cannot be discharged into surface waters, which are drawn upon for domestic and agricultural use. It is therefore reinjected into the steam formation through injection wells.

The low reservoir pressure of 500 psig and the high permeability of the formation allow very high injection rates to be sustained, over 1,000 gpm in some cases. Reinjection is fairly new to The Geysers operation, the first reinjection beginning in April 1969, but no plugging or decrease in injectivity has been observed to date. Two other idle steam wells have since been converted to reinjection wells, and all plant-condensate effluent is now returned to the formation. This condensate will probably be revaporized and produced again, and should therefore extend the producing life of the field.

ACKNOWLEDGMENTS

The author wishes to express his appreciation to his many associates in the Magma Power Company, Thermal Power Company, and Union Oil Company who have contributed either directly or indirectly to the writing of this paper, and in particular to Mr. Delbert E. Pyle, Division Manager of Operations, Union Oil Company of California.

REFERENCES

Craft, B. C., and M. F. Hawkins, Jr. 1959. Applied petroleum reservoir engineering. New York, Prentice-Hall, pp. 326–27.

Cullender, M. H. 1955. The isochronal performance method of determining the flow characteristics of gas wells. Trans. AIME, v. 204, p. 137.

Garrison, L. E. 1972. Geothermal steam in The Geysers–Clear Lake region, California. Geological Society of America Bull., v. 83, pp. 1449–68.

Koenig, J. B. 1972. The world-wide status of geothermal exploration and development. Compendium of first-day papers presented at the first conference of the Geothermal Resources Council, pp. 1–4.

Ramey, H. J., Jr. 1968. Petitioners vs. Commissioner of Internal Revenue. 1969 Tax Court of The United States, 52 T.C. no. 74, 1970.

7. Design and Operation of The Geysers Power Plant

JOHN P. FINNEY

The Pacific Gas and Electric Company (PG&E) has for 15 years been engaged in the design, construction, and operation of dry-steam geothermal electric-power generators at The Geysers, California. This paper discusses our research at The Geysers and the design evolution of the several power units now in operation there.

The Geysers Power Plant is located in the rugged mountains of northeastern Sonoma County, some 80 miles north of San Francisco. Since the facility came on the line in 1960, PG&E and the group of geothermal-steam suppliers headed by Union Oil Company of California, from whom steam for the plant is purchased, have continued its expansion. The plant's eight turbine-generator units produce a net output of 290,000 kw. Two more 53,000-kw units are scheduled to go into operation in 1973. And in 1974, after a single unit rated at 106,000 kw goes into service, the total net output of The Geysers Power Plant will be 502,000 kw, making it the largest geothermal-power installation in the world.

The first attempt to develop geothermal energy to generate electricity at The Geysers was made in the 1920s (Allen and Day, 1927). Two small steam-engine generators drawing steam from shallow wells were used to light The Geysers Resort. Some 30 years later, the first large-scale development was undertaken. In 1955 and 1956, the Magma Power Company and the Thermal Power Company obtained leases around the natural steam vents near the resort and began a drilling program. In 1957, PG&E tested the six wells drilled and found that they could power a small turbine-generator unit economically if steam could be obtained for about 2.5 mill/kwh (Bruce and Albritton, 1959).

John P. Finney is with the Pacific Gas and Electric Company, San Francisco.

The following year PG&E signed a contract with Magma and Thermal providing for the installation of an initial 11,000-kw unit to be followed by additional units if the project proved successful. Unit 1 went into operation in 1960 (Bruce, 1961), and unit 2, rated 13,000 kw, began operation in 1963. Units 3 and 4, each rated 27,000 kw, went into operation in 1967 and 1968, respectively (Barton, 1970).

In 1967, Union Oil of California, which had also acquired geothermal land holdings, joined Magma and Thermal in an expanded joint venture that now holds some 15,000 acres. By 1970, PG&E had signed a new contract with the joint venture providing for the increase of geothermal capacity at an orderly pace. Units 5 and 6 went into service in 1971, units 7 and 8 in 1972. About 100 Mw a year will be installed for as long as the steam suppliers prove up additional steam reserves in what is currently the world's largest known dry-steam geothermal reservoir (Finney, 1972).

It could be said that the prototype unit 1 was a research and development project. Although there was some information available on the experience of the Italian and New Zealand geothermal plants, it was felt that only by building and operating a geothermal unit could this energy source be proved as a competitive power producer. During its design, extensive investigations were made on power-plant material, the configuration of the power cycle, and the power-plant equipment (Bruce and Albritton, 1959). And material and environmental research is continuing.

Plant Cycle and Configuration Studies

The initial investigations were concerned with establishing the adequacy of the steam source. At that time, the steam suppliers were required to demonstrate sufficient flow from existing wells to supply the first turbine-generator unit. The pressure-flow characteristics of the steam wells vary, but in general the flowing wellhead pressures decrease with increased production rates. Shut-in pressure on the wells is 450–500 psi (gauge pressure); and the steam has a constant enthalpy of 1,200–1,205 Btu/lb.

To fully utilize the energy of the steam, it was felt necessary to employ condensing steam turbines exhausting below atmospheric pressure. Since this area has no source of condenser cooling water, cooling towers had to be incorporated into the cycle as a heat sink. Studies established that the most economic utilization of the resource resulted with a turbine-inlet pressure of 100 psig and an exhaust pressure of 4 inches Hg absolute. The electric-power output of the wells is at about its maximum under these turbine conditions. At higher pressures, the increased energy available from the steam is more than offset by the diminished steam-production rates from the wells. At lower pressures, the increased production rate from the wells is offset by the decreased available energy of the steam. Moreover, the size and cost of steam-turbine generators and their auxiliary equipment increase rapidly at lower pressures. It was found that under all atmospheric and operating conditions, the rate of steam flow to the turbines exceeded the cooling-tower evaporation rate. At the operating conditions selected, about 80 percent of the turbine steam flow is evaporated in the cooling tower. This finding meant that the condensed exhaust from the turbine could be used as cooling-tower makeup water.

It followed that a direct-contact turbine condenser could be used in this power-plant design. This type of condenser is less expensive than the surface type used in conventional fossil-fuel power plants, where high-purity condensate is required for return to the boiler. In a direct-contact condenser, steam condenses by mixing with cooling water as it cascades over a series of trays. Two designs of the direct-contact condenser are available: the barometric type and the low-level type. The barometric design was used in the first four units at The Geysers Power Plant (Barton, 1970). In this design, the combined condensed steam and cooling water leaves the condenser by gravity down a barometric leg that is sealed beneath a water surface. Two advantages accrue to this design: the turbine-generator unit can be mounted at ground level; and the unit is therefore inherently fail-safe for the reason that water cannot enter the turbine. Offsetting these advantages are the substantial costs of mounting the large, heavy condenser on a supporting steel structure well above the turbine level and running the large duct from the turbine exhaust up to the top of the condenser. Unit 5 and subsequent units use the low-level type condenser. With this design, the turbine is mounted on a concrete pedestal and the condenser is placed at floor level directly beneath the turbine exhaust. Owing to the lower condenser elevation, pumps are not required for the condenser cooling water. If the proper relative height between condenser and cooling tower is selected, the vacuum head in the condenser induces the cooling water to flow from the cooling tower basin to the condenser (Finney, 1972).

Many investigations were conducted on the composition of the geothermal steam from producing wells. It was found that the steam has a noncondensable fraction that varies from well to well but averages less than 1 percent by weight. For purposes of equipment design, the fraction is taken as 1 percent. The range of concentration of noncondensable gases is shown in Table 1. By comparison, the noncondensable gases in the steam in a conventional fossil-fuel steam-power unit amount to less than 0.01 percent by weight, these in fact owing mainly to air leakage into the condenser. Because of its smaller cost and trouble-free operation, it was decided to use steam-jet gas-removal equipment rather than mechanical vacuum pumps. Two-stage jet ejectors are used; these have been found to make efficient use of the motive steam, which is at turbine-inlet pressure. The first stage compresses the gases from the exhaust pressure of 4 inches Hg absolute to about 5 psia. The second stage further compresses the gases to about 14.5 psia (1 psi above atmospheric pressure) for discharge out of a stack extending above the power building. The inter and after condensers required to condense the ejector steam are barometric-type direct-contact condensers. The ejectors consume about 5 percent of the full-load steam flow to a unit.

The flow diagram for unit 1 is shown in Fig. 1. Except for flow rates and minor differences in arrangement of equipment, it is typical of the first four units. Several steam wells are connected to a single generating unit. Their combined production characteristics closely match the required full-load turbine-inlet flow and pressure requirements. Centrifugal separators installed in the steam lines from each well remove small rocks and dust entrained in the steam from the wellbores. The steam lines from several individual wells are manifolded into a main line leading to one unit. Should a turbine-generator unit trip out of service,

TABLE 1
Range of Concentrations of Noncondensable Gases in Supplied Steam

Gas	Low	High	Design
Carbon dioxide	0.0884%	1.90%	0.79%
Hydrogen sulfide	0.0005	0.160	0.05
Methane	0.0056	0.132	0.05
Ammonia	0.0056	0.106	0.07
Nitrogen	0.0016	0.0638	0.03
Hydrogen	0.0018	0.0190	0.01
Ethane	0.0003	0.0019	—
TOTAL	0.120%	2.19%	1.00%

pressure-control or -relief valves on the main steam line will limit the pressure to 150 psig by exhausting steam to atmosphere through mufflers.

A wye-type strainer is installed in the main steam line just ahead of the turbine to remove any large particles that may pass through the centrifugal separators. Between the strainer and the turbine are one or more turbine stop valves, which shut off steam flow to the turbine quickly in emergency situations. These are followed by one or more trip-check valves. These valves, check valves installed backward, with the flapper or disks latched open against the steam flow, are not used in fossil-fuel turbine installations, but are used in this configuration as a second line of turbine protection because dust in the steam tends to reduce reliability of valve operation. Steam enters the turbine through one or more control valves that regulate the steam flow to control turbine load. These are butterfly valves, which are less prone to malfunction resulting from the suspension of material in the steam; where more than one control valve is used, the valves are operated in sequence to give better control characteristics.

The turbine exhausts to the barometric condenser at about 4 inches Hg absolute. The combined condensed steam and cooling water are pumped from the condenser hotwell to the cooling tower by the circulating water pump. The cooling tower is an induced-mechanical-draft tower of conventional design. Cold water from the tower basin is pumped to the condenser by another circulating water pump to complete the cycle.

Increasingly larger units have been installed at The Geysers Power Plant. Experience gained in operating the earlier units has assured the reliability of the power source, justifying the installation of additional capacity in larger blocks. With the costs of power-plant equipment, construction, operation, and steam increasing, the one way to reduce production costs is through the economy of larger installations. Concern about transporting heavy power-plant equipment on steep, narrow mountain roads has been eased somewhat by the continuing improvement of public roads to the area from the railhead and of the field roads of the steam suppliers. In 1974, a single unit of 106,000 kw (Unit 11) will be placed in operation. Beyond this, no substantial increase in the size of the units is anticipated, owing mainly to transportation difficulties and the length of steam-supply lines.

The cycle diagram for the 53,000-kw units is shown in Fig. 2. The

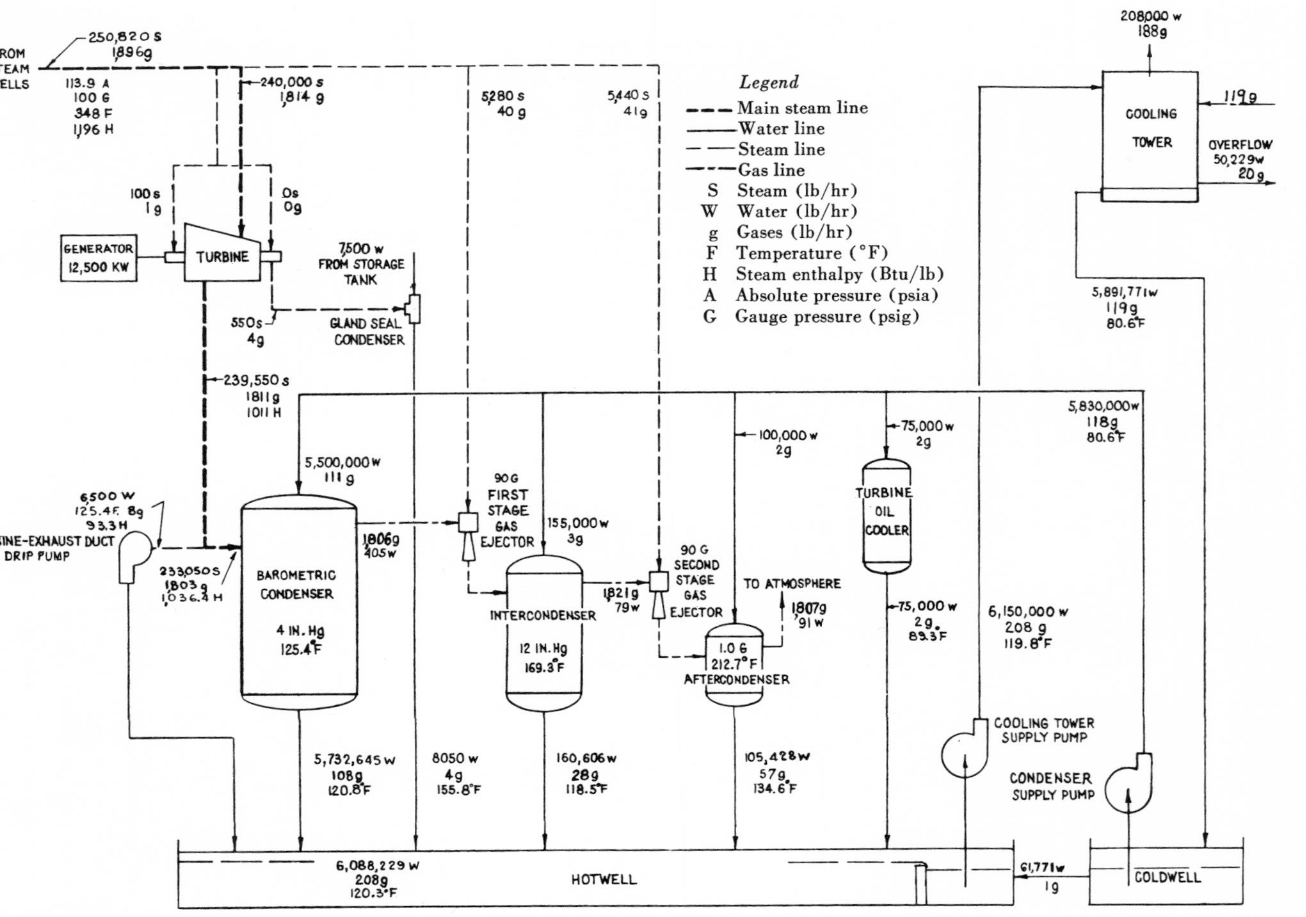

Fig. 1. Heat balance diagram at design load, unit 1, The Geysers Power Plant

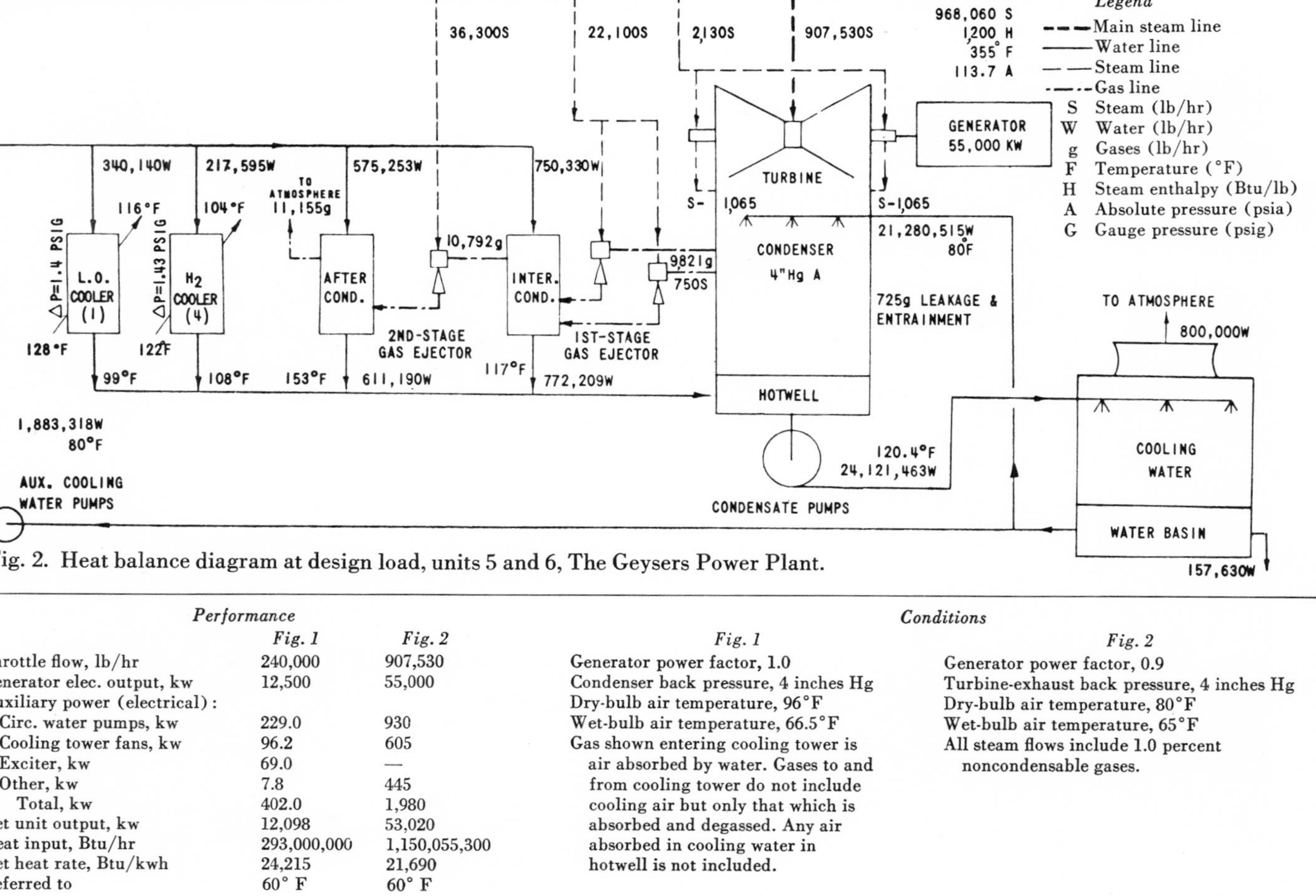

Fig. 2. Heat balance diagram at design load, units 5 and 6, The Geysers Power Plant.

Performance

	Fig. 1	*Fig. 2*
Throttle flow, lb/hr	240,000	907,530
Generator elec. output, kw	12,500	55,000
Auxiliary power (electrical):		
Circ. water pumps, kw	229.0	930
Cooling tower fans, kw	96.2	605
Exciter, kw	69.0	—
Other, kw	7.8	445
Total, kw	402.0	1,980
Net unit output, kw	12,098	53,020
Heat input, Btu/hr	293,000,000	1,150,055,300
Net heat rate, Btu/kwh	24,215	21,690
Referred to	60° F	60° F

Conditions

Fig. 1

Generator power factor, 1.0
Condenser back pressure, 4 inches Hg
Dry-bulb air temperature, 96°F
Wet-bulb air temperature, 66.5°F
Gas shown entering cooling tower is air absorbed by water. Gases to and from cooling tower do not include cooling air but only that which is absorbed and degassed. Any air absorbed in cooling water in hotwell is not included.

Fig. 2

Generator power factor, 0.9
Turbine-exhaust back pressure, 4 inches Hg
Dry-bulb air temperature, 80°F
Wet-bulb air temperature, 65°F
All steam flows include 1.0 percent noncondensable gases.

Fig. 3. The Geysers Power Plant, units 1 and 2.

turbine steam conditions are 100 psig at 355°F (179°C) and 4 inches Hg absolute exhaust. The turbine is a single-shell, double-flow design with 23-inch last-stage blades. Steam enters the turbine through two 24-inch lines, each of which has the same strainer and turbine-valve provisions as the earlier units. The design of the 106,000-kw unit has not been completed. The turbine, a two-shell, four-flow, 23-inch last-stage blade machine, is essentially two of the 53,000-kw turbines in tandem. It is connected to a single large generator. The cycle conditions and plant arrangement will be similar to the 53,000-kw units. The distinguishing feature of these larger units is the use of the low-level-design direct-contact condenser. Substantial plant-cost savings result from the use of this type of condenser.

Fig. 3 shows photographs of units 1 and 2; Fig. 4 shows units 3 and 4; and Figs. 5 and 6 show units 5 and 6.

Materials Investigations

Prior to the construction of the first unit, we investigated the suitability of materials for plant equipment in the environments anticipated in service. Because of the constraints of the construction schedule, exposure tests were necessarily brief, and some accelerated testing techniques were used (Bruce and Albritton, 1959). But although there have been some unanticipated long-term service effects, these tests were, in general, quite successful. As development at The Geysers has proceeded, we have come to a better understanding of the service life obtainable from various materials.

We recently completed a 4-year corrosion test on materials for the condensate systems, in cooperation with the International Nickel Company. Two sets of test coupons were installed at two different locations to allow us to analyze corrosion effects for 2 and 4 years' exposure. One set was installed in the condensing zone in the top of a condenser and the other was submerged in turbulent flow in a condenser hot well. Both sets of coupons, when removed after 2 years' exposure, had accumulated a buildup of condensate precipitate. The corrosive attack, which was quite similar, may be summarized as follows: (1) austenitic stainless steels showed minimal deterioration; (2) aluminum alloys and chromium stainless steels had slight overall attack, but were susceptible to pitting (many samples were perforated); (3) copper alloys, plain carbon steel, and cast irons had excessive overall loss. The 4-year exposure coupons have recently been removed and are now being analyzed. Addi-

Fig. 4. The Geysers Power Plant, units 3 and 4. The expansion loops in the foreground allow the steam pipes to contract when the plant is shut down. The steam rising from five low stacks on the line at the left marks the blowdown valves; when the plant is shut down, the steam escapes through these valves.

Fig. 5. The Geysers Power Plant, units 5 and 6, nearing completion. The main steam lines from the producing wells had not yet been completed.

Fig. 6. The Geysers Power Plant, units 5 and 6, in operation. The main steam lines enter the building at the right. Units 3 and 4 can be seen beyond the power-transmission lines at the right.

tional corrosion testing is planned for the condensate system. The materials to be evaluated will be selected after the 4-year test data are fully analyzed. It is anticipated that these tests will investigate not only compositions, but fabrication alternatives and the effects of heat treatment on various compositions.

Atmospheric-corrosion racks have been placed at several locations adjacent to the power building at one of the units. Various aluminum alloys and stainless steels have now been exposed for about one year. Exposure tests of the different plating methods and of several additional metals are planned.

Sulfur bacteria are known to exist in The Geysers environment, but there have been no corrosion failures attributable to their presence. Further research will be undertaken on their possible effects.

The main steam lines, valves, and strainers are made of carbon steel that is not appreciably corroded by the dry geothermal steam. The turbines are of conventional design for low-temperature and -pressure steam service. However, particular attention has been directed to smoothing steam-flow paths. Turbine blading of 13-percent chromium steel has been used in order to minimize erosion from the particulate matter in the steam. In the lower-pressure stages, where the moisture content of the expanded steam increases, conventional moisture-removal provisions are employed.

After the steam condenses and is exposed to air in the condenser and cooling tower, some of the hydrogen sulfide is oxidized to sulfate, which, in combination with other compounds in the condensate, becomes highly corrosive. All wetted parts of the condenser, including the condenser internals and all wetted areas of condensate pumps, are made of austenitic stainless steel. The condenser shells are carbon steel clad with austenitic stainless steel. The piping for the condensers is austenitic stainless steel and aluminum.

The cooling towers are of conventional design, but special materials are used. All wetted metal parts are Type 304 stainless steel or aluminum. In those cases where cast iron is the only acceptable material, it is protected by a thick coating of coal-tar epoxy resin or other plastic material. The cooling-tower fill material is redwood, polystyrene, polypropylene, or polyvinyl chloride. Of these, polyvinyl chloride is preferred because of its strength and fire-retardant qualities. All condensate and cooling-water piping is made of stainless steel, aluminum, or plastic-lined material. Heat exchangers are fabricated of aluminum or austenitic stainless steel. Concrete surfaces in contact with water are coated with coal-tar epoxy compounds or synthetic rubber. Several new coating materials are currently being exposed in one of the cooling towers.

The hydrogen sulfide in the plant atmosphere has caused some problems with electrical equipment because of its corrosivity to copper alloys and silver. Transmission lines and hardware are therefore primarily aluminum. Tin-alloy coatings, although somewhat effective in resisting corrosion, have not been satisfactory. Electrical contacts are particularly susceptible to corrosion. Platinum and gold inserts and plating appear to be a good solution to contact-corrosion problems. In the newer units, the electrical relays most important to unit protection have been placed in a "clean-room" environment. This room is maintained at a slightly positive pressure, with clean air supplied through activated-charcoal filters.

Environmental Investigations

The small amount of hydrogen sulfide contained in the steam is partially oxidized to elemental sulfur and sulfate in the power cycle, but most of it is released to the atmosphere from the cooling towers, the balance from the gas-ejector vents. About two-thirds of the release comes from the cooling towers, the balance from the gas ejectors. The hydrogen-sulfide concentration around the units poses no health hazard, but its unpleasant odor can be detected in some areas and it does cause corrosion of equipment and material at the plant. The Geysers area has a history of natural hydrogen-sulfide emission from a number of natural steam vents; the explorers who discovered The Geysers in 1847 noted the odor and named the stream that traverses the area "Big Sulfur Creek." It is interesting that the concentrations of hydrogen sulfide and other noncondensable gases from wells that have been in production for several years are decreasing with time. This suggests that the noncondensable gases have accumulated in the higher zones of the producing strata and are being depleted. PG&E has a research project under way to develop a method of mitigating the hydrogen-sulfide release from the units, but it is too early to predict success.

In the power cycle, about 80 percent of the steam flow to the turbines is evaporated in the cooling towers and the remainder is effluent water containing boron, ammonia, and bicarbonate, in addition to sulfate and finely suspended sulfur. When the early units were in operation, this effluent water was discharged into Big Sulfur Creek. After unit 4 was placed in service, the quantity of effluent water was too large for adequate dilution by the creek, and a number of chemical processes were investigated for removal of these undesirable substances but were found to be prohibitively expensive. The steam suppliers carried out a parallel research project on reinjection of the effluent back into the less-productive steam wells. This procedure was found to be a practical way of handling the effluent water and is now being used routinely.

Operation

Since there are no boilers with their attendant complex auxiliaries and controls, The Geysers units are simpler to operate than fossil-fuel plants. The units in fact operate more like a hydroelectric plant than a steam power plant. Power-plant operators are not in full-time attendance. Sixteen men operate and maintain units 1 through 6. One or two roving operators work the day shift and at least one is on duty afternoons and nights.

The importance of clean steam for sustained operation of the turbine-generator units cannot be minimized. Experience in recent years has shown that close coordination with the steam suppliers is required for economical operation of both the resource and the generating facilities. Sudden changes in steam requirement cause changes in the flow of steam from the wells, which tend to jar small rock or sand particles into the wellbores, which in turn might be carried into the steam piping and turbines. Therefore, startup, loading, and shutdown of turbine-generator units must be closely coordinated with the suppliers, who need 24 hours' notice of startup to clear shut-in wells of water and debris, warm up the steam-collection piping system and drain it of condensate, and get the wells up to rated flow. Experience has shown that certain mechanical equipment (Budd, this volume) is needed at the wellheads and in the steam-collection piping systems to deliver steam of good quality.

PG&E is advised each time a well is put into service. Steam temperatures and steam-line drains are then closely monitored by the supplier for water and dirt; at the first indication of contaminants, the turbine-generator unit is tripped to prevent damage. New steam-monitoring systems are now being installed on the main steam-supply lines near the turbines. Isokinetic steam samplers will take a representative steam sample across the diameter of the main steam line. The sample will be condensed and analyzed for dirt by the use of a nephelometer, which measures turbidity by light scattering. In steady-load operation, the impurities in the steam are as low as 1 ppm; in transient operation, much higher. Because of the long, complicated startup procedures, geothermal units should be operated as base-load units (i.e. not as peak-load units, which must operate without interruption).

The buildup of dustlike material on the turbine blades has led to a number of problems. This material can cause a loss in turbine efficiency and a reduction of steam flow through the turbine by closing up the nozzle areas. More seriously, the material can build up on the shrouds of the turbine blades to such an extent that the centrifugal load of the buildup can cause the shrouds to loosen, resulting in extensive turbine-blade damage. The material appears to be fine quartz particles embedded in a matrix of glassy borates. Since it is partially water-soluble, the heavier deposits are found in the drier high-pressure turbine stages. Water-washing facilities are now being installed on the turbines in the hope that the material can be removed on-line, i.e. without shutdown and overhaul. It is too early to predict the success of this program.

Experience is still being gained on the proper operating period between turbine overhauls. It is now believed that an operating period in excess of 24 months may be achieved. It is yet to be determined, however, what proportion of the former turbine problems is due to continuous operation and what to improper procedures. It is hoped that after well-established starting and loading procedures for the units have been in effect for several years, the overhaul interval can be extended to 5 to 6 years, as is now routine with our fossil-fuel units.

Conclusions

At The Geysers Power Plant, the generation of electric power from geothermal energy has proved both an engineering and a commercial success. Power-production costs compare favorably with those of PG&E's fossil-fuel steam-generating units. Problems associated with particulate matter in the steam and with the corrosivity of the condensate have been mitigated by attention to design, maintenance, and operation. Several processes for the reduction of hydrogen-sulfide emission from the units are now under study. And significant improvements have been made in overall noise abatement by both the steam suppliers and PG&E.

Additional generating capacity is scheduled each year through 1974, at which time the total capacity will be 502,000 kw, making The Geysers Power Plant the largest geothermal power installation in the world. Beyond that, additional capacity will be installed as further steam reserves are developed by the steam suppliers and permits are obtained from the California Public Utilities Commission.

REFERENCES

Allen, E. T., and A. L. Day. 1927. Carnegie Institute of Washington, publication no. 378.

Barton, D. B. 1970. Current status of geothermal power plants at The Geysers, Sonoma County, California, U.S.A. Paper for the United Nations Symposium on the Development and Utilization of Geothermal Resources, Pisa, Italy, agenda item IX/8.

Bruce, A. W. 1961. Experience generating geothermal power at The Geysers power plant, Sonoma County, California, U.S.A. Paper for the United Nations Conference on New Sources of Energy, Rome, Italy, agenda item II.A.2(b).

Bruce, A. W., and B. C. Albritton. 1959. Power from geothermal steam at The Geysers power plant. Paper 2287. J. Power Div., Proc. Am. Soc. Civil Engineers, v. 85, no. PO 6.

Finney, J. P. 1972. The Geysers geothermal power plant. Chem. Eng. Progr., v. 68, no. 7, pp. 83–86.

8. The Vapor-Turbine Cycle for Geothermal Power Generation

J. HILBERT ANDERSON

For thousands of years, man has known that the inside of the Earth is hotter than the surface. But only during the past few centuries have we learned that mechanical energy can be generated by transferring heat from a warm body to a cold body. This is a heat engine. Knowing these two principles, we are led logically to consider the possibilities of producing power by transferring heat from the hot interior of the Earth to the cooler exterior. This is geothermal power. There is more than enough heat available in the Earth to generate all the power we need for millions of years to come at the present rate of worldwide power consumption; assuming sole dependence on geothermal heat, it would take some 41 million years for man to reduce the temperature of the Earth 1°F. Really, the major questions are: how accessible is the heat at an economic temperature level and how do we convert it to power competitively with other heat engines?

The thermodynamic efficiency of a perfect heat engine is

$$\text{Eff}_{t} = (T_1 - T_2)/T_1$$

where T_1 is the absolute temperature of the hot body and T_2 is the absolute temperature of the cold body, or heat sink. From this equation it is obvious that more work is produced per unit of heat as the hot-body temperature, T_1, increases. There are many places in the world where fairly high temperatures occur within drilling distance of the Earth's surface. Drilling costs, however, increase greatly with depth: for example, an Oklahoma drilling crew recently reported temperatures of 360°F at a depth of 24,500 ft and a drilling cost of $6 million (*Pitts-*

J. Hilbert Anderson is Consulting Engineer, York, Pennsylvania.

burgh Press, 1971). By contrast, a recent well in the Imperial Valley recorded 390° at a depth of only 2,600 ft at a small fraction of that cost. Figure 1 shows the increase of well costs with depth. The data, taken from technical journals (Bowen and Groh, 1971; Barnea, 1972) and newspaper articles, show that although costs vary widely, they increase dramatically at depths of more than 10 or 15 thousand feet. Because of the limitations imposed upon productivities by casing size and well depth (Budd, this volume), it is unlikely that well depths of more than 15,000 ft would be found economically feasible in geothermal-power applications. For if the curve is a valid indication of cost, a well 15,000 ft deep costs $800,000; if the heat from such a well is sufficient to produce 10,000 kw, then the well cost is $80 per kw, which may not be economically acceptable. At higher well outputs, however, costs per kw would improve. Thus wells of maximum output and at shallow depths offer the best prospects for power production.

Since our cold heat sink is at the Earth's surface and the hot body is deep under the surface, we need some means of transporting the heat from the source to the surface, where it can be converted to power. The

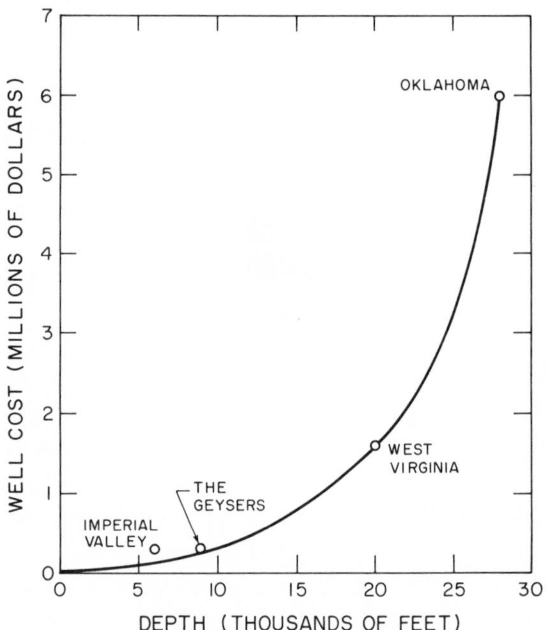

Fig. 1. Drilling cost vs. depth.

obvious means is fluid flow; and since water is our most abundant and cheapest fluid, it is the obvious fluid choice. Moreover, nature has supplied the water under the ground, already heated by hot rock, at countless points around the world. There are, for example, more than a thousand hot springs already mapped in the continental United States.

There are several means available to us to employ water as a heat-transport fluid:

1. The water may occur in a porous hot zone and vaporize into steam, which can be passed through a wellbore to the surface and expanded directly through a steam turbine to produce power. This is what is being done at The Geysers, in California, where hot water in the formations is overlaid by an immense steam reservoir that is trapped by the impermeable overlying rocks. For this condition to exist, fluid pressures in the reservoir must be below hydrostatic. In this case the steam becomes superheated. The dry steam, which is only mildly corrosive, works satisfactorily in a rather conventional steam turbine. But such underpressured fluid reservoirs require unique geological conditions (White, this volume), and natural-steam geothermal fields are a rare phenomenon.

2. In many places the water table is high enough that there is a layer of cold water above a hot zone. The shallow cold water maintains hydrostatic pressure above the hot water beneath, which is highly superheated. If a well is drilled through the cool-water zone into the hot-water zone and if the pressure in the wellbore is lowered, the superheated water can be brought to the surface and flashed into steam. Since the average density of the steam/water mixture is lower than that of the water alone, the mixture is lifted to the surface. The mixture, moreover, is always at saturation temperature with respect to pressure. It follows that the temperature of the mixture drops as it flows upward, and there is a consequent loss in available energy when it reaches the surface. At the surface, the steam can be separated from the water and expanded through a turbine. But using steam to lift the water is inefficient and wastes a good deal of the available energy. In addition, the heat remaining in the hot water at turbine-inlet pressure is wasted. Part of this heat can be recovered by flashing to steam at a lower pressure, but because low-pressure steam turbines are costly the economics are often doubtful.

In some cases, the steam flashed from the hot water is so corrosive or contains so much other gas that maintenance of a steam turbine is not practical. In such cases the steam can be used to boil water in a heat exchanger, and the clean steam can be expanded through a steam turbine. This has the disadvantages of adding heat-exchanger costs to the system, and of further reducing the available energy in the steam. As a result, the steam-turbine cost per kw generated is also higher.

3. A new and quite different geothermal-power process, the vapor-turbine cycle, has been developed to eliminate the disadvantages inherent in the methods described above. In this process the hot water brought to the surface is passed through a heat exchanger, where it gives up heat to boil and superheat a high-density vapor. The vapor then expands through a turbine to produce power and is recycled to the heat exchanger. The practical and economic advantages of this process may be great enough to turn the vast potential of geothermal energy into reality.

The vapor-turbine cycle is shown in simple form in Fig. 2.* Water is pumped to the surface by a deep-well pump. And to ensure that no steam is produced and no dissolved gases escape from the water, the water is kept above saturation pressure throughout the process. The water gives up its heat in a heat exchanger to heat a liquid, boil it, and superheat it. The cooled water is then injected into a well nearby, carrying all dissolved gases and most of the solids in solution with it. The water thus injected maintains the underground reservoir pressures, and very likely finds its way back through the hot porous zone to be reheated. The water thus becomes simply a means of transporting heat from the hot rocks beneath to the power plant above, in a continuous cycle.

Various fluids can be used in the power cycle. Because isobutane has shown the most favorable overall economics for developing power from water of about 325°F (the water-inlet temperature of the first plant we designed), its use has been assumed throughout this paper. Freon is being used in a small power plant built in the Soviet Union (Moskvicheva, 1971).

The cycle diagram shows how isobutane vapor, expanding through a turbine, yields its energy to drive a generator. The exhaust gas is condensed back to the liquid state in a water-cooled condenser. From there it is pumped back through the isobutane liquid heater and boiler to repeat the cycle. Since the boiler feedpump draws quite appreciable power, it is advantageous to drive it with an isobutane turbine, as shown in the diagram.

* Patents have been applied for. The cycle has been named the "Magmamax" power process by Magma Energy, Inc., of Los Angeles.

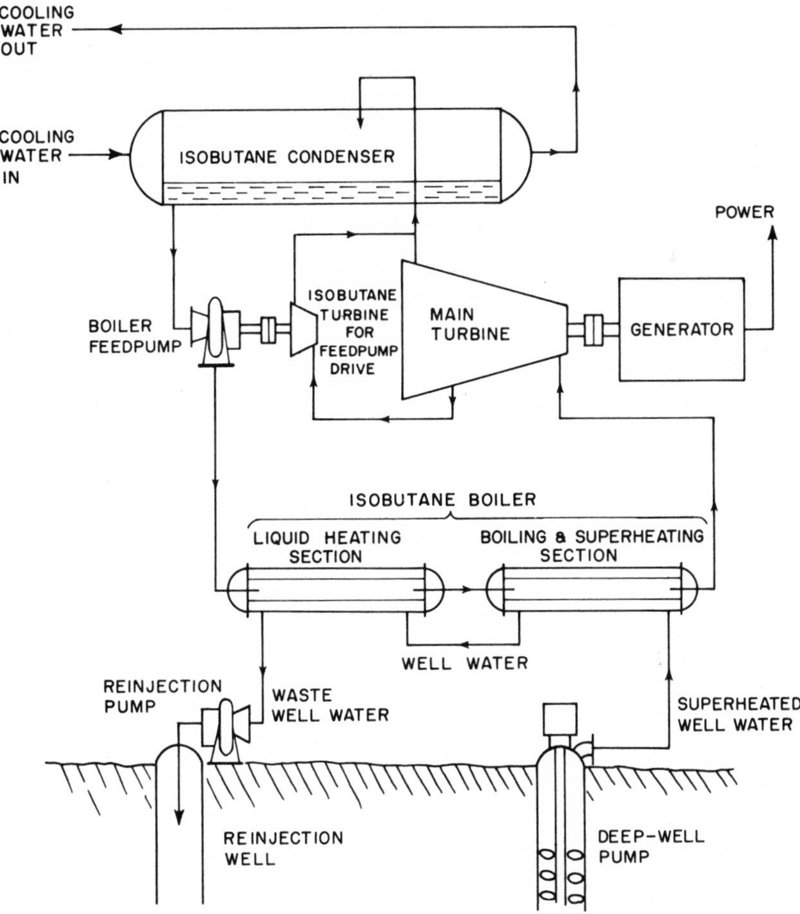

Fig. 2. Vapor-turbine cycle diagram.

In a vapor-turbine cycle, many parameters must be evaluated to optimize power cost. Choice of fluid, turbine-inlet temperature and pressure, condensing temperature, type of cooling system, well depth, and other factors affect the overall cost of the plant and the production system. For example, lowering the turbine-inlet temperature increases the temperature difference between water and isobutane and reduces the required heat-transfer surface for a given heat flow. However, lowering the turbine-inlet temperature also decreases cycle efficiency, and therefore increases the required heat flow per kwh generated. There are many other such factors that must be brought into balance for optimum cost design.

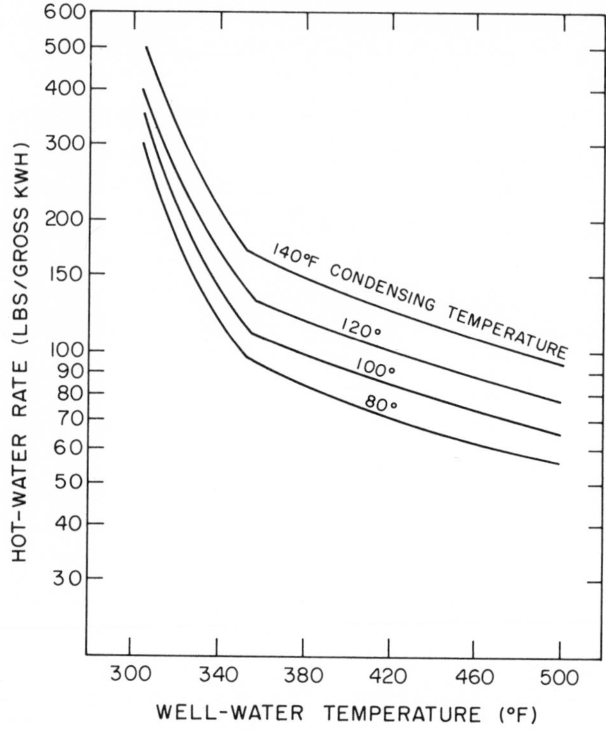

Fig. 3. Water rate vs. well-water temperature for the isobutane cycle.

Figure 3 shows typical performance curves for an isobutane cycle (Anderson, 1972), with water flow per kwh plotted against water temperature from the well. It is interesting to note that condensing temperature is more important than well-water temperature in determining cycle efficiency. For example, at a well temperature of 400°F and a condensing temperature of 100°, the water rate is 91 lb/kwh. If condensing temperature is reduced to 80°, the water rate drops to 76 lb/kwh. To achieve a water rate of 76 lb/kwh at 100° condensing temperature, we would need to raise the well temperature to 450°, an increase of 50°. A geothermal plant of this type thus benefits greatly from the use of a condensing system designed to take full advantage of lowest available atmospheric temperatures.

One of the important advantages of a vapor-turbine plant is that lower condensing temperatures can be used than are economically possible with a steam-turbine plant. The physical size of any turbine is determined mainly by the diameter of the last stage wheel and by the size of the exhaust connection. The physical size is also a major factor in the cost. Both the size of the last stage wheel and the exhaust connection are determined by the volume rate of flow. At 80°F condensing temperature the specific volume of steam is 633 ft^3/lb, whereas that of isobutane is only 1.68 ft^3/lb. This is the basic reason why an isobutane turbine is so much smaller and cheaper to build for operation at low condensing temperatures. Steam at 100°F has a specific volume of 350 ft^3/lb. This means that a steam turbine designed for 80° must have almost twice the exhaust end flow area of one designed for 100°.

Although the water rate per kwh decreases with increasing well temperature, it is not necessarily true that the combined cost per kw of production system and power plant decreases with increasing well temperature. Generally speaking, higher temperatures require deeper drilling, and deeper drilling becomes expensive. Deeper wells also require more pumping power and costlier pumps, and higher-temperature water may be more corrosive and may require costlier materials for its pumps, pipes, and heat exchangers. For these reasons, each application must be carefully analyzed to determine its best cycle. We have made much progress in this work, and engineering analysis of optimum cycles for each location can now be made in a relatively short time.

The importance of designing a power plant at high thermal efficiency is demonstrated by Fig. 4, which plots heat rejected from the plant against useful energy output. Note that the heat-rejection load increases tremendously at low efficiencies. For example, a nuclear power plant at 30 percent efficiency must reject 2.33 kwh of heat energy for each kwh generated. For a geothermal plant at 10 percent efficiency, 9 kwh must be rejected for each kwh generated, almost four times as much. Heat-rejection equipment is expensive and is a major cost factor in geothermal plants. To compound the problem, many geothermal areas occur where there is an inadequate supply of cheap cooling water. Thus, one of the most serious problems in large-scale geothermal power development is cooling.

The vapor-turbine cycle using isobutane or other suitable fluids has many advantages over the flashed-steam or indirect steam-heating cycles:

1. Water pumped from the reservoir at pressures above saturation reaches the surface at nearly maximum well temperature, whereas water lifted by steam suffers severe temperature losses.
2. Since water at full pressure retains its gases in solution, the gases

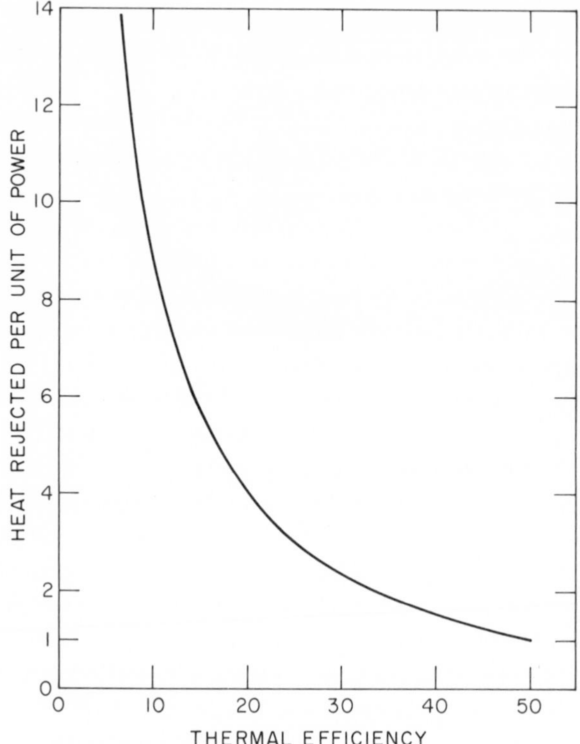

Fig. 4. Heat rejection vs. thermal efficiency.

can be returned to the ground without danger of atmospheric pollution.

3. If steam and dissolved gases were permitted to escape from the water, the chemical composition of the water would change, very likely causing precipitation of solids out of solution, and plugging of wells.

4. Keeping water at high pressure throughout the heat exchangers helps to minimize heat-exchanger tube stresses. The possibility of stress corrosion, which is often the chief cause of failure in high-temperature hot-water heat exchangers, is thereby reduced.

5. Because the vapor turbine incorporates fewer stages, and because the vapor volume change through the turbine is not as great, the vapor turbine generally offers more efficiency than the steam turbine.

6. Because wheel speed in vapor turbines is lower than that in steam turbines, design problems are simpler and blade stresses are much less severe.

7. The vapor-turbine cycles can be relatively quiet; flashed-steam cycles require costly noise-abatement measures.

8. Isobutane turbines operate above atmospheric pressure throughout the cycle. The possibility of air and oxygen getting into the turbine and causing corrosion or explosive mixtures is eliminated. Air entering the system under vacuum is a major cause of corrosion in steam systems.

9. Isobutane and other such fluids are relatively simple and inexpensive.

10. Isobutane turbines are much smaller and, therefore, less costly than steam turbines of the same power output.

11. Isobutane remains dry throughout its expansion through the turbine, thus eliminating the erosion of blades by water droplets that is so common in steam turbines.

12. Because isobutane is compatible with oil, internal bearings can be used in the turbine, yielding a turbine much more rugged and lower in cost, and requiring only a single shaft seal at the coupling end of the turbine, thus eliminating the long, complicated, leaking shaft seals required in a steam turbine.

13. Since isobutane is noncorrosive, there should be no need for the expensive stainless steels often required in various parts of the turbine in the flashed-steam cycle.

14. Because the condensed isobutane has lower density and lower latent heat than steam, cavitation damage should not occur in the boiler feedpump.

15. Because corrosion problems are less severe, the isobutane boiler feedpump can be made from cheaper materials than a water feedpump.

16. Because isobutane turbines have much lower rotating inertia than steam turbines of the same power, the short-circuit torque problem from the drive couplings is virtually eliminated.

17. Because isobutane turbines can be designed to utilize lower condensing temperatures, cycle efficiencies are improved and water rates are reduced below those of steam turbines.

18. With no air or noncondensable gas in the condensers, isobutane condensers can be made 100 percent effective; with steam condensers, gas in the system reduces condenser efficiency by increasing condensing pressure.

19. Steam turbines can require substantial gas-removal equipment; isobutane turbines, none.

20. The isobutane cycle permits efficient transfer of heat from well water down to quite low temperatures; water can be discharged from the generating plant at temperatures as low as 120°F. By contrast, steam

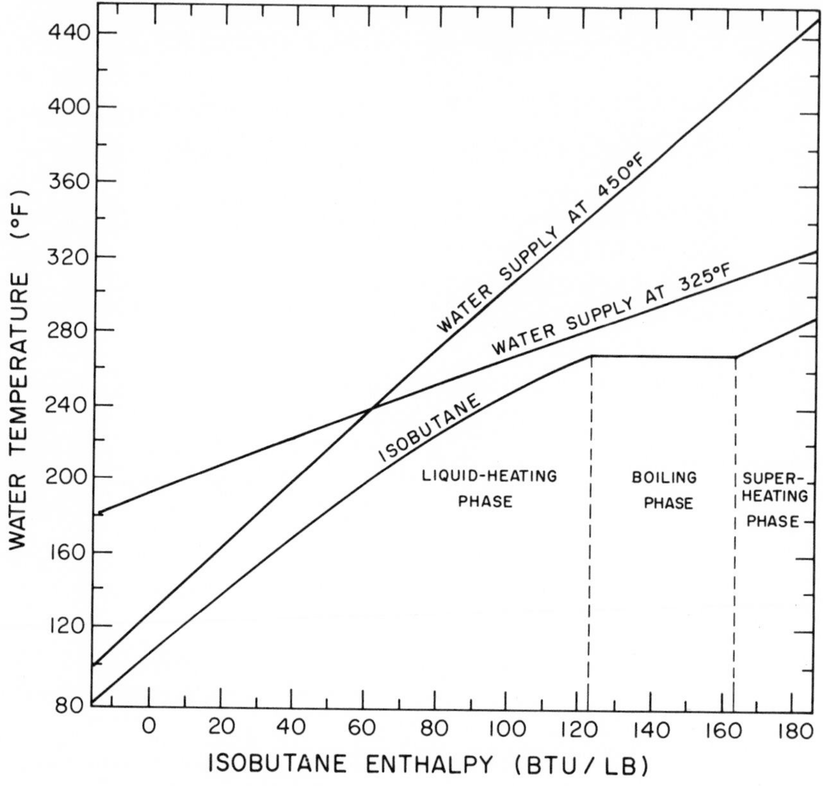

Fig. 5. Water temperature vs. isobutane temperature.

cycles could rarely be economic at water-discharge temperatures below 212°.

Figure 5 shows the heat-transfer relationship between well water and isobutane (Anderson, 1972). Since the specific heat of water is nearly constant, the temperature varies almost linearly with heat release to the isobutane. But the specific heat of isobutane is *not* constant through its liquid, boiling, and superheat phases (*Handbook of Fundamentals*, 1967). Therefore, with a constant ratio of water flow to isobutane flow, the temperature difference cannot be constant. For water at 325°F the temperature leaving the system must be quite high, because the temperature difference becomes too small near the liquid boiling point. For water at 450° the temperature difference becomes smallest at the outlet, and there is a wastefully high temperature difference in the boiling and superheat phases. The effect of these slope changes is a break in the

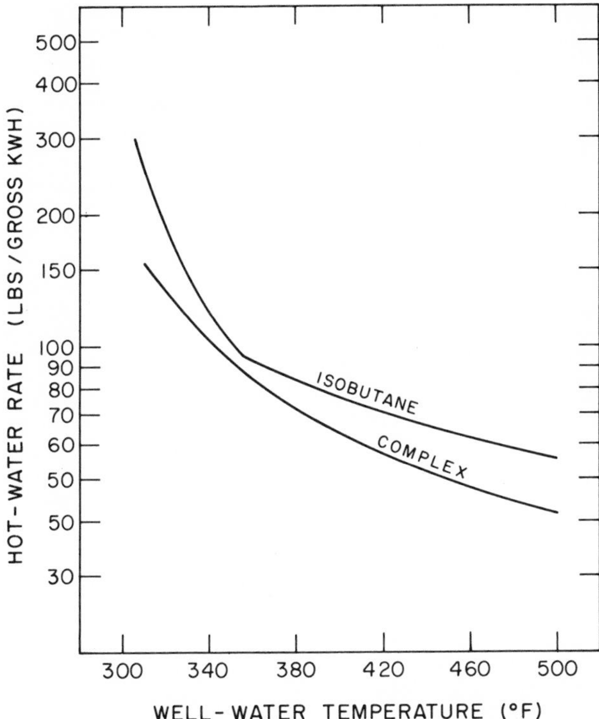

Fig. 6. Water rate vs. well-water temperature.

water-rate curves for isobutane, as shown in Fig. 3. By the use of more complex cycles, we can improve the water rate somewhat, as shown in Fig. 6, in which the water rate for the more complex cycle is plotted against that for the typical curve from Fig. 3.

The use of pumps to lift the water from the wells adds some cost and complexity. But their use is amply justified by increased water production per well, greatly increased power output per well, and all of the other advantages listed above. Shaft-driven hot-water pumps have been used quite successfully in Iceland since 1961 (Bodvarsson, 1972).

At higher temperatures and greater depths it becomes advisable to use turbopumps, which are smaller and simpler than shaft-driven pumps, much easier to install or remove for maintenance, and operable at much higher temperatures. And although they require more power than shaft-driven pumps, this increased power is still a small fraction of total system output.

Fig. 7. A 9,000-kw isobutane turbine on test at York Division of Borg-Warner Corporation.

The first isobutane vapor turbine manufactured for Magma Energy, Inc., is shown on test in Fig. 7. This turbine is a three-stage, radial-flow turbine. It will turn at approximately 7,000 rpm, and is expected to deliver approximately 9,000 kw at the generator terminals, depending on cycle conditions. The turbine has internal bearings, requires only one external shaft seal, and should be virtually maintenance-free. It has a casing bore of only 38 in.

With the development of efficient vapor-turbine cycles, better pumping equipment, better heat exchangers, and more efficient cooling systems, we now feel confident that low-cost power can be produced from many available sources of geothermal heat.

REFERENCES

Anderson, J. H. 1972. Geothermal heat, our next major source of power. Paper presented at meeting of the American Institute of Chemical Engineers.

Barnea, J. 1972. Geothermal power. Scientific American, v. 226, no. 1.

Bodvarsson, G. 1972. Private correspondence, Reykjavik, Iceland.

Bowen, R. G., and E. A. Groh. 1971. Geothermal-earth's primordial energy. Technology Review.

Canjar, L. N., and F. N. Manning. 1967. Thermodynamic properties and reduced correlations for gases (Houston: Gulf Publishing Co.).

Handbook of Fundamentals. 1967. American Society of Heating, Refrigerating, and Air Conditioning Engineers.

Moskvicheva, V. N. 1971. Geopower plant on the Paratun'ka River. USSR Academy of Sciences.

Pittsburgh Press, 1971. December 12.

9. Water from Geothermal Resources

ALAN D. K. LAIRD

Averting projected water shortages has become a serious problem in many places. Importation, reclamation, and desalination systems are proposed as solutions, but all take large amounts of power, and power shortages are also projected. The situation is further complicated by increasingly stringent antipollution requirements. Interest has therefore focused recently on geothermal energy, since it appears to be a potentially clean source of large amounts of power. Estimates of the magnitude of geothermal-energy resources (Rex, this volume) vary according to the assumptions made, but all indicate that very large amounts of water are available from the resources. Relatively small-scale geothermal power production and some water production have been carried on in several parts of the world (Koenig, this volume). In these systems water carries thermal energy from the reservoirs to the surface: in some cases, condensate from steam turbines may be available as freshwater supplies; and since liquid water produced from geothermal wells is generally saline and already hot, distillation comes immediately to mind as a means of making more fresh water available. Consequently, several groups have studied large-scale geothermal distillation as a component of energy-production schemes.

In certain circumstances, moreover, it may be advantageous to employ the electricity thus generated to desalinate sea water or brackish water at some more or less remote location. For such situations, membrane desalting is more appropriate than distillation. Both classes of desalting process will be discussed.

Alan D. K. Laird is Director of the Sea Water Conversion Laboratory and Professor of Mechanical Engineering at the University of California, Berkeley.

Because of the wide range of geothermal-reservoir conditions, water-production processes could take many forms. Geothermal reservoirs are classified (White, this volume) as hot-water, or liquid-dominated, system if they produce principally liquid, and as steam, or vapor-dominated, systems where they contain mostly steam. Liquid-dominated reservoirs may be particularly valuable as water supplies, since they often contain large amounts of water. Reservoirs of this type are in use at a number of locations throughout the world (Koenig, this volume). Vapor-dominated reservoirs, such as that at The Geysers (Budd, this volume), tend to be more limited as sources of water. There are also vast bodies of hot, dry rock that may in the future be artificially converted into liquid- or vapor-dominated reservoirs if current research and development projects are successful. These may or may not be made to supply usable water, depending on the methods employed and the availability of suitable sources of water. But whatever the reservoir type, if it can be made to produce power it can contribute to making fresh water available.

For geothermal-resource development, liquid-dominated reservoirs appear at present to be the most important; they are the most readily available and offer the potential for water and power production, regional water-inventory management and salinity control, and mineral production. Consequently, the examples of geothermal desalination taken up in this chapter will be based principally on liquid-dominated reservoirs.

Idealized Liquid-Dominated Reservoirs

One quadrant of an idealized liquid-dominated reservoir is shown schematically in Fig. 1. The block of strata depicted may be taken as a mile or more on a side. There may be an impervious layer overlying the reservoir, as suggested by the shaded bands. The edge in the foreground represents the centerline of the hottest part of the reservoir. The permeability of the reservoir strata and the differences in liquid density, owing to differences in temperature, result in circulation of the geothermal liquid, as described by White (this volume) and as demonstrated by Schrock, Fernandez, and Kesavan (1970) on a laboratory scale for a somewhat similar system. This very slow circulation, indicated by the curved arrows in the diagram, brings hot liquid relatively close to the surface, where it can be tapped by producing wells, as

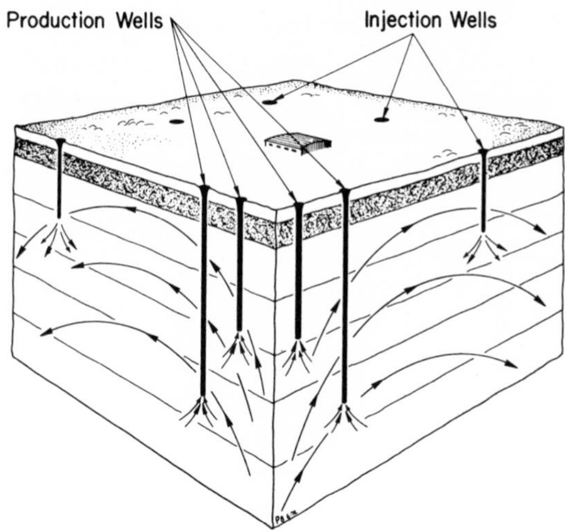

Fig. 1. Geothermal reservoir (schematic).

shown. Cooler liquid water to replace the hot fluid removed from the reservoir is injected into the system on the flanks of the reservoir, where it can follow the natural convection currents away from the hot region near the surface. Eventually, the injected water may enter the hot part of the reservoir at a lower level and be withdrawn as hot fluid by the production wells.

In such a system the hot geothermal fluid from the production wells would be piped to the power/water plants. Electricity and fresh water not used locally would be exported. Reject brine could be disposed of locally or at some distance by releasing it to surface waters, allowing it to dry in an evaporation pond, or returning it underground by seepage or deep-well injection. If a net loss of fluid in the reservoir is to be avoided, local or imported low-quality water would have to be provided for injection.

Figure 2 is a generalized schematic diagram of one such system. The geothermal fluid is separated into a vapor stream and a brine stream; this is known as an "open" system. The vapor, or steam, is fed through a turbogenerator to produce electric power. The residual thermal energy may be disposed of to a body of water, or to the atmosphere via evaporative coolers or dry-air coolers. And the condensate from the turbine exhaust may be used as a source of fresh water. If the brine, too, is needed as a source of fresh water, the brine stream could be

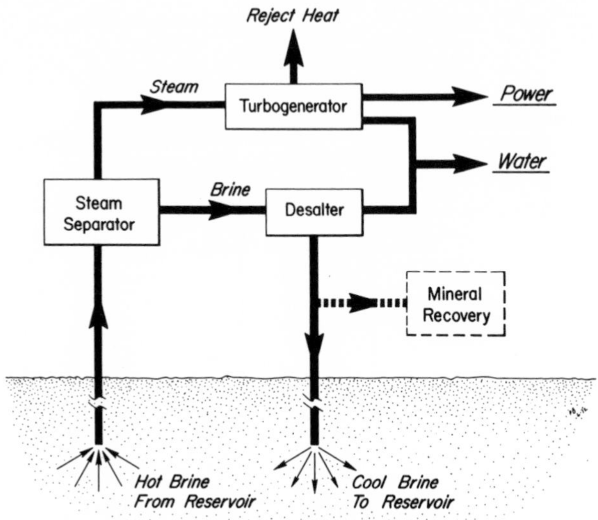

Fig. 2. Open-system geothermal power/water plant.

desalinated. If economically feasible, minerals could be recovered directly or as a byproduct of some water-treatment process. The open system is used for power production in New Zealand and is the basis of the development at Cerro Prieto, Mexico (Koenig, this volume).

Under certain conditions, however, the open system has disadvantages. If undesirable precipitation of minerals, release of gases, or changes in chemical balance leading to severe corrosion accompany the phase change of a particular geothermal brine, the "closed" system, in which phase change is suppressed and no fluid is allowed to escape, may be more desirable. In this system, shown schematically in Fig. 3, a pump near the producing zone in the well maintains sufficient pressure in the brine to suppress phase change as the brine comes to the surface, passes through a heat exchanger (which cools it while heating a secondary fluid), and returns to the reservoir. The secondary fluid is used to drive a turbogenerator in a closed cycle. Brine treatment may be necessary to make the brine suitable for reinjection, as in the open system. And mineral production, again, may be a byproduct of such treatment. Since no water is produced directly by this system, only the open system will be considered here. The closed system (the vapor-turbine cycle) is described in detail by Anderson (this volume).

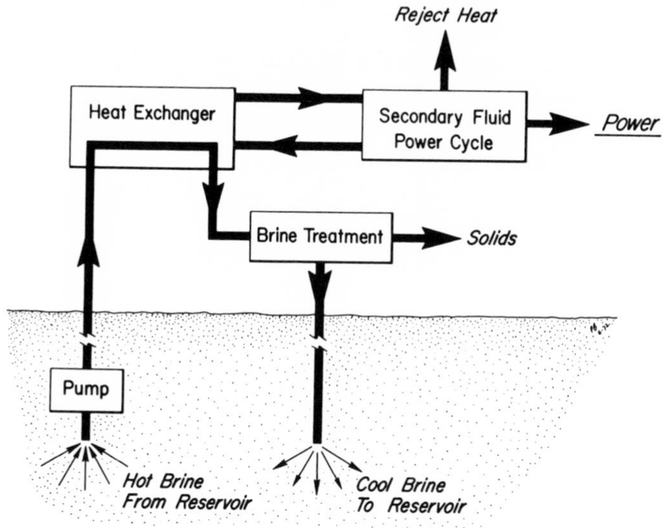

Fig. 3. Closed-system geothermal power plant.

Water and Power Generation: Effects of Some System Parameters

The most efficient recovery of water from brines requires complex plants, and the choice of plant to be used is influenced by brine characteristics as well as other considerations. A few simple examples (see Table 1) will demonstrate the effects of various system parameters and choices of process. These examples are all based on a production rate of 1 million pounds per hour of well fluid.

A simplified flow diagram and thermodynamic-process diagram for Case 1 are given in Fig. 4. Assumed state points (sp) and calculated values are listed in Table 2. The brine and steam mixture entering the system from the wellhead (sp 1) is separated into liquid (sp 2) and vapor (sp 7). The liquid is passed through a mixed-phase turbine producing useful work, brine (sp 4), and vapor (sp. 9). The brine from this turbine is partially evaporated by throttling to a lower pressure and temperature (i.e. flashed) in the flash chamber to provide cooling for condensation of the vapors from the turbines. The mixture (sp 5) resulting from this flashing separates into a reject-brine stream (sp 6) and a vapor stream (sp 10). The vapor is condensed in the air condenser to form fresh-water product (sp 11).

The vapor (sp 7) from the separator is passed through the steam

TABLE 1
Approximate Calculated Water and Power Production for Six Cases and Various Assumed Conditions

Production parameter	Case 1	2	3	4	5	6
Reservoir temperature, °F	600	600	600	600	400	300
Wellhead						
Temperature, °F	444.6	444.6	444.6	444.6	300	200
Pressure, psia	400	400	400	400	67.0	11.5
Lowest temperature, °F	212	160	160	160	160	160
Fresh water/well fluid, ratio	0.79	0.83	0.79	0.87	0.43	0.26
Water production rate[a]						
Million gallons/day	2.04	2.15	2.05	2.26	1.13	0.62
Thousand acre-ft/yr	2.29	2.41	2.30	2.55	1.27	0.70
Power production rate, Mw	17.7	24.3	24.3	24.1	6.7	1.2
Percentage of Case 2						
Water rate	95	100	95	105	53	29
Power	73	100	100	99	28	5

[a] At 90-percent plant availability per million lb/hr of fluid supplied.

TABLE 2
State Points and Calculated Values for Case 1 (Fig. 4)

State Point[a]	Mass (lbs)	Pressure (psia)	Temperature (°F)	Enthalpy (Btu/lb)	Salinity ratio[b]
1	1.000	400.0	445	617	1.00
2	0.753	400.0	445	424	1.33
3	0.753	18.6	224	398	1.33
4	0.592	18.6	224	192	1.69
5	0.592	14.7	212	192	1.69
6	0.212	14.7	212	180	4.72
7	0.247	400.0	445	1205	0.00
8	0.247	18.6	224	1027	0.00
9	0.161	18.6	224	1155	0.00
10	0.380	14.7	212	1151	0.00
11	0.380	14.7	212	180	0.00
12	0.408	18.6	224	192	0.00

[a] Brine properties are approximated by water properties.
[b] Salinity ratio is relative to state point 1.

turbine, producing useful work and an effluent mixture of liquid and vapor (sp 8). This stream, along with the vapor from the mixed-phase turbine, is condensed inside coils in the flash chamber to produce fresh water (sp 12).

Nearly four-fifths of the well fluid is converted to fresh water. The

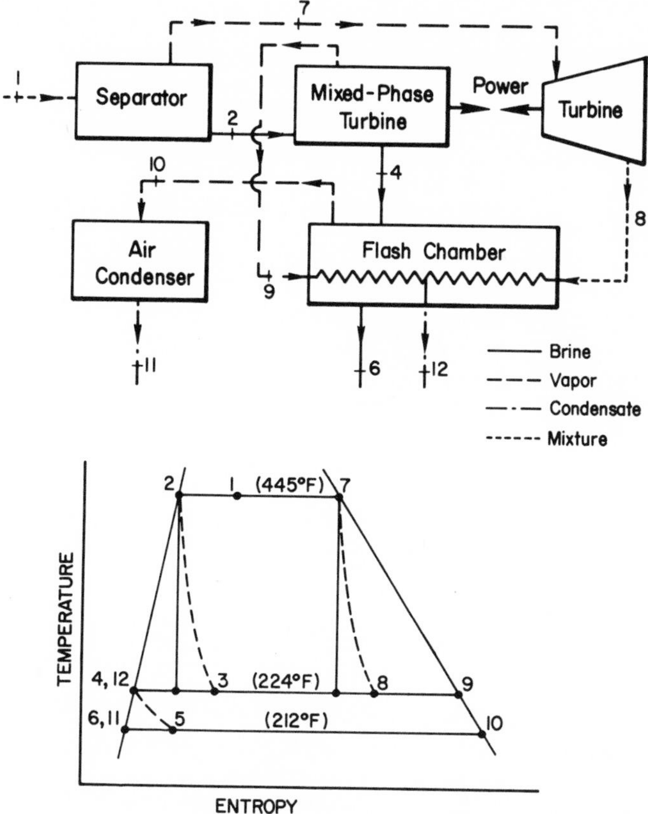

Fig. 4. Sample open-system geothermal power/water plant (Case 1): flow diagram (above); thermodynamic-process diagram (below).

brine returned to the reservoir is thus 4.7 times more concentrated than the well fluid. The calculated power-production rate is 17.7 Mw and the water-production rate at 90-percent availability (i.e., operating 0.9 year each year) is about 2 million gallons per day (mgd) or 2,290 acre-feet per year. Since the fresh-water product is above its normal boiling point, further cooling would probably be necessary. This might be accomplished by evaporation, by mixing with cooler receiving water, or by heat loss from distribution piping. But such high discharge temperatures for the water and brine constitute a thermodynamic loss that wastes geothermal energy.

The effects of reducing the temperature at which thermal energy is rejected are indicated by Case 2, in which the turbine-discharge tem-

TABLE 3
State Points and Calculated Values for Case 4 (Fig. 5)

State Point[a]	Mass (lbs)	Pressure (psia)	Temperature (°F)	Enthalpy (Btu/lb)	Salinity ratio[b]
1	1.000	400.0	445	617	1.00
2	0.753	400.0	445	424	1.33
4	0.501	381.5	440	419	2.00
6	0.380	6.0	170	128	2.63
8	0.125	4.7	160	128	8.00
9	0.247	400.0	445	1205	0.00
10	0.247	400.0	445	424	0.00
12	0.245	381.5	400	419	0.00
14	0.186	6.0	170	138	0.00
15	0.059	6.0	170	1134	0.00
16	0.252	381.5	440	1204	0.00
18	0.255	4.7	160	1130	0.00
19	0.421	4.7	160	128	0.00
20	0.266	6.0	170	138	0.00
21	0.002	381.5	440	1204	0.00

[a] Brine properties are approximated by water properties.
[b] Salinity ratio is relative to state point 1.

perature is taken as 170°F (instead of the 224° of Case 1) and the heat-rejection temperature as 160° (instead of 212°). Other parameters are held constant, as shown in Table 1. Compared to Case 1, the fraction of well fluid converted to fresh water is increased from 79 to 83 percent, water production from 2,290 to 2,410 acre-feet per year, and power by about 37 percent from 17.7 to 24.3 Mw.

Case 3 is a simplified version of Case 2. The steam turbine is eliminated and the wellhead mixture (sp 1) is passed directly through a mixed-phase turbine. Other conditions remain the same. Water production is 5 percent less, and power, at 24.3 Mw, is about the same as for Case 2.

These three systems have the disadvantage that the noncondensable gases pass through much of the equipment, perhaps causing corrosion unless these gases are removed early in the succession of processes. Moreover, if it were found necessary to collect noxious gases for disposal, their larger volume at low pressure would require larger equipment and their disposal might cost more than at high pressure. Indeed, noncondensables at high pressure might be disposed of rather simply by mixing them with some of the liquid stream being injected into the reservoir; and high-pressure removal of noncondensables also avoids

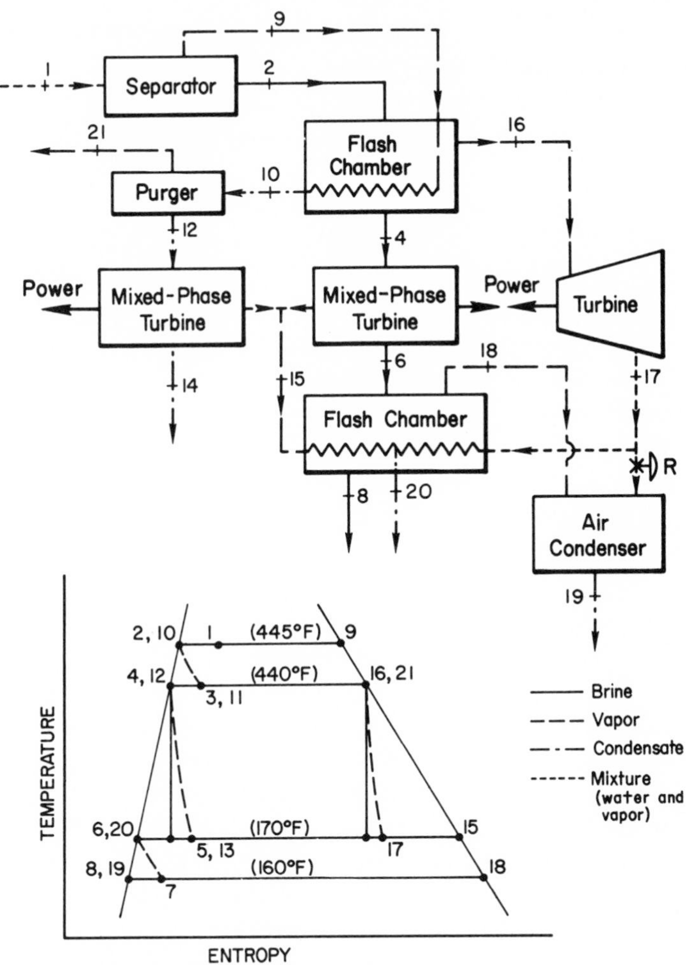

Fig. 5. Sample open-system geothermal power/water plant (Case 4): flow diagram (above); thermodynamic-process diagram (below).

the need for large vacuum pumps that arises when condensation occurs below atmospheric pressure.

Such difficulties can be overcome by more complex systems that may also provide greater flexibility of operation. Case 4, shown in Fig. 5 and Table 3, is a variation of Case 2, differing from the previous cases chiefly in disposing of the noncondensables at high pressure, and in bypassing some of the discharge from the steam turbine to the air condenser, as a means of regulating the fraction of well water that is converted to fresh water. Two mixed-phase turbines are needed to keep the fresh-water and brine streams separate.

The brine stream from the separator at sp 2 is flashed to sp 3 to provide cooling for condensation of the wellhead steam, permitting capture of the noncondensables. The brine from this flash chamber is passed through a mixed-phase turbine and second flash chamber (through state points 6, 7, and 8), as in the previous cases. The residual brine (sp 8) is pumped out and disposed of.

The steam from the separator (sp 9) condenses inside the coils of the first flash chamber and enters the purger as fresh water (sp 10) to be stripped of noncondensable gases. There are several procedures for removing noncondensables, each causing the loss of a different amount of water and leaving the liquid in a different thermodynamic condition. For this analysis it is assumed that the process would be equivalent to the flashing of the liquid from sp 10 to sp 11, followed by the expelling (with the noncondensables) of the vapor thus produced (sp 21). The remaining water (sp 12) is then passed through a mixed-phase turbine, and the effluent fresh water (sp 14) is pumped out as product water. The vapor at sp 15 passes through the condenser in the second brine-flash chamber (to sp 20).

The vapor from the first brine-flash chamber (sp 16) passes through the steam turbine, as in Cases 1 and 2. By means of control valve R (Fig. 5), the correct amount of the effluent mixture (sp 17) is passed through the condenser in the second brine-flash chamber to evaporate the desired amount of brine, thus bringing the concentration of the reject brine to the desired level (sp 8), which is equivalent to regulating the fraction of well water that is converted to fresh water (in this case, seven-eighths). The vapor (sp 18) from the second flash chamber is condensed in the air condenser, along with the effluent at sp 17 from the regulating valve, to sp 19.

The calculated fresh-water production of this system is about 0.87 lb per lb of well fluid between 160° and 170°F, or 2,550 acre-feet per year per million pounds per hour of well fluid supplied (at 90-percent plant availability). The gain in water production is about 5 percent over Case 2. Power production, at 24.1 Mw, is about 1 percent less than for Case 2, largely because Case 4 wastes some of the available energy, particularly from the steam-turbine exhaust, to control the salinity of the residual brine.

Although several wells (see, for example, Residencia, 1968; Resources Agency, 1970) have produced water with the high thermal content assumed in the above cases, water with substantially lower thermal content is much more likely to be encountered, particularly at shallow depths. It is generally considered more economical to produce power from these lower-temperature waters by using systems employing secondary working fluids, as described by Anderson (this volume). However, the thermodynamic effectiveness of the open systems discussed above is about the same as for those using a secondary fluid. Consequently, a measure of the effect of reduced reservoir temperature, and thus of thermal content of the wellhead fluid, can be furnished by two more examples, which are in other respects similar to Case 3.

For Case 5, it has been assumed that the thermal content of the reservoir is equivalent to that of liquid water at 400°F and that the fluid enters the system at a wellhead temperature of 300°. For Case 6, the corresponding temperatures are 300° and 200°. (Production parameters for these two cases are also given in Table 1.) The heavy penalty paid for lower reservoir temperatures is apparent. The differences between Cases 1 and 2 show that there is also a severe penalty for using higher waste-heat rejection temperatures. The performance of such systems may be much improved by employing evaporative cooling to yield significantly reduced lowest temperature (i.e., lowering of line 10–11 in Fig. 4). It may also be concluded from these examples that even a modest reuse of the heat of condensation of the vapors results in significantly improved water production.

Multiple-Effect and Multistage Distillation Processes

Multiple-effect (ME) distillation, in which each effect takes place at a lower temperature than the preceding one, is one of the oldest and best desalination methods. Modern high-performance, ME sea-water-conversion plants may employ 10 to 25 effects to produce from 8 to 20 pounds of water for every pound of steam condensed to heat the first effect. Extensive reuse of the heat of condensation to evaporate an equal amount of brine on each reuse (or "effect") is an application of ME distillation. (A general discussion of desalination methods is given by Spiegler, 1966.)

A particularly interesting application of this principle has been suggested by the Bureau of Reclamation, U.S. Department of the Interior, in a recent report (BuRec, 1972). It consists essentially of a three-effect vertical-tube evaporator (VTE) operating within about 26°F of the assumed wellhead temperature of 398°. The incoming mixture is assumed to be 20 percent steam. The first effect is similar to that of Case 4, including the removal of noncondensable gases. The two additional effects are similar except that with little noncondensable gases, no provision is made for their removal. The brine from the third effect is five times as concentrated as the wellhead mixture; i.e., 80 percent of the wellhead fluid is converted to fresh water. The production rate is therefore about 2,500 acre-feet per year per million pounds per hour of well fluid at 90-percent availability. The steam from the third effect is passed through a turbogenerator to produce some 10 Mw. Both these rates are comparable to those of Case 4 when allowance is made for the lower wellhead temperature in the Bureau's example.

Another distillation method that has been proposed for desalting geothermal brines is the multistage-flash (MSF) process in which hot brine is passed through a series of chambers, each chamber at a lower pressure than the one preceding. One step of the process is accomplished in each stage. Some of the brine is flashed as it enters each chamber. Since evaporation is not instantaneous, each chamber is several feet long to allow time for the vapor to disengage itself as the brine flows along the bottom of the chamber. Within each chamber are tubes inside which cooler brine flows toward the hot end of the series of chambers; the vapor produced is condensed on these tubes. The condensate drains into a trough below the tubes. This distillate, as well, is passed through the chambers in a manner similar to that for the hot brine; its vapors also condense on the condenser tubing, thereby providing more heat for the cool brine. Commercial MSF sea-water distillation plants have from 10 to 30 or more stages. Spent thermal energy is normally rejected to the ocean.

Some of the advantages of the MSF distiller for sea-water conversion carry over to geothermal-brine distillation. It is probably the best-developed desalination technique and dominates the field—over 90 percent of all desalination in the world is performed by this process. Its hot-brine channel can be made sufficiently free of obstructions that precipitating solid materials (such as silica, which is apt to be encountered in geothermal distillation) will not pile up and reduce the plant's effectiveness. In addition, deposits that may adhere to channel surfaces can be scraped off economically. The MSF process also shares with other geothermal-distillation systems the advantage that the feedwater needs no additional heating; in the sea-water version of MSF distillation, feedwater must be heated—by a heater introduced between the discharge end of the condenser tubes and the entrance to the series of flash chambers. But this advantage of the geothermal system carries with it a serious disadvantage: in sea-water-desalting applications, the cool brine inside the condenser tubes is incoming feedwater being heated; for geothermal-desalination applications, no good substitute has been found for this expeditious method of cooling the condenser, which would otherwise allow a reasonably high fraction of geothermal brine to be desalted. Consequently, the ME process, which can convert as much of the geothermal brine as may be desired, at present has the edge on the MSF process. In fact, because of recent advances in heat-transfer technology, the ME process is beginning to challenge MSF distillation for sea-water conversion, and the MSF process is likely to be relegated to the role of the feedwater heater.

At the present time, large distillation plants (MSF and VTE plants of about 5 mgd capacity per unit) can produce water at a cost of about $0.60 per 1,000 gallons. Within a decade the cost should be halved; and if a large market for distillation plants develops, costs should be halved again by the year 2000.

Energy-Budget Management

Cooling-water problems in MSF geothermal distillation are part of the general problem of handling the energy, water, and solids budgets in geothermal systems. None of these is destroyed by the various procedures; they are merely redistributed or converted to different forms. Thermal energy cannot be stored, but may be converted to electrical, chemical, or potential energies that can be stored. But not all the thermal energy taken from a geothermal reservoir can be converted to other forms—it must flow into materials at lower temperatures or be radiated away. Unconverted thermal energy is known as spent, reject, or waste heat. Its entry into the environment may be a matter of indifference, or it may be approved or disapproved depending on whether the local heating it causes is considered as thermal enrichment (beneficial) or as thermal pollution (detrimental). Regardless of people's attitudes, some thermal energy must be rejected to the environment as a consequence of any activity involving energy. Consequently, the manner of disposing of waste heat is important in geothermal-resource utilization.

The amount of thermal energy that would be transformed to chemical energy simply in removing salt from saline solutions by some perfect process—about 4 kwh per 1,000 gallons of water converted from sea water—is generally small compared to the geothermal energy that can be converted to electricity, returned to the reservoir in the reject brine, or otherwise disposed of. In Case 2, for example, the approximate relative amounts of energy available above 80°F in these four categories are 1 for desalting, 32 for electricity, 5 for injection, and 184 for rejection to the surface environment and radiation to outer space. If cooling water is available at 80°, heated to 160° (the lowest temperature for Case 2), and injected into the reservoir with the brine to replace the amount of reservoir fluid extracted, these relative values become 1:32:31:150, respectively. Such a distribution of energy is typical of proposed geothermal water/power schemes. If only water is produced, and if little attempt is made to return thermal energy to the reservoir, practically all of the geothermal energy extracted becomes waste heat.

Exporting electricity from the region of the geothermal field not only expands the area in which the thermal energy must eventually be rejected, but also allows desalting to be decentralized. A high-performance sea-water-conversion plant, for example, consumes about 30 kwh per 1,000 gallons of fresh water produced. Consequently, the 24.3 Mw of power shown for Case 2 would produce some 19 mgd more fresh water if no power were lost in transmission. This is an impressive water-producing potential that would be even greater if brackish water were to be desalted.

If the electricity were not transmitted over long distances, but were sold locally, with the proceeds used to buy fresh water or the means of producing it, where needed, desalination far from the geothermal energy source would be more economical. An analogous exchange might occur between two users on the same river, seeking augmentation of water in short supply: if one could be economically supplied with desalted water and the other could not, the former might trade his water rights with the latter for money to produce desalted water.

Processes for Decentralized Desalination

The principal desalination methods that operate on electric power are vapor-compression distillation, reverse osmosis, and electrodialysis.

All are well-established commercial processes, and each has compelling advantages under particular conditions.

The vapor-compression (VC) distillation process is based on the phenomenon that raising the pressure of water vapor by means of a compressor raises its temperature, thereby making a temperature difference available that can be used to cause condensation of the compressed vapor on one side of a heat-transfer surface and evaporation of an equal amount of water from brine on the opposite side of the surface. This evaporand (vapor from the brine) is fed to the compressor. The best designs usually incorporate a vertical-tube evaporator that has little resistance to heat flow, so that the temperature difference—and consequently pressure difference—across its tubes can be small, to minimize the compressor power. The staging principle can be incorporated into VC plant designs (Tleimat, 1969). And for big plants, an MSF distiller would probably be used to heat feedwater. Many other refinements to the basic process can be made, and technological improvements are to be expected (Laird, 1971a).

The VC process can be made very economical. It is highly competitive for small- (10,000 gpd) to medium-capacity (500,000 gpd) sea-water-conversion service and for use with other relatively concentrated brines. It is not suitable, however, for feedwaters (such as geothermal brines) containing appreciable silica or other salts that become less soluble as the temperature decreases. Nor is it competitive with reverse osmosis or electrodialysis for desalting brackish waters.

Reverse osmosis (RO), or hyperfiltration, desalination depends on membranes that pass water but hold back most of the salts when a pressure difference higher than the osmotic pressure is maintained across the membranes. Because the osmotic pressure increases with concentration difference across the membrane, both the energy required and the cost of product water increase with the salinity of the feedwater.

There are many successful variants of this process. It is most competitive for saline waters in the concentration range from 2,000 to 5,000 parts per million (ppm). In this range of concentration, water-production cost for RO should be about one-half that for desalting sea water by distillation.

Electrodialysis (ED) desalination depends on the use of two kinds of membrane that are selectively permeable to ions. One kind of membrane passes positive ions (cations) better than the other, which passes negative ions (anions) more readily. Alternate cation and anion membranes are arranged as the walls of parallel channels through which saline water is flowed, and across which electric current is passed. The electric current carries ions of opposite sign through the membranes on opposite sides of each channel. Consequently, the concentration in every second channel is reduced, but in alternate channels it is increased. The channels are made long enough to obtain the required reduction in concentration of the product water. Under favorable conditions, the electric current flowing across the channels is proportional to the number of ions it transports, and therefore to the required reduction of salinity. Consequently, more concentrated saline waters require proportionately more electrical energy for their desalination.

Electrodialysis is generally favored for freshening brackish waters with concentrations up to 3,000 ppm. Since this range of salinities is relatively low, ED costs should be about one-third that for desalting sea water by distillation.

Water-Budget Management

A number of matters should be considered in handling the regional water budget in geothermal developments. The operator of a geothermal plant on the seacoast or near other large supplies of water, particularly one near a large water market, would probably have little trouble in keeping his plant's water system in balance. Far from the sea, however, careful manipulation might be necessary.

In arid regions, liquid-dominated reservoirs should be expected to occur in low-lying basins where there is little or no natural drainage. Water brought in by precipitation would be balanced by evaporative losses. To ensure long-term productivity of a reservoir supplied by a large heat source, maintaining pressure in the reservoir might have to be accomplished at the expense of reinjecting all the water produced. This would require heat rejection to the atmosphere via dry-air heat exchangers and other water-saving measures.

If the water content of a reservoir were large relative to its thermal-energy content, it might be permissible to expend the water inventory to provide evaporative cooling for heat rejection. Irrigation with desalted water might also be feasible if evapotranspiration and other losses were within safe limits.

Reservoir overdraft should not be allowed if intolerable surface subsidence might result from the shrinkage of supporting strata. Such shrinkage is commonly associated with the reduction in reservoir pressure caused by fluid withdrawal. Drying the strata or exposing them to air may also cause shrinkage.

Seismic activity is also a possible result of altering reservoir conditions, especially since geothermal areas tend to be in areas of crustal deformation. It is generally believed (Bowen, this volume) that maintenance of fluid pressures in geothermal reservoirs is one way to avoid inducing earthquakes.

For very large systems, it may be economical to import water for pressure maintenance and cooling. The Imperial Valley in southern California is an example of a geothermal area in an arid region some distance from the coast. The large-scale development contemplated (BuRec, 1972) includes the export of as much as 2.5 million acre-feet per year of desalted water from the Valley to the Colorado River. Rex (1970) and others have suggested importing sea water as makeup water for the Valley. It was presumed that an amount of sea water equal to the net brine withdrawal from geothermal reservoirs, plus the amount used for cooling, would be imported from the Gulf of California or the southern coast of California, over 100 miles away.

For medium-sized developments, the makeup, or replacement, water may be locally available. Large amounts of saline groundwater occur in many places, and waste waters can be important sources. An interesting possible use of irrigation-drainage water has been proposed (BuRec, 1972; Rex, 1970) in connection with development of geothermal resources in the Imperial Valley. This is an arid basin below sea level, with no drainage outlet. Agricultural-drainage water maintains the Sal ton Sea, which essentially constitutes the drainage sump for the Valley ("Salton Sea Project," 1969). Water inflows of about 1.3 million acre-feet per year are balanced by evaporation from the 360-square-mile surface of this salt lake. It has been estimated (Rex, 1970) that 150,000 acre-feet of water could be withdrawn annually and used for pressure maintenance in geothermal reservoirs. This withdrawal would lower the level of the Salton Sea, reducing its surface area until the reduction in evaporation (about 12 percent) balances the withdrawal. The Bureau of Reclamation estimates that 125,000 acre-feet of water would have to be withdrawn from the Salton Sea to supply replacement and cooling water for their demonstration facility, which is designed to produce about 100,000 acre-feet per year of desalted water and 420 Mw of power. Since the fresh water is to be contributed to the Valley's irrigation system, this withdrawal may have less effect on the Salton Sea's level than the export of water from the Valley recommended in the Bureau's large-scale scheme.

Solids-Budget Management

The use of local waters as replacement for reservoir fluid is an important aspect of managing regional salt budgets. It may also provide a means of disposing of other solids. And since desalination concentrates the salts, it, too, is a useful salt-management tool. The net effect of the fresh-water production is to make room underground for the injection of brine, which removes salts from the surface environment. If other underground void space is available, it could be used for disposal of saline waters in an independent salt-control procedure. The salts might thereby be controlled within the region. An alternative method of salinity control would be salt exportation. The economies to be gained by combining water and power production, salinity control, and possibly other activities may make an entire system economically feasible and provide multiple benefits, whereas each activity may not by itself be viable.

Such may be the case in the Imperial Valley. Stabilization of the salinity and level of the Salton Sea, reduction of irrigation-water salinity, provision of cooling for geothermal water and power plants, prevention of surface subsidence, and augmentation of the Colorado River flow are all among the benefits to be expected from the proposed development of geothermal resources in the Imperial Valley. A subreconnaissance-level cost estimate (BuRec, 1972) indicates that the benefits from the whole scheme would be considerably greater than the costs.

Cost Projections

In the Bureau's study, the unit cost of water delivered at the rate of 100,000 acre-feet per year was estimated at $85 to $130 per acre-foot for transportation distances less than 80 miles. Power from the associated 420-Mw electric-power system would cost 3 or 5 mills per kwh at the plant boundary, depending on whether Federal or private funding, respectively, were used. The capital cost of the water plant would be about $0.69 per daily gallon of capacity, which is about that for the

MSF sea-water-conversion plant of about the same unit size now operating at Rosarita Beach, Mexico. Electric-plant cost would be about $130 per kw of capacity.

For this large-scale development in the Imperial Valley, the estimated unit cost for 2.5 million acre-feet per year of distilled water is about $100 to $150 per acre-foot, delivered between 100 and 250 miles away, with sea water imported from 100 miles away. The unit costs of 10,000 Mw of electric power are given as 3 or 5 mills per kwh, on the same basis as for the 420-Mw plant. The estimated water cost at the plant site is about $70 per acre-foot. For converted sea water from a plant of 10 to 50 mgd capacity using the latest technology, this would be a low price.

If the above water prices were subsidized by the 2 mills per kwh difference between the estimated Federal and private costs, the price of each acre-foot could be reduced by about $87. A subsidy based on the difference between estimated production costs and local power costs of about 7 to 10 mills per kwh could result in water being delivered free to the water customers at the expense of the power customers and the public.

It is unlikely that any form of desalination will compete with large systems able to supply good water at prices reflecting only pumping costs from a nearby river. However, several desalination methods available for upgrading otherwise unusable waters (waste waters, saline groundwaters, sea water) are already competitive with long-distance importation schemes. In the next few decades, desalination costs should be reduced by a factor of 3 or 4. Consequently, desalination, in many regions, will become the most economical water-supply method. Limitations on desalination might prove to be in the supply of energy or in the capital needed to build the plants. Successful geothermal-resource development, however, should improve the supply of energy directly. And the increasing growth in the country's economy should provide the financial basis for large-scale, desalinated water-supply systems. Geothermal desalination itself is now very promising and should contribute significantly to future water-management systems.

As technology improves and associated problems (Anderson and Axtell, 1972; Laird, 1971b) are resolved, the costs of developing geothermal resources will decrease, and more and more geothermal water/power plants will become economically attractive.

REFERENCES

Anderson, D. N., and L. H. Axtell, compilers. 1972. El Centro Conference, Geothermal Resources Council. Davis, California.

Bureau of Reclamation, U.S. Department of the Interior. 1972. Geothermal resource investigations, Imperial Valley, California.

Laird, A. D. K. 1971a. Desalting technology. *In* D. Seckler, ed., California water: A study in resource management. Berkeley: University of California Press, pp. 127–60.

——— 1971b. Ranking research problems in geothermal development. U.S. Department of the Interior, Office of Saline Water Research and Development, Progress Report 711. Washington, D.C., U.S. Govt. Printing Office.

Residencia de Obras Geotermicas del Valle de Mexicali. 1968. Campo geotermico, Cerro Prieto, B.C. (Table 1).

Resources Agency. 1970. Compendium of papers, Imperial Valley: Salton Sea Area. Section Y, Geothermal Hearing, Sacramento, California, October 22–23.

Rex, R. W. 1970. Investigation of geothermal resources in the Imperial Valley and their potential value for desalination of water and electricity production. Institute of Geophysics and Planetary Physics, University of California, Riverside.

"Salton Sea Project, California." 1969. Federal-State Reconnaissance Report. U.S. Department of the Interior and The Resources Agency of California.

Schrock, V. E., R. T. Fernandez, and K. Kesavan. 1970. Heat transfer from cylinders embedded in a liquid-filled porous medium. Paper CT 3.6, 4th International Heat Transfer Conference, Paris-Versailles. *In* Heat transfer 1970. Amsterdam: Elsevier.

Spiegler, K. S., ed. 1966. Principles of Desalination. New York: Academic Press.

Tleimat, B. W. 1969. Novel approach to desalination by vapor compression distillation. ASME Publication 69-WA/PID-1, Amer. Soc. Mech. Engineers Annual Meeting.

10. Environmental Impact of Geothermal Development

RICHARD G. BOWEN

The motive force of our industrial-technological society is the use of stored energy. Within the United States, where about 6 percent of the world's population uses 35 percent of the world's energy, many are beginning to question whether all of this expenditure of nonrenewable resources is necessary. For although it would be catastrophic to prohibit the use of energy stored in fossil fuel or in fissionable material, it would be equally catastrophic to use energy at its projected potential rate of increase, which finds electric-power production doubling every 10 years.

A running confrontation has ensued between those whose projections call for more production and consumption of electricity and those who insist that past values and practices have brought us to the brink of disaster, that the price for "more" is too high. Arguments have also revolved around claims and counterclaims of the proponents and opponents of the various methods of producing electricity, each faction claiming its method is the "cleanest." Because the power plants proper are salient in the public eye, the controversy has centered on them, almost to the exclusion of the other steps of the fuel cycle, some of which have much greater environmental impact. To render valid judgment on the environmental effects of the several means of producing electric power, it is necessary to look beyond the power plant to the total fuel cycle.

The geothermal plant is unique in that all of the steps in the fuel cycle are localized at the site of the power-production facilities. At the other end of the power spectrum is the nuclear-reactor plant, in which the actual power-production facilities are a small fraction of a cycle requir-

---

Richard G. Bowen is Economic Geologist, State of Oregon Department of Geology and Mineral Industries, Portland.

ing a complex industrial-support system for each reactor. Intermediate in complexity, and varying somewhat in rank with the type of fuel used, are the fossil-fuel plants. Thus in all instances except geothermal the environmental impact of the fuel cycle extends far beyond the bounds of the power-generating plant.

Characteristics of Geothermal-Energy Production

Like other thermal power plants, the geothermal plant involves the production and use of steam, expanding it through a turbine and condensing it at the turbine exhaust. The geothermal plant differs from the conventional-fuel or nuclear plant in its method of steam production and the quality of its steam. At the geothermal plant the steam is produced in nature's own boiler by the natural circulation of water coming into contact with hot rocks in the depths of the earth. Depending upon conditions within the reservoir, the geothermal fluids may be in the form either of slightly superheated dry steam or of pressurized hot water. The condition of the fluid in turn controls the method of utilization and the potential environmental impact from its production.

The most desirable type of geothermal field, and the only kind that has proved to be economically viable with existing technology in the United States, is the dry-steam field, with its sole example that at The Geysers, California. But projects are under way that will utilize the more prevalent hot-water fields in other parts of the country, and the environmental effects of the various power-production cycles employed must be examined.

Dry steam and hot water differ chiefly in the quantity of geothermal fluid that must be brought to the surface to produce a given amount of electric power. At The Geysers, it takes about 20 pounds of steam to produce 1 kwh of electricity. In a dry-steam field the total well production can be utilized in the power cycle; of this 20 pounds, approximately 15 is evaporated in the cooling system and the remaining 5 is disposed of by returning it to the production reservoir. Aside from the aesthetic impact of the simple presence of the geothermal power plant, the only release of products is the venting of modest quantities of noncondensable gases entrained in the steam. More will be said about the nature and quantity of these gases.

There are two possible methods of producing power from hot-water fields. One is to flash water to steam at reduced pressures and then treat the steam in the same manner as the dry-steam power plant. This is the method successfully used in New Zealand and Mexico. The other method, the vapor-turbine cycle, described by Anderson (this volume), uses heat exchangers and a turbine with a separate working fluid.

Because of the lower enthalpy of hot water, both of these methods bring much greater quantities of fluid to the surface than dry-steam fields produce, per kwh of electricity produced. The actual amount is dependent upon the water temperature and the flashing pressure, but in actual conditions it ranges normally between 75 and 150 pounds per kwh of electricity (Hansen, 1964). With the simple flashing method, utilizing the steam condensate for cooling, the disposal of 60 to 135 pounds of geothermal water is required. This can be done in many ways, but in the United States injection into the producing reservoir is probably most desirable. The vapor-turbine cycle requires less water than the flashing method per kwh of electricity, but all of the geothermal fluid must be disposed of, since it is not used for cooling.

Except in their manner of returning fluids to the geothermal reservoir, both the simple flashing system and the vapor-turbine cycle have proved successful in basic concept. The problem with reinjection lies in the fact that most geothermal hot waters contain some dissolved solids, and the lowered pressures and temperatures may cause salt precipitation, which in turn might reduce porosity or plug fractures in the reservoir. The net result could be a decrease in permeability and capacity for accepting further reinjection fluids—to the inevitable detriment of productivity. In dry-steam fields, where reservoir steam is of high purity, injection is practiced successfully. In this case the injected steam condensate is essentially distilled water and contains only a few parts per million of salts.

An advantage of the vapor-turbine cycle is that it entails no release of noncondensable gases; the geothermal fluid is contained under pressure and not allowed to expand at any time. Without the need for expansion, the noise level at the field is much lower than that of the hot-water flashing system, which can generate considerable noise during the test phase (though during the production phase everything is contained and inaudible). A commensurate disadvantage in the vapor-turbine cycle is the need for a supplementary source of cooling water, since steam condensate is not directly available.

To gain proper perspective on the environmental impact of producing electricity from a geothermal plant, it is necessary to understand the basic character of its various manifestations and to compare the relative impact of other thermal power cycles—nuclear and fossil-fuel—since each produces its own effects. The kinds of effects produced may be categorized by their impact on the land, on the air, and on the water.

Impact on the Land

Natural steam is produced by drilling wells to a depth of 300–2,700 m (1,000–9,000 ft) until a productive steam aquifer is tapped, as is done in the production of natural gas. The pressure of the steam causes it to flow to the surface, where it is collected in insulated pipes and delivered to turbines. At The Geysers field, where individual wells have an average production of about 7 Mw, about 150 wells are required for a 1,000-Mw plant. With the present spacing at The Geysers this would amount to about 12 square miles of land. And additional acreage must be set aside for new wells, to maintain the needed steam supply as production from existing wells declines (see Budd, this volume).

The wells, pipelines, and power plants of the producing geothermal field, such as that at The Geysers, modify the existing terrain. This aspect of geothermal development is one of the main objections voiced by environmental groups. But the development of a geothermal field need not be out of harmony with the surroundings. The geothermal field at Larderello, Italy, has been compatible with many other land uses during its 60 years of development and production. Because the wells, gathering lines, and power plant use only small patches and strips of the field, most of the land is being used for varied agricultural industry, with many farms, vineyards, and orchards interspersed among the pipelines and wells (see Figs. 1 and 2).

Another example of multipurpose utilization is The Geysers field (Fig. 3). Prior to its development as a producing power resource, The Geysers area was a wilderness, with much of the land owned by private hunting clubs that devoted the land to forage for deer. This use, along with cattle grazing, continues at The Geysers field.

The impact of the construction of wells, pipelines, and power plants is most evident during the development period, and for a large field this could extend over several years.

Drilling operations in a geothermal field are comparable to construction activities in noise impact and are equally episodic in nature. The

Fig. 1. The Larderello geothermal field in Italy, showing the steam-gathering lines. Note the compatibility of the steam-extraction facilities with other land uses. (Photo by Ira E. Klein, U.S. Bureau of Reclamation.)

noise problem is associated mainly with drilling operations and steam escape during testing. Once the field is in production the noise level declines to that of other power plants. The drilling of dry steam wells requires special techniques; at the present time, drilling into the production zone uses air, rather than mud, for the circulating fluid that removes the drill cuttings from the hole. This results in a "controlled blow-out" of the well during the time the steam zone is penetrated, amounting to only a few days out of the total drilling time. There is no danger involved because the pressures are relatively low and the blow-out can be quenched at any time by pumping water down the drill string. When the well is completed it must again be allowed to blow until the accumulated dust and rocks are removed from the bore hole. This constitutes the clean-out period, and until it is completed the wells cannot be completely muffled. Mufflers now in use at The Geysers field during drilling and testing operations have significantly lowered the noise level in the field, and new developments promise further decline of the noise levels.

Land subsidence and seismic effects are other potential effects of geo-

Fig. 2. Larderello power plant with steam-gathering lines passing through an orchard in the foreground. This site has been used for electric-power generation for over 60 years with minimal environmental impact. (Photo by Ira E. Klein, U.S. Bureau of Reclamation.)

thermal development. The possibilities of these two environmental hazards were raised by the Department of the Interior in its Environmental Impact Statement for the Geothermal Leasing Program (1972), and by the Sierra Club. But neither at the Larderello field, where production has been carried on for 60 years and on a relatively large scale for 30 years, nor at The Geysers, with its 12 years of operating experience, have subsidence or seismic effects been observed. Although these phenomena have been noted under special circumstances in certain oil fields, there is no reason to relate such problems to the dry-steam geothermal field, where the geologic conditions are entirely different. Hot-water fields, however, could present a problem, as we shall see.

Subsidence can occur whenever support is removed from beneath the ground. It has been noted in oil fields, in mines, and from the pumping of subsurface waters. In most cases where subsidence is caused by the removal of ground waters the pumping is from a relatively shallow depth. In oil fields the fluids have come from greater depths and subsi-

Fig. 3. The Geysers steam field, in the mountainous terrain of northern Sonoma County, California. Unit 4 is on the hill at the left, not yet in operation. Steam plumes are evident; the wells producing them are being bled to keep them open prior to adding them to the system. Today, with the system on the utility company's line, steam rises from the cooling towers rather than from the hills.

dence occurs only under special conditions, i.e., when the fluids being removed are at greater than normal pressures for the depth of the reservoir. These conditions constitute an "over-pressured" reservoir, the fluids providing support to the overlying column of rock. Removal of this support may lead to subsidence. Injection of water around the periphery of the field replaces the petroleum with water, thus alleviating the problem.

Because of the geologic circumstances under which dry-steam fields develop, subsidence should not be expected to occur. The production reservoir of a dry-steam geothermal field consists of fracture zones, solution channels, or other permeable cavities filled with vapor, possibly from the "boil-off" of a deeper hot-water reservoir. A unique characteristic of the dry-steam geothermal fields is the near-constant pressure of the vapor wherever measurements are made throughout the vertical section of the reservoir. At The Geysers and the other dry-steam fields so far discovered in the world, the steam temperatures and pressures are about 240°C (465°F) and 34 kg/cm^2 (480 psia). White, Muffler, and Truesdell (1971) discuss the reservoir thermodynamics that explain this phenomenon. This near constancy of pressure, even at depths greater than 2,500 m (8,200 ft), where hydrostatic pressures would normally be about 280 kg/cm^2 (4,000 psia), indicates that for a dry-steam field to exist the host rocks must be competent and therefore not subject to subsidence from the removal of vapor.

Hot-water fields, by contrast, could behave more like an unconsolidated petroleum reservoir, and unless pressures are maintained by fluid return there may be subsidence. Indeed, this has occurred in Wairakei, New Zealand (Hatton, 1970), where the water is not returned to the reservoir. Much has been learned about subsidence from the exploitation of petroleum reservoirs, and with the proper understanding and practices, any geothermal area where this could be a problem can be stabilized.

In relating the exploitation of geothermal resources to seismic hazards it must be considered that the unstable conditions in the Earth's crust leading to the presence of geothermal phenomena are also those conditions producing faults and earthquakes. Thus geothermal and seismic phenomena are geographically inseparable. In fact, the presence of high seismic incidence is one of the exploration clues used in the search for geothermal reservoirs (Clacy, 1968). However, the intensity of individual shocks within the thermal areas and associated with volcanic activity (the source of geothermal heat) is usually of a relatively low order, much lower than that associated with major crustal movements along faults (Ward, 1972). There is much to be learned about the interrelationship of thermal and seismic phenomena, and the drilling and exploitation of geothermal fields should add new information to this field of knowledge. But there is no evidence that geothermal production has increased the seismicity of an area.

Concern over seismic hazards arises in part from the process of reinjection of the spent geothermal fluids. Incidents of seismic activity relating to the injection of fluids in waste-disposal operations, such as that at the Rocky Mountain Arsenal near Denver (Evans, 1966), and to water-flooding operations to repressurize declining oil fields (Raleigh et al., 1970, 1971), have involved injection at pressures exceeding hydrostatic. In such instances, reinjection could open and lubricate preexisting fractures and zones of weakness or extend the fracture pattern, causing increased seismic activity and perhaps structural damage. But geothermal reservoirs are at subnormal pressures and the return of fluids merely maintains preexisting pressures in the reservoir and would not cause the increasing seismicity noted in other conditions. The low pressure existing in geothermal reservoirs facilitates the reinjection of fluids into the field in two ways: first, because reservoir pressure is less than hydrostatic, the water's weight produces sufficient head to ensure its entry into the formation without pumping; second, the returning water seeks the area of lowest pressure, thus minimizing the chances of geothermal fluids' migrating into other aquifers.

In order to compare the impact on the land from geothermal operations with the effects occasioned by the nuclear-fuel and the fossil-fuel cycles, all of the steps in the production of fuels should be considered. Mining, milling, refining, enrichment, conversion, and fabrication must be performed before the nuclear-fuel cells enter the reactor. Mining is the first major step. Over its 30-year active life, a 1,000-Mw reactor will need about 1,000 tons of enriched uranium (Westinghouse Electric Corporation, 1968, p. 45). Using the enrichment ratio required by the present generation of pressurized water reactors, this would require the production of 6,000 tons of natural uranium. The current grade of uranium ore mined in the United States, and the figure usually used for reserve projections, amounts to about 4 pounds of uranium per ton, but over the life of the plant the grade of ore is expected to decline and will probably average 3 pounds per ton. This would require the mining of 4,000,000 tons of ore over the life of the plant, or an average of 133,333 tons per year.

The U.S. Atomic Energy Commission reports (1972, p. 51) that the uranium-mining industry currently holds more than 19 million acres of land for mining and exploration. Not all of this land will be mined out, but the considerable amount of uranium ore that must be extracted to supply projected needs will constitute a major impact on the land.

The milling of uranium ore also creates a substantial impact on the land. Of the many millions of tons of ore that have been milled in the United States to date, most of it is still standing in large waste dumps adjacent to the mills. These waste dumps not only are an eyesore but in some cases represent a landslide hazard, as do the waste dumps from other types of mines. More seriously, they carry the threat of radionuclide contamination to the environment.

Another impact of the nuclear-fuel cycle is the massive construction required by the various steps in the conversion cycle. Of particular significance are the gaseous-diffusion enrichment plants. Three of the plants currently in operation were built originally to supply the uranium needed for weapons, but they are now being used to process nuclear fuels for commercial reactors. These three plants, built at a cost of over 2 billion dollars (Hogerton, 1964, p. 14), consume tremendous amounts of electricity in their operation; in 1962, for example, they consumed 47 billion kwh of electricity or about 5 percent of the total amount of electric power generated in the United States (Westinghouse Electric Corporation, 1968, p. 14). The increasing demand for nuclear fuels will make necessary the construction and operation of new enrichment plants, and the consumption of large blocks of electricity for this purpose.

The transportation and handling of nuclear fuels, especially the spent fuels, is a potential environmental hazard. The isolation and storage of the high-level fission wastes from the several reprocessing plants, whose volume is estimated by the AEC to be about 60 million gallons by the year 2000, requires large, guarded disposal sites. In addition to these high-level wastes, there are large volumes of low-level wastes, such as tailings and various wastes from other steps in the fuel cycle, that must be isolated or diluted and dispersed into the environment. Each of these exigencies uses or compromises occupied land. The high-level wastes may indeed require permanent protection from entry into the environment. And although it is not possible to estimate the amount of subsidiary land that may be required by each reprocessing plant, a considerable amount of surface and/or underground storage facilities may be needed.

Fossil-fuel generating plants, particularly those fired by coal, require a vast acreage of land for mining, railroad yards, and fuel handling. A coal-fired plant of 1,000-Mw capacity would require about 70 million tons of coal over its 30-year life (U.S. Congress, 1969, p. 125). With a ratio of 2:1 overburden to coal, this would amount to the movement of about 200 million tons over the life of the plant. Moreover, land is required to accommodate the washing and shipping of the coal and to dispose of the fly ash and clinkers. Coal-fired electric power plants usually require more land than do nuclear plants for the plant site proper, but because of the simplicity of the fossil-fuel cycle and because its wastes do not require guarded isolation, the total land requirements are less than those for the nuclear plant.

Oil- and gas-fired thermal plants generally create less local impact than the coal plant because the fuel is most often delivered by pipeline or barge, and little area is required for fuel storage. All of the problems created by combustion are present, but generally to a lesser extent than with the coal-fired plant, since oil and gas contain fewer deleterious elements. Natural gas is the cleanest fuel available, but because supplies are diminishing rapidly and new sources appear to be extremely expensive it will probably not be considered for base-load plants, but only for peaking purposes. Again, as in the case of coal and nuclear plants, land in other areas must be devoted to producing, processing, and transporting the fuel.

In summary, a geothermal power plant, particularly during its developmental period, appears to incur more impact on the land than do other thermal plants; but all of the components of the total geothermal system are at a single site. With nuclear power, the thermal reactor and power-generating facilities are a small part of the power cycle—the top of the iceberg. The fossil-fuel cycle is intermediate in simplicity and land use between the geothermal system and the nuclear cycle.

Impact on the Air

Gases are rejected to the air from each type of thermal power plant. But because the geothermal plant operates without combustion, the volume of noxious gases produced is far less and is of a different nature than that from a fossil-fuel plant. The natural steam is predominantly water vapor; that at The Geysers, for example, yields 99.5 percent water vapor. The noncondensable gases are about 80 percent carbon dioxide, with lesser amounts of methane, hydrogen, nitrogen, ammonia, and hydrogen sulfide (Bruce, 1971). Of these, hydrogen sulfide presents the most serious environmental problem. At The Geysers, hydrogen sulfide runs about 2 to 6 percent and averages 4.5 percent of the noncondensable gases from the producing wells (Goldsmith, 1971, p. 31), or about 225 parts per million of the steam.

Because of the remoteness and the relatively small size of the power plants at The Geysers, and because of the lower release per unit of power than from fossil-fuel plants, the hydrogen-sulfide emission has not caused the producers much concern. However, the expansion of the field and the increasing awareness of the necessity for minimizing all releases have caused the power company and the steam producers to begin studies to lower the hydrogen-sulfide emission. Their studies show that most of the noncondensable gases are drawn from the direct-contact condenser. A part of the hydrogen sulfide, however, goes into solution in the condensate, where it is converted to sulfates and elemental sulfur. Materials-balance calculations (McCluer, 1972) indicate that 30 percent of the hydrogen sulfide is oxidized and retained at the cooling towers or injected with the condensate as sulfates and elemental sulfur. Laboratory tests have shown that by altering the chemistry of the condensate by the addition of sulfur dioxide the oxidation of hydrogen sulfide to sulfur and water can be accelerated. If the field tests of this process are successful, it may be possible to overcome this environmental problem (Barton, 1972, p. 33).

To place the release of hydrogen sulfide from geothermal plants in its proper perspective, the release should be compared to that of fossil-fuel plants. Using for comparison a 1,000-Mw plant fired by coal with 1 percent sulfur and the steam conditions at The Geysers, the fossil-fuel plant would release 140 tons of sulfur dioxide per day (U.S. Congress, 1969, p. 115). By comparison, the geothermal plant with a flow of 430 million pounds of steam per day containing 0.0225 percent hydrogen sulfide would bring to the surface 48.4 tons of hydrogen sulfide per day. If 30 percent is returned to the reservoir with the steam condensate, as seems to be the ratio now, the total release would be

33.9 tons or about one-fourth the sulfur dioxide from the coal plant. This constitutes the release without pre-treatment. If the method described by Barton (1972, p. 33) is successful, the hydrogen-sulfide release can be lower.

Carbon dioxide, the major component of the noncondensable gases in the natural steam, would total about 860 tons a day from a 1,000-Mw plant. The fossil-fuel plant of the same electrical capacity produces about 20,000 tons of carbon dioxide a day (Holdren and Herrera, 1972), or more than twenty times that of the geothermal plant. And the geothermal plant releases no oxides of nitrogen, smoke, fly ash, or other aerosols.

Radioactivity of the gases and steam is at or very near natural background. Tests have shown that The Geysers steam has an alpha radiation level of 0.015×10^{-7}mCi/ml, well below the U.S. Public Health Service permissible concentration for drinking water (Bruce and Albritton, 1959).

A nuclear reactor has less total release to the air than a geothermal power plant, but when the complete nuclear-fuel cycle is considered the total impact on the air is many times that of the geothermal plant. Dust, smoke, and radionuclides are released from the mining operations at the outset of the cycle, and each step of the cycle produces various new releases. Although most of the individual releases are minor, the sum total of their effects is considerable, especially that from the fuel-reprocessing plants.

Another factor of air pollution to be weighed in the nuclear-fuel cycle is the total gases produced from hydrocarbon fuels by the machinery necessary to mine, mill, and process the uranium, and to transport it. Currently, the ore is mined and milled in the Rocky Mountains and shipped to the Midwest and South for refining, enrichment, and conversion; the fuel units are fabricated in California, shipped to a reactor in, for example, Oregon, then to a fuel-reprocessing plant in New York; and the wastes are shipped to a storage site in, say, Washington or South Carolina.

Fossil-fuel plants employing the combustion of coal, oil, or natural gas produce large amounts of carbon dioxide, nitrogen oxides, sulfur oxides, and, especially with coal, fly ash. These products create visible air pollution as well as other effects that have been the object of most of the complaints against fossil-fuel plants.

Impact on the Waters

The natural thermodynamic constraints placed on any steam cycle require the rejection of 60 to 70 percent of the total energy produced. This is normally done by circulating cooling waters through the condenser to pick up this reject heat and dissipate it into a larger body of water such as a river, lake, or ocean. Rejection of heat into the body of water can cause local environmental degradation, or at least a change in the biota, and is generally described as thermal pollution. It is inherent in the energy conversion of the thermal-electric plant, and is present in all of the new types of power-production methods being used and considered, including fusion and magnetohydrodynamics. One way to alleviate thermal pollution is to reject the waste heat directly to the atmosphere and not through an intermediate body of water, as is now the practice. Cooling towers will accomplish this transfer, but require large quantities of low-cost water, or, in the case of the dry cooling tower, an extremely large capital investment. With the present system of producing electricity from the dry-steam fields, which is technologically feasible for the flashed hot-water fields as well, all of this waste heat is either returned to the producing reservoir or rejected directly to the atmosphere via the cooling towers, thus creating no thermal pollution. The closed-cycle vapor-turbine system, as described by Anderson (this volume) will require the same heat-rejection system as conventional thermal plants.

The necessity for large quantities of water is becoming one of the limiting factors in the location of thermal-generating plants. In the Rocky Mountains, where there are large coal resources, there is already a shortage of surface and ground water for other uses. Adding the load of several new thermal plants will cause a severe strain on the available water resource. So great are the requirements for cooling water that at a recent national AAAS symposium on "Power Generation and Environmental Change" it was estimated that by 1980 one-sixth of the freshwater runoff in the United States will be used to cool power plants, increasing to one-third by the year 2000 (Holcomb, 1970). Dry cooling towers and condensers are a partial answer to the problem, but they add significantly to the capital costs of the plants and lower their efficiency.

The geothermal plant, which relies on natural steam at lower temperatures and pressures (and therefore bearing less usable heat) than those of the manufactured steam of the fossil-fuel or nuclear plant using the same cooling system, will evaporate more water than the other types of plants. With cooling towers, a 1,000-Mw geothermal plant evaporates 30 to 35 million gallons of water a day; a nuclear plant, 25 to 30 million; and a fossil-fuel plant, 15 to 20 million gallons a day. These volumes are closely related to the respective thermal efficiencies of the plants, which are about 14 to 16 percent for the geothermal plant, 32 to 34 percent for the nuclear plant, and 36 to 40 percent for the fossil-fuel plant.

But thermal efficiency as so measured is a characteristic of the power plant, not of the total cycle. Indeed, the thermal efficiency of the nuclear-fuel cycle should be based on more than just the conversion of fission energy to steam energy; it should consider as well the energy requirements of each step of the conversion of uranium ore to enriched reactor fuel, and the energy required for transporting, handling, and guarding the wastes!

By contrast, geothermal plants do not require a supplementary source of cooling water when using natural steam or the flashed cycle. The natural steam, after passing through the turbine, is condensed, piped to the cooling towers, and then recirculated back to cool the condenser. By this method the field at The Geysers produces about 20 percent more condensate than is evaporated. This surplus is then returned to the reservoir where it originated, thus prolonging the useful life of the field. A geothermal plant is the only type of thermal power plant that does not compete with other uses for our dwindling supplies of water.

Hazards to Ground-Water Aquifers

One of the questions raised about geothermal power development is its potential to contaminate surface and ground waters. True, in the early days of the exploration and development of geothermal resources in this country, several improperly cased wells blew out during drilling, allowing geothermal fluids to enter shallower aquifers or nearby streams. Also used to illustrate the danger from thermal waters are the wells drilled in the Salton Sea region. Here the extremely saline brines, which contain about 33 percent dissolved solids after flashing, constitute a hazard if allowed to enter and mix with the irrigation waters in the region. But hypersaline brines are probably restricted to the Salton Sink proper and are not present elsewhere in the Imperial Valley; they are in fact found in only a few places in the world. Geothermal waters generally carry higher percentages of dissolved solids than do nonthermal waters, because their higher temperatures have increased the rate of dissolution of the more volatile chemicals of the host rocks. But in many cases the thermal waters are of sufficient purity to be used for agricultural and industrial purposes. For example, in Klamath Falls, Oregon, the geothermal waters are used directly for stock watering (Peterson and Groh, 1967), and in Boise, Idaho, the geothermal waters are used for domestic hot waters (Wells, 1971). In Iceland there is a long history of geothermal-water utilization for both heating and domestic use.

The hazard of surface-water contamination has delayed the development of hot-water geothermal fields in the United States. Although this type of geothermal field has been developed successfully elsewhere—notably at Wairakei, New Zealand, and Cerro Prieto, Mexico, where the effluent is rejected into the surface streams—attempts to develop a field in the Imperial Valley have thus far been slowed. The main deterrents are the high salinity of the geothermal fluids found in the area, the extremely precarious water situation existing in the Valley, and the high cost of the farmland. A development planned by the Bureau of Reclamation is outlined by Laird (this volume). This plan would allow multiple use of the geothermal resources of the Imperial Valley and answer many of the questions raised concerning the environmental hazards of hot-water fields.

Although the development of the hot-water fields has been delayed in the United States, mainly because of the problem of disposing of the large volumes of water, there is every reason to believe that hot water will be utilized in the future, for it does have several advantages over dry-steam systems. Primarily, it appears to be much more abundant and is producible from shallower depths, which would make it particularly useful for space heating and for industrial and agricultural purposes. In fact, the hot waters are already used extensively for heating purposes in Hungary, the Soviet Union, Iceland, New Zealand, and Japan, as well as in the western United States and in the multipurpose development in the Imperial Valley.

Hot-water wells can be drilled by conventional drilling techniques using mud as the circulating fluid rather than air. This cuts down the noise level and the escape of steam and dust from drilling that characterize the dry-steam wells. In areas where the pressures are not excessive it is common practice to drill hot-water wells even within cities (Klamath Falls, Oregon; Boise, Idaho; Rotorua, New Zealand; and Budapest, Hungary). This would be extremely difficult, if not impossible, to do with a dry-steam well.

Dry-steam fields, such as those at The Geysers and Larderello, do not pose the problem of saline-water disposal, since these salts are not transported in the steam phase. Most of the foreign material in natural steam is in the gaseous state, in the form of noncondensable gases, as discussed above. However, a certain amount of deleterious material is present, usually amounting to a few parts per million of boron and ammonia. These form salts that persist in the condensate and are injected back into the producing reservoir, along with that fraction of condensed cooling water that is surplus to the needs of the plant. Consequently there is no release of either thermal waters or chemical contaminates into the surface waters or other usable water sources from the present production of geothermal energy.

Most of the potential hazards to the waters from the dry-steam geothermal operation occur during the development of the field, when drilling muds are required and construction upsets the normal water pattern of the area. With proper care these operations do not present an environmental hazard. And in any event, the hazard is brief and negligible compared to the hazards created by the extensive construction required by the competing power sources and more dramatically from mining, which must continue over the entire life of the nuclear or fossil-fuel power plant.

Conclusions

The environmental impact of any power-production system is reflected in the number and complexity of the steps in the fuel and production cycle. Because geothermal power plants utilize naturally occurring steam, they need no complex steam-generating equipment or extensive mining, processing, storage, or transportation facilities, as do other thermal power plants.

The chief impact from the use of geothermal power occurs during the period of development of the field and construction of the steam-gathering lines and power plants, but the impact is limited to the area of the field and poses nothing like the vast disruptions of the landscape concomitant with mining the fuels for other thermal power plants. During the productive lifetime of the geothermal field, which can extend over many decades, most of the area can be used for other purposes. At Larderello, for example, where natural steam has been used to produce electricity for 60 years, farms, orchards, and vineyards cover much of the land surface.

Natural steam does contain a small percentage of noncondensable gases that are vented to the air. But compared to the amounts dissipated by fossil-fuel plants, these gases—mostly carbon dioxide but also nitrogen, hydrogen, methane, and hydrogen sulfide—are minor. Compared to the total gaseous release from all steps in the nuclear-fuel cycle, the overall volume and toxicity of gases from the geothermal plant is, again, minor.

Dry-steam geothermal developments pose no hazard to water supplies. Moreover, dry-steam and flashed-steam power plants supply their own cooling water by condensing their steam, and are therefore independent of the sources of condenser cooling water that are needed by other types of thermal plants. Hot-water geothermal systems will have an effect on the waters, but in most cases it will be to bring into use waters that are below the economic drilling depths of water that is currently in use, or to upgrade the quality of currently unusable waters, thus making these waters themselves a valuable resource. The multipurpose development planned in the Salton Sea region is an imaginative scheme that could well be duplicated in other areas.

The simplicity of the geothermal-steam cycle enhances its reliability, another factor that needs to be considered when assigning priorities of development. Because the geothermal-power cycle is self-contained, it needs no outside support to maintain the production of electricity; there are no railroads or mines or complex processing facilities to be put out of service by a strike or natural catastrophe; and the reliability of nature's own boiler is paramount.

REFERENCES

Barton, D. B. 1972. The Geysers power plant: A dry steam geothermal facility. *In* Compendium of First Day Papers, presented at the First Conference of the Geothermal Resources Council, El Centro, Calif. Davis, Calif.: Geothermal Resources Council, P.O. Box 1033, pp. 27–38.

Bruce, A. W. 1971. Geothermal power: On line. *In* Papers presented at the First Northwest Conference on Geothermal Power, May 21, 1971. Olympia, Wash.: Washington State Department of Natural Resources.

Bruce, A. W., and B. C. Albritton. 1959. Power from geothermal steam at The

Geysers power plant. *In* Power Division, Proc. Amer. Soc. Civ. Eng., v. 85, no. P06, part 1, p. 34.

Clacy, G. R. T. 1968. Geothermal ground noise amplitude and frequency spectra in the New Zealand volcanic region. J. Geophys. Res., v. 73, p. 5377.

Evans, D. M. 1966. Man-made earthquakes in Denver. Geotimes, v. 10, no. 9, pp. 11–18.

Goldsmith, M. 1971. Geothermal resources in California: Potentials and problems. California Institute of Technology, Environmental Quality Laboratory, 45 pp.

Hansen, A. 1964. Thermal cycles for geothermal sites and turbine installation at The Geysers power plant, California. *In* New sources of energy; United Nations Conf. on New Sources of Energy. New York: United Nations, v. 3 (Geothermal Energy: 11), pp. 365–76.

Hatton, J. W. 1970. Ground subsidence of a geothermal field during exploitation. *In* United Nations Symp. on Util. of Geothermal Resources, Pisa, Italy. Elmsford, N.Y.: Maxwell Scientific International, Inc.

Hogerton, J. 1964. Atomic Fuel: U.S. Atomic Energy Commission, Division of Technical Information, Understanding the Atom Series, 40 p. Illus.

Holcomb, R. W. 1970. Power generation: The next 30 years. Science, v. 167, pp. 159–60.

Holdren, J., and P. Herrera. 1972. Energy: a Crisis in Power. San Francisco: Sierra Club Battlebook Series, 4.

McCluer, H. K. 1972. Pacific Gas & Electric Company, oral communication.

Peterson, N. V., and E. A. Groh. 1967. Geothermal potential of the Klamath Falls area, Oregon: A preliminary study. Ore Bin, v. 29, no. 11, pp. 209–31.

Raleigh, C. B., J. Bredehoeft, J. H. Healy, and J. P. Bohn. 1970. Earthquakes and water flooding in the Rangely Oil Field. *In* Abstracts with programs, Geol. Soc. Amer., v. 2, p. 660.

Raleigh, C. B., J. H. Healy, J. D. Bredehoeft, and J. P. Bohn. 1971. Earthquake control at Rangely, Colorado. Trans. Amer. Geophys. Union, v. 52, p. 344.

U.S. Atomic Energy Commission. 1972. Statistical data of the uranium industry, GJO–100. Grand Junction, Colo.: Grand Junction Office, U.S. Atomic Energy Commission.

U.S. Congress, Joint Committee on Atomic Energy. 1969. Selected materials on environmental effects of producing electric power. Washington, D.C.: U.S. Govt. Printing Office, 553 pp.

U.S. Department of the Interior. 1972. Supplement to draft: Environmental impact statement for the geothermal leasing program, revised Chapter IV, Sec. C—Alternatives to proposed action; Appendix G—Energy alternatives; Appendix H—Proposed unit plan regulations. 175 pp.

Ward, P. L. 1972. Microearthquakes: Prospecting tool and possible hazard in the development of geothermal resources. Geothermics, v. 1, no. 1, pp. 3–12.

Wells, M. W. 1971. Heat from the earth's surface. J. West, v. 10, no. 1, pp. 53–71.

Westinghouse Electric Corporation. 1968. Nuclear fuel. Pittsburgh, Pa.: Nuclear Fuel Division, Westinghouse Electric Corporation, 47 pp.

White, D. E., L. J. P. Muffler, and A. H. Truesdell. 1971. Vapor-dominated hydrothermal systems compared with hot-water systems. Econ. Geol., v. 66, no. 1, pp. 75–97.

11. Stimulation of Geothermal Systems

ANTHONY H. EWING

The development of geothermal energy as an important national energy resource depends entirely on the extent to which geothermal heat in its more abundant forms can be located and extracted economically. The existence of a vast potential resource of geothermal heat is well established. But it is also well established that only a small fraction of this heat is recoverable with existing technology; and if geothermal energy is to become a significant source of electric power it thus becomes essential that we develop technologies that will increase this extractable fraction.

Geothermal energy exists in the Earth's crust in several forms, as described by White (this volume). Some of these forms present few problems. Where nature has provided high thermal gradients, sufficient heat-transporting fluids, adequate formation permeability, suitable chemical composition, and suitable reservoir configuration, the ideal conditions for productive steam fields exist. Only a few such systems have been found. Currently, production of dry steam for the generation of electricity is important at only a few locations, such as The Geysers, California, Larderello, Italy, and Matsukawa, Japan.

More abundant are geothermal systems containing superheated liquid water that can flash to steam under reduced hydrostatic pressure as it is brought to the surface. In such systems only 10 to 20 percent of the produced fluid results in usable steam. Hot-water/steam systems are known to exist in Iceland, Chile, Turkey, New Zealand, Mexico, the

Anthony H. Ewing is with the Division of Applied Technology, U.S. Atomic Energy Commission, Washington, D.C.

United States, and elsewhere. Their relative importance is discussed by Koenig (this volume).

Still more abundant are hot-water systems that produce liquids at temperatures below 180°C. The conversion of such fluids to steam for electricity production by the steam cycle is not considered economical with current technology.

In addition to these geothermal-fluid systems, there are formations covering hundreds of square miles that contain immense volumes of hot rock of abnormally high thermal gradients but do not contain sufficient fluids to produce steam or hot water. Some of these formations are believed to exist at temperatures between 300° and 600°C at relatively shallow depths of less than 3 km. In most areas of the United States, in fact, perhaps with the exception of the north-central states, the average formation temperature is in excess of 150°C within a depth of 10 km. These formations can be made productive by the introduction of surface water under pressure.

Geothermal formations of one type or another are widespread throughout the United States, but are most abundant in the western states, as shown in Fig. 1, a map of known geothermal areas released by the U.S. Department of the Interior in 1967.

Potential Capacity

At present there are only about ten geothermal areas in the world producing electricity, with a combined capacity of less than 1,000 Mw. This output represents the equivalent of one modern power plant and is insignificant in terms of the nation's present energy requirements. Estimates of the contribution that geothermal energy may ultimately make to our total energy reserve differ widely. Considering only naturally occurring geothermal systems lying at depths of less than 3 km, and assuming that 1 percent of the heat energy in such systems is recoverable by present technology near present-day energy-generation costs, White estimates that 2×10^{19} calories are available, an amount roughly equivalent to the U.S. energy consumption for 1972. And at depths up to 10 km, the total energy available in such systems is estimated by White to be about 500 times greater, though its exploitation would require improvements in the technology of drilling and stimulating marginal systems. Much more energy would become accessible if methods were developed for the recovery of energy from hot, dry rocks. Estimates put the energy recoverable from this source at 100 to 1,000 times

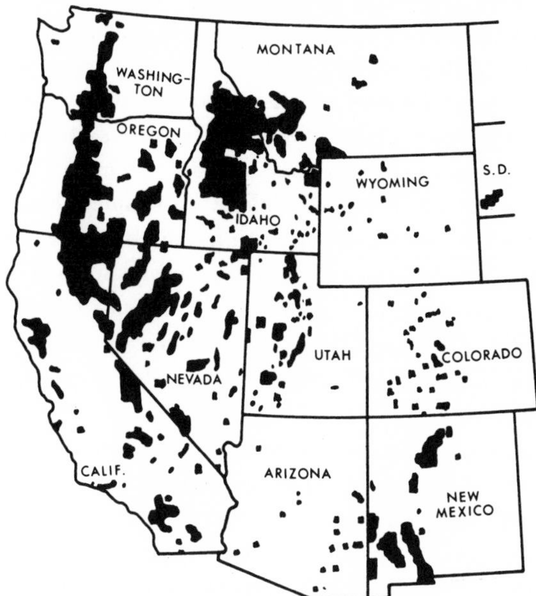

Fig. 1. Geothermal areas in the western United States (from the U.S. Geological Survey, 1967).

that available in hydrothermal systems, or 10^{24} to 10^{25} calories. Systems exploiting hot, dry rock are thus potentially capable of supplying U.S. energy needs for 50,000 to 500,000 years at the present rate of consumption. But in both hydrothermal and dry-rock systems the upper estimates assume technological improvements, including principally the development of the vapor-turbine cycle (for heat-extraction from lower-temperature resources) and stimulation methods for increased well productivity. The vapor-turbine cycle discussed by Anderson (this volume) holds promise for utilizing the thermal energy of fluids at relatively low temperatures (less than 180°C). Prospects for artificially stimulating the productivity of economically submarginal geothermal-steam fields, hot-water aquifers, or hot, dry rock formations also hold great promise.

Stimulation of Geothermal Steam and Hot-Water Systems

Although dry-steam geothermal systems are not abundant, they produce electricity more efficiently than other geothermal systems. Production problems are relatively minor, and the discovery of vapor-dominated systems comparable to that at The Geysers would lead to rapid development and production by private enterprise. But both the Larderello and The Geysers dry-steam fields have experienced a decline in steam pressure from producing wells through the years. Reinjection of condensate from the turbine exhausts, already in practice at The Geysers, or the injection of surface waters into the system should retard depletion, but for a variety of reasons production can be expected nonetheless to decline with time. It is difficult to estimate how effective the stimulation of dry-steam systems might be, but it is reasonable to assume that a geothermal well that has been known to produce but has declined to subeconomic levels may be amenable to stimulation by techniques similar to those in use for gas and oil wells. Thus, studies and field experiments to determine appropriate stimulation techniques need to be undertaken.

The greater abundance of hot-water systems over dry-steam systems suggests the need for a correspondingly greater research and development effort, although they pose more serious development and utilization problems. For example, since superheated water flashed at the surface results in only about 10 to 20 percent of usable steam, it is evident that means should be found to extract heat from the formation rock, as well. The use of low-yield nuclear explosives to create artificial geothermal-fluid reservoirs in natural aquifers offers the greatest promise for improved exploitation of this resource. Raghavan, Ramey, and Kruger (1971; this volume) have suggested several modes of extracting production from hot-water systems that have been raised to viability in this manner.

Increased formation permeability might also be achieved by the use of small-charge, properly spaced, high-energy chemical explosives; the concept is described in detail by Austin and Leonard (this volume).

Some areas of investigation that may hold promise for the stimulation of steam or hot-water geothermal systems are:

1. Increasing the steam-to-liquid production ratio of geothermal wells by creating a void for steam separation within the underground system. The liquid might remain underground to be reheated and revaporized, eliminating costly surface separation and the handling of residual hot liquids.

2. Increasing the effective wellbore diameter by explosive stimulation, thereby preventing the deposition of solids from impeding the flow of geothermal fluids.

3. Increasing the formation permeability in high-temperature steam or hot-water formations (see below) where the permeability is otherwise too low to allow economical production.

4. Providing adequate volume and permeability in geothermal formations to allow economical reinjection of condensed steam or cooled liquids, thereby utilizing more of the available heat of the formation.

Stimulation of Dry Geothermal Formations

The development and production problems attendant upon the use of dry, hot geological formations are even less known than those encountered with low-temperature geothermal aquifers, and it is in this area that research needs are most evident. The stimulation of geothermal resources has been under consideration since the early speculations of Carlson (1959) and Kennedy (1964), and several techniques of stimulation methods have since been proposed.

One method proposes using injected water to produce steam from dry, hot geothermal deposits previously fractured by a suitable array of multiple large-yield nuclear explosives. The feasibility of the approach was examined by Burnham and Stewart (1970) and by a joint Government/industry study (American Oil Shale Corp. et al., 1971). These studies analyzed detonation configurations yielding an optimum volume of fractured rock for the extraction of heat from a closed-loop cycle of surface water. The analysis of Burnham and Stewart (this volume) forecasts operation of a 200-Mw power plant for 30 years.

Smith et al. (also this volume) propose using hydraulic thermal-stress fracturing for the stimulation of geothermal formations having insufficient natural permeability. Cold surface waters would be used to pressurize natural or stimulated fractures; the resulting shrinkage of the rock would cause additional cracks, perhaps sufficient in extent to propagate increased permeability over an extensive volume of the formation. A second hole would then be drilled to intersect the upper reaches of the hydraulically fractured region. The cold water initially pumped into the deeper well would establish closed-loop circulation through the cracked zone into the new, higher well, and a vapor-turbine cycle at the surface would be used to produce electricity from the heated water emerging.

Each of these potential stimulation methods warrants further investigation.

Conclusions

One of the important problems in the development of geothermal resources is the necessity for proving up reserves sufficient to justify the expense of constructing a power plant directly at the site of a reservoir. The potential investment in the power plant may be large compared to the investment already made in the exploration phase. But the increased well productivity promised by artificial stimulation may make such decisions easier. The main objectives of research on the development of geothermal resources are (1) the ability to utilize the resource however it manifests itself in nature, (2) the technology necessary for efficient extraction of energy from the resource, (3) the ability to predict the productivity and longevity of the resource, and (4) the ability to develop the resource in an environmentally tolerable manner.

REFERENCES

American Oil Shale Corporation, Battelle-Northwest, Westinghouse Electric Corporation, U.S. Atomic Energy Commission, Lawrence Livermore Laboratory, and Nevada Operations Office–AEC. 1971. A feasibility study of a geothermal power plant. PNE-1550.

Burnham, J. B., and D. H. Stewart. 1970. The economics of Plowshare geothermal power. Proc. Am. Nucl. Soc. Symp. on Engineering with Nuclear Explosives, CONF-700101.

Carlson, R. H. 1959. Utilizing nuclear explosives in the construction of geothermal power plants. Proc. Second Plowshare Symposium, UCRL-5677.

Kennedy, G. C. 1964. A proposal for a nuclear power program. Proc. Third Plowshare Symposium, TID-7695.

Raghavan, R., H. J. Ramey, Jr., and P. Kruger. 1971. Calculation of steam extraction from nuclear-explosion fractured geothermal aquifers. Trans. Am. Nucl. Soc., v. 14, p. 695.

12. Recovery of Geothermal Energy from Hot, Dry Rock with Nuclear Explosives

JOHN B. BURNHAM AND DONALD H. STEWART

Almost every application of underground nuclear explosives relies on high fracturing efficiency. The Plowshare geothermal concept described by the American Oil Shale Corporation et al. (1971) entails the generation of power from the energy contained in deposits of hot, dry rock. The rock is fractured by an array of sequentially fired, fully contained nuclear explosives; when fracturing is complete, water is injected through fill pipes and steam is withdrawn to the surface facility. Figure 1 illustrates a plant designed to recover the energy of a pluton (a hot, intrusive formation) utilizing 1,000-kt nuclear explosives. The temperature, depth, and nuclear-cavity diameters (R_C) are representative of what would be yielded in a granitic intrusive with an assumed thermal gradient of 125°C/km.

The economics of most mining operations are sensitive to the costs of drilling and blasting. Costs range from over $2/ton for drilling and blasting in the lead-zinc industry (*Economics*, 1959) to as low as $0.03/ton in the open-pit copper mines (*Surface Mining*, 1968, p. 885). The economics of energy produced by the Plowshare geothermal concept are no different from those of the mining industry. The low thermal conductivity of rock requires very large heat-transfer areas if meaningful amounts of energy are to be extracted at a useful rate. Figure 2 illustrates the economic potential of hot, dry rock when used in this manner. One cubic mile of rock at 350°C, when cooled to 150°, yields a usable energy equivalent to that of 300 million barrels of oil—an amount, at present oil prices, worth approximately $1 billion. Many

J. B. Burnham is Program Director for Economic Analysis, and D. H. Stewart is Manager of Geothermal Programs, at Battelle-Northwest, Richland, Washington.

geologists, moreover, believe that the incidence of dry geothermal sites will be two to three orders of magnitude greater than that of fluid-bearing geothermal sites.

The American Oil Shale Corporation et al. (1971) discuss the sensitivity of Plowshare-style energy production to the fracturing efficiency of nuclear explosives. Their report furnishes the details of plant design and the technical and economic assumptions made. The relationship between the number of nuclear explosives required, the energy recovered from the nuclear explosives (that portion of the energy from the explosion that remains locally in the form of heat), and the energy extracted from the hot rock is given by

$$N = \frac{E - NJW}{\varepsilon\tau V_f} \qquad (1)$$

where N is the number of nuclear explosives required; E is the energy required to run a 200-Mw plant over a 30-year lifetime, or 1.91×10^{11} kwh (for an 80 percent plant factor and a 22 percent conversion efficiency); J is the energy recovered per kt yield, or 1.05×10^6 kwh/kt; W is the nuclear-explosive yield, in kt; ε is the fraction of energy recov-

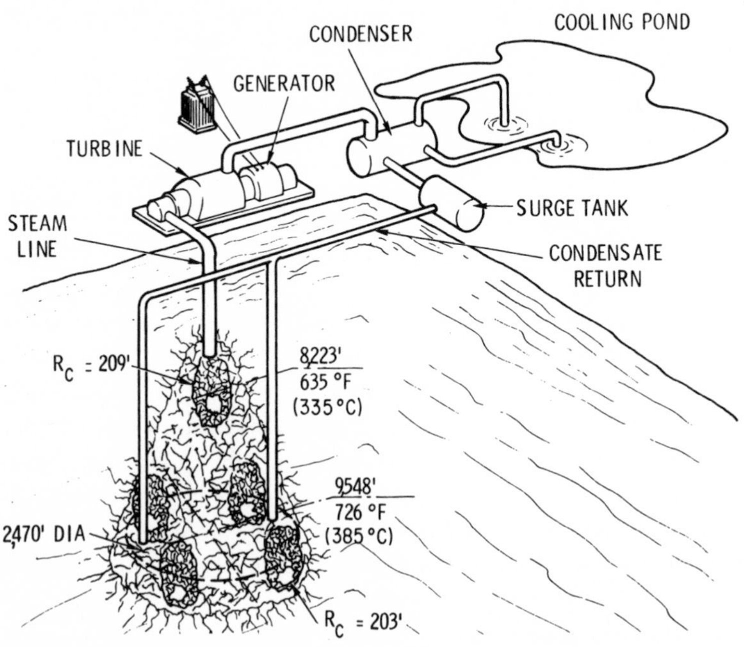

Fig. 1. A power plant using the Plowshare geothermal concept.

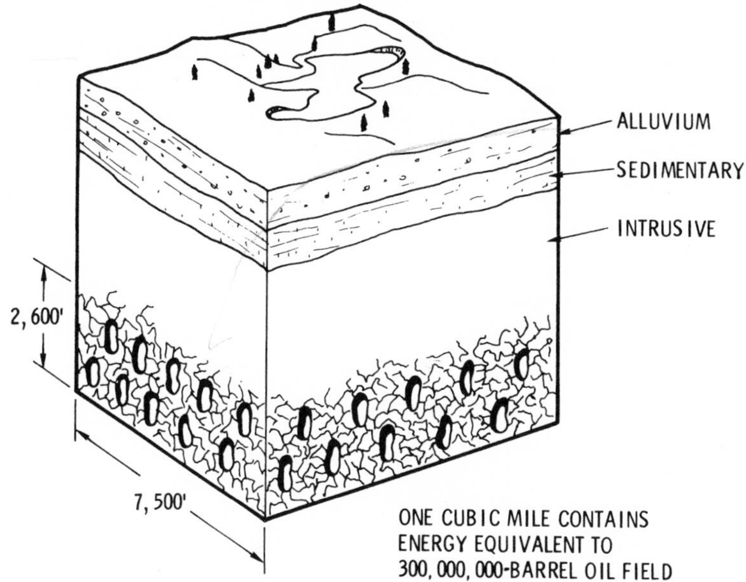

Fig. 2. A hypothetical dry geothermal reserve (one cubic mile) developed with nuclear explosives.

ered from rock, assumed to be 0.9; τ is the sensible heat, 180 kwh/m^3; and V_f is the fractured-rock volume in m^3. Basically, the energy required to run the plant for its lifetime must equal the sum of the energy recovered from the nuclear explosives and the energy recovered from the fractured hot rock. Of particular interest is the quantity V_f, the fractured-rock volume, which equals the cavity volume times a factor called "M," or fracturing-efficiency coefficient. This coefficient, in turn, is made up of two parts: one is the ratio of fractured-rock volume to cavity volume; the second is a factor called the enhancement factor. The relationship is given by

$$M = \left(\frac{r_f}{r_c}\right) e_h \qquad (2)$$

where r_f is the fracture radius; r_c is the cavity radius; and e_h is the enhancement factor.

Although an enhancement factor has never been measured for multiple nuclear explosions, it is possible to derive a reasonable range of values for this coefficient. Atchinson (*Surface Mining*, 1968, p. 357) reports data from experiments in Lithonia granite in which fractured volumes, first established by a completely contained explosion, in-

TABLE 1
Number of Nuclear Explosives in a Dual-Emplacement Array Needed to Support a 200-Mw Plowshare Geothermal Plant

Fracturing-efficiency coefficient, M	Number of explosives of given size		
	200 kt	500 kt	1,000 kt
20	238	96	48
54	106	42	22
129	48	20	10

NOTE: Conditions: rock temperature, 350°C; placement depth of device, 3 km.

creased by a factor of 36 when blasting was done to a free surface. It must be realized, of course, that a preexisting cavity does not react precisely as a free surface. In the first place, the intercepted geometry will be limited. Second, the change in density between the base rock and the preexisting cavity will be less than in free-surface blasting. But at the same time, in a multiple array, the shock waves will be intercepting more and more preexisting cavities.

Kutter and Fairhurst (1971) have performed laboratory experiments that simulate totally enclosed blasts in rock. These experiments have shown that the stresses needed to extend an existing crack are lower than those required to initiate a new one. Their work has also shown that radial cracks tend to enlarge with subsequent detonations, and that the radial cracks pointing normal to a free surface have the greatest tendency to enlarge. It is these radial cracks extending toward preexisting cavities that will, when enlarged, increase the interconnection and the effective fractured volume of an array. These same experimenters demonstrated the increase in fracturing produced by gas pressures that work with radial cracking to enhance fractured volume.

Table 1 illustrates the number of nuclear explosives that are required to operate a 200-Mw geothermal-electric power plant for 30 years at an 80-percent plant factor, calculated from Eqs. (1) and (2). It was assumed that rock temperatures of 350°C would be reached at a depth of 3 km with fracturing efficiencies, M, over a range of 20 to 129. The fracturing efficiency of 20 corresponds to a ratio of fractured radius to cavity radius of 2.5 and an enhancement factor of about 1.3. A fracturing efficiency of 129 corresponds to a value of r_f/r_c of 3.5 and an enhancement factor of 3. There are of course many such combinations of enhancement factors and fracture radii, yielding a great range of M values. But a

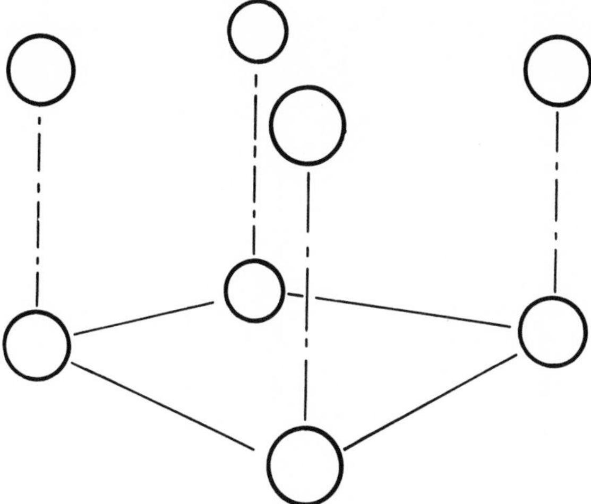

Fig. 3. Schematic diagram of dual-emplacement array.

fracturing efficiency of 50–100 does not seem particularly optimistic. The fact that all of the numbers in the table are even results from the particular array selected for these calculations. The array is based on the dual-emplacement concept illustrated in Fig. 3. The calculated number of explosives in this concept is always arbitrarily rounded upward to the next even number.

The dual-emplacement array is designed to take advantage of an explosive-emplacement concept that allows two devices to be emplaced at approximately the same cost as one, though the cost of the explosives themselves is of course doubled. A second array, designed for close packing, is oriented in such a way that each shot interacts maximally with its nearest neighbors, as shown in Fig. 4. The conical geometry is designed to facilitate the injection of water into the array. We recognize that developing the water-injection system for such a large underground array will be a major problem; hydraulic flow through this large interconnected-fracture volume must be designed for maximum energy extraction. It is quite possible that as experimental data become available, some combination of these designs will evolve—one perhaps in which a body-centered cubic structure results from introducing another explosive pair (one-half the cube size displaced) into the simple cubic array shown in Fig. 3.

Recent field data indicate that some of the values used in these cal-

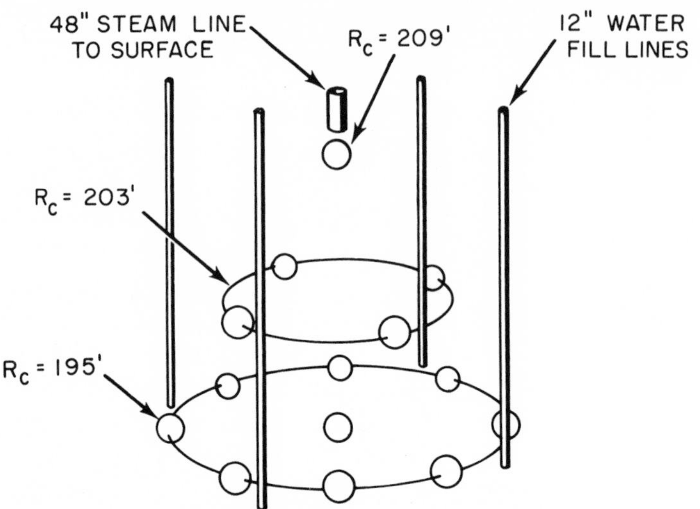

Fig. 4. Schematic diagram of close-packed array of nuclear explosives.

culations may have been overconservative. For example, Blackwell (1972) has shown that a thermal gradient of about 300°C/km exists in a geothermal area near Marysville, Montana. This is more than twice the gradient of 125°/km used in our calculations. Blackwell's data indicate that so long as the gradient remains constant with depth a temperature of 700° might occur at 3 km in this area. By cooling this rock from 700° to 200°, the average thermal efficiency will increase from 22 to 30.5 percent. The combination of the increase in thermal-conversion efficiency and the increase in available sensible heat in the rock may decrease the number of devices required by a factor greater than 2.5.

Other factors are involved, one of which—the diameter of the emplacement holes—is important in the economics of this concept. If the explosives could be emplaced in smaller-diameter holes, rather than the 24-inch-diameter holes assumed in these calculations, substantial savings on each hole could be realized. Another factor, not considered here, is a recent court decision, not a precedent but in fact confirming an earlier decision, that geothermal-energy resources are depletable. This decision will reduce the cost of this type of energy by about 0.3 mill/kwh.

Kutter and Fairhurst (1971) also indicate that the increase in radial cracking with subsequent nuclear explosions may significantly increase the enhancement factor beyond the range of 1 to 3 used in these calculations. Moreover, the possibility of thermal fracturing has not been accommodated in the calculations; though its effect is unknown at this time, it could be significant in increasing the fractured volume.

A number of unanswered questions in the economics of a Plowshare geothermal plant remain. Undoubtedly, considerable work will have to be done on the design of the array to solve the hydraulic-flow problem. It will be important to drive the water and steam through areas of low permeability into areas of high permeability without excessive channeling. The dissolution of silica may be an important problem requiring research. With high temperatures and pressures, an appreciable amount of silica may be carried up with the steam into the turbine system. If this becomes severe enough, an intermediate heat exchanger may be required at the surface. Even though most of the radioactive nuclides are frozen in the molten rock, some will have enough solubility in the steam to come to the surface (Charlot et al., 1971). Though the problem does not seem severe at this time, the available data are still inadequate. Finally, the size of nuclear explosives required to make the concept economically sound may be a problem. Although recent discoveries of higher-temperature rock make costs look much more favorable, the use of larger devices will of course decrease power costs in any configuration of the concept. But the limit of device size that can be used in the continental United States is uncertain. The chemical and physical effects of underground nuclear explosions are discussed in more detail in the papers by Sandquist and Whan and by Krikorian (this volume).

Figure 5 illustrates the overall economics of a 200-Mw plant as a function of the fracturing efficiency of the array and the size of the nuclear explosive. Low-efficiency arrays are not attractive even with large explosive yields. With intermediate fracturing efficiency, though, the power from this plant concept would seem to be competitive at device sizes in the range of 500 kt and up. For higher fracturing efficiency, the plant seems to be competitive economically at nuclear-explosive yields of 300 kt and greater. The use of a hotter intrusive, such as the one near Marysville, Montana, would significantly decrease these costs. They would be decreased, for example, to about 5 mill/kwh for 1,000-kt devices with a fracturing efficiency of 54. With the same fracturing efficiency, power costs of 5.8 and 8.8 mill/kwh could be expected from 500-kt and 200-kt nuclear explosives, respectively.

Geothermal energy does indeed constitute a significant energy source, one that can be recovered with the use of nuclear explosives. The

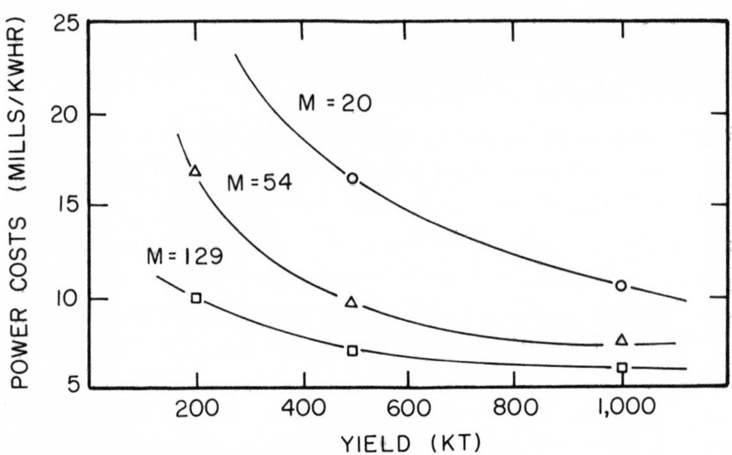

Fig. 5. Power costs from 200-Mw Plowshare geothermal plant as a function of nuclear-explosive yield and fracturing efficiency.

economic potential of this energy source is comparable to that of most mining operations that require the fracturing of rock: the economics are highly dependent on the efficiency with which the rock can be fractured.

ACKNOWLEDGMENTS

The work reported here is concerned with the fracturing of hot rock as an outcome of a study of Plowshare geothermal energy jointly sponsored by the American Oil Shale Corporation and the Atomic Energy Commission. Much of this work was performed by that study team. We would like to acknowledge the contributions of many individuals at Lawrence Livermore Laboratory, Westinghouse Electric Corporation, and Battelle-Northwest. We would like to especially acknowledge the contributions of C. E. Chapin of Lawrence Livermore Laboratory.

REFERENCES

American Oil Shale Corporation–U.S. Atomic Energy Commission. 1971. A feasibility study of a geothermal power plant. PNE-1550.

Blackwell, David. 1972. Exploration and development in dry-hot systems. ARPA Conference, Lake Arrowhead, California, May 30–June 2.

Charlot, L. A., et al. 1971. Plowshare geothermal steam chemistry. BNWL-1614.

Economics of mineral industry. 1959. AIME, p. 189.

Kutter, H. K., and C. Fairhurst. 1971. On the fracture process in blasting. Int. Jour. Rock Mechanics and Mining Sciences, v. 8, no. 3.

Surface mining. 1968. AIME.

13. Explosive Stimulation of Hydrothermal Reservoirs

HENRY J. RAMEY, JR., PAUL KRUGER & RAJ RAGHAVAN

Geothermal-energy resources exist in several forms, notably vapor-dominated (dry-steam) systems, liquid-dominated (hot-water) systems, and dry (hot-rock) geothermal formations. Each of these forms has a potential for exploitation. The vapor-dominated system offers the best features for production of electricity; it is the form most sought after and the rarest. The production problems of vapor-dominated systems are described by Budd (this volume). If other dry-steam fields as economically attractive as that at The Geysers are discovered, their development will be rapidly undertaken by private enterprise. Hot-water systems, which are also being seriously investigated for power production, pose greater exploration, development, and production problems. And the utilization of dry geothermal formations for power generation is even less understood than that of hot-water systems; the stimulation of hot-rock formations is discussed in other papers in this volume.

White (this volume) estimates that hot-water geothermal systems are many times more abundant than vapor-dominated systems. Moreover, if the promise of two infant technologies is realized, the geothermal aquifer can become an important source of electrical power. The first of these technologies, the vapor-turbine power cycle, is described by Anderson (this volume). The second, the explosive stimulation of geothermal-fluid production, which creates an open reservoir in or adjacent to a fluid-bearing aquifer, is the subject of this paper; for convenience we may divide the subject into five topics: (1) energy aspects

---

Henry J. Ramey, Jr., is Professor of Petroleum Engineering and Paul Kruger is Professor of Nuclear Civil Engineering at Stanford University, Stanford, California. Raj Raghavan, formerly Acting Assistant Professor of Petroleum Engineering at Stanford, is now Senior Research Engineer at Amoco Production Company, Tulsa, Oklahoma.

of hydrothermal reservoirs, (2) explosive stimulation, (3) environmental effects, (4) production methods, and (5) performance forecasting.

Energy Aspects of Hydrothermal Reservoirs

The energy content of hydrothermal reservoirs can be evaluated with respect to steam-production requirements: steam turbines currently operating in geothermal fields require about 100 psia minimum inlet steam pressure for efficient operation. The energy content of a hypothetical system—one having, say, a reservoir temperature of 500°F and a reservoir porosity of 25 percent—can be estimated (Table 1) for a pressure reduction across the useful life of the system, from an initial saturation pressure of 681 psia (at 500°) down to the abandonment pressure of 100 psia (at 327.8°). The table summarizes the available-energy content for two types of geothermal system: the dry-steam reservoir and the hot-water reservoir. (Neither is subject to liquid recharge, i.e. reinjection, for purposes of this discussion.) The available energy above 327.8° at 100 psia saturation pressure is expressed in Btu and as fuel-oil equivalent (17,000 Btu/lb oil). The difference in available energy in the two reservoir types is apparent: the hot-water reservoir contains substantially more than the steam-filled reservoir. Although the latent heat of vaporization is high for steam with respect to the sensible heat in liquid at the same temperature, the much larger mass of water in the hot-water aquifer provides greater heat availability. But in both systems the mass of rock is large compared to the mass of fluid. Thus most of the heat available is present in the rock rather than in the fluid.

For either system, the total available energy is less than an equivalent

TABLE 1
Available Energy Content for Hypothetical Geothermal Reservoirs

	Steam reservoir			Hot-water reservoir		
	Mass	Available energy		Mass	Available energy	
Component	(10^6 lb)	(10^6 Btu)	(bbl oil)	(10^6 lb)	(10^6 Btu)	(bbl oil)
Aquifer rock	5.410	233	40	5.410	233	40
Fluid	0.016	15	3	0.534	101	18
TOTAL	5.426	248	43	5.944	334	58

NOTE: Data based on a geothermal reservoir of 25 percent porosity 1 acre-foot in extent, initially at 500°F, with production at delivery pressure of 100 psia (327.8°). Oil density 340 lb/bbl.

TABLE 2
Production from Hypothetical Geothermal Reservoirs

Production parameter	Steam reservoir	Hot-water reservoir
Initial fluid content (lb)	16,160	533,800
Content remaining at abandonment (lb)	1,950	62,300
Production:		
As steam (lb)	14,210	372,000
As water (lb)	0	99,500
TOTAL	14,210	471,500
Recovery:		
As percent of initial fluid mass	87.9	88.3
As percent of total available energy	5.6	99.1

NOTE: Data based on a geothermal reservoir of 25 percent porosity 1 acre-foot in extent, initially at 500°F, with production at delivery pressure of 100 psia (327.8°). Mean produced fluid enthalpy 1,000 Btu/lb.

58 bbl oil/acre-foot, a value low compared to the energy content in conventional oil reservoirs. Since most of this energy, 40 bbl oil/acre-foot, resides in the rock, it is clear that production methods permitting the recovery of the heat in the rock, as well as the heat in the fluid, are worth investigating. Heat may be recovered from the rock itself by one of two processes: (1) flashing the existing liquid to steam within the pore space (which of course is effective only while the liquid lasts) or (2) recycling colder fluid back into the formation.

Steam reservoirs. Although the data in Table 1 indicate that hot-water reservoirs have considerably higher initial available-energy content than steam reservoirs, the steam reservoir has been more highly sought after as a particularly clean energy source with few production problems. Production from steam reservoirs is comparable to production from natural-gas reservoirs, in the sense that pressure is reduced by expansion of the in-place gas. Table 2 summarizes results for the production of saturated steam per acre-foot of reservoir to an abandonment pressure of 100 psia under the reservoir conditions given. About 14,000 of the in-place 16,000 lbs of steam would be produced before abandonment, for a mass recovery of 87.9 percent. However, the recovery of energy would be only 5.6 percent of the total energy in the system, since the expansion of steam in the rock pore space is essentially isothermal. Energy can be recovered from the rock only by a production process that significantly reduces the temperature of the rock. Although two of the world's most important geothermal fields are of the dry-steam type, it does not appear that energy recovery from the rock itself will be important in these fields unless a liquid-reinjection program is conducted.

Hot-water reservoirs. Table 2 also shows the production results for hot-water reservoirs under the same initial conditions—500°F saturated fluid. Of the 534,000 lbs of in-place water in the 1 acre-foot of the hot-water system, about 372,000 lbs of steam and 100,000 lbs of liquid water would be produced before pressure depletion. Some 88.3 percent of the initial mass in the system would have been produced. About 70 percent of the in-place water would have been flashed to steam by pressure reduction. Since this process is *not* isothermal, the temperature reduction on boiling permits recovery of thermal energy from the rock. The steam produced by flashing from the hot-water reservoir represents almost all of the total available energy in the reservoir system (fluid and rock). The usable steam available for electric-power generation, 372,000 lbs, would be more than 20 times the total steam content in the steam-reservoir case. These figures represent an ideal or optimum result, in the sense that all of the fluid and rock have been assumed maintained in equilibrium down to the endpoint conditions (pressure below 100 psia). In practice, such a system could be produced in a variety of ways, each leading to poorer recovery of fluid and energy. For example, production of liquid only during the initial stages would lead to a smaller fraction being produced as steam.

The recovery of energy from the flashing of hot pressurized water in surface equipment is strongly dependent on water temperature. At 500°, about 32 percent of the mass of liquid water can be flashed to steam by reduction of pressure from 681 psia at saturation to atmospheric. At 400°, where saturation pressure is 249 psia, only about 20 percent can be flashed. And at 250°, only about 4 percent can be flashed. The recovery of heat from the host rock in a geothermal reservoir requires the flashing of the hot water in the pore space of the aquifer or the reinjection of cold water. Thus, in-place boiling or cold-water injection become key factors in the recovery of geothermal energy.

Factors affecting boiling in porous media. Little attention has been given to date to the phenomenon of boiling within porous media, and a number of technical problems must be examined. For example, flat-surface vapor-pressure data may not be appropriate within pore spaces where curved liquid surfaces would be expected, owing to capillarity.

Although water-vapor pressure measurements have been made at 94°, within the pores of a sandstone, by Calhoun, Lewis, and Newman (1949), data at the temperatures expected in geothermal reservoirs are not available. Their data, which assume 760 md permeability, 27.9 percent porosity, and a temperature of 94°F, may be summarized as follows, where vapor-pressure ratio is the ratio of flat-surface vapor pressure to vapor pressure in the porous media:

Fraction of pore space occupied by water	0.103	0.058	0.0325
Vapor-pressure ratio	1.101	7.692	37.313

The figures indicate a remarkable lowering of the flat-surface vapor pressure with decreased pore occupancy. By contrast, Cady (1969) and Bilhartz (1971) did not observe significant vapor-pressure lowering for unconsolidated sands for temperature levels of 300° to 400°. Why the capillary-pressure effect is lacking in this case is not known, though capillary-pressure phenomena are known to be quite different for unconsolidated sands and consolidated rocks. Work is continuing on this phenomenon. The two-phase flow of liquid water and steam through porous media is another problem. Donaldson (1968) has discussed boiling and flow of steam/water mixtures in porous media. Although specific heats and heat of vaporization for boiling water are known to be much greater than those for vaporized liquid hydrocarbons, most available information on two-phase flow is for hydrocarbon systems under isothermal conditions.

Another factor affecting boiling in porous media is the salt content of the liquid phase. The effect of salt content for liquid at 400° may be summarized as follows, where vapor-pressure ratio is the ratio of pure-water vapor pressure to salt-solution vapor pressure:

Salt concentration (ppm)	1,000	10,000	100,000
Vapor-pressure ratio	1.0003	1.003	1.033

It appears that vapor pressure is lowered about 3 percent for a salt content of 100,000 ppm. Wells drilled near the Salton Sea in California have produced geothermal brines of salt content in excess of 300,000 ppm. Information is sparse concerning the combined effects of salt content and capillarity upon vapor pressure within pore space. And perhaps a more important consideration is the effect of salt precipitation upon the pore volume and permeability of the porous medium where boiling takes place. This has long been a concern in the consideration of recharge (reinjection) of salt waters in a natural geothermal-steam production. But the Salton Sea geothermal brines are exceptional; most known geothermal liquids contain less salt than sea water, and it is not expected that salt in solution will be a major factor in production.

Explosive Stimulation

The greater abundance and greater potential energy recovery of hot-water geothermal systems with respect to dry-steam systems suggests a need to find methods for stimulating the production of hot-water systems. Explosive stimulation is one of the more promising methods. The use of chemical explosives for stimulation of geothermal reservoirs is discussed by Austin and Leonard (this volume). For surface excavations with shallow-depth burial, chemical explosives have proved useful in yields of up to around 5 kt. In deeper formations, the emplacement of large volumes of explosive and the higher formation temperatures limit the practical explosive yield.

Nuclear explosives, which become more efficient in size and cost for yields above about 3 kt, have proved useful for large-volume fracturing of underground formations. And special nuclear explosives are being designed for use in deep-emplacement wells of diameters nominally those employed by the oil and gas industry (Kruger, 1972). Nuclear-explosion stimulation of natural-gas production in deep, low-permeability gas fields is currently under study in the United States with the 29-kt Gasbuggy experiment, the 40-kt Rulison experiment, and several other experiments now in the planning stages. In the Soviet Union, nuclear fracturing has been successfully demonstrated for the stimulation of oil production. The use of low-yield (5–100 kt) nuclear explosives also appears promising for the recovery of geothermal energy.

The concept of nuclear-explosive fracturing to stimulate production from marginal geothermal resources was first proposed by Carlson (1959), who considered the use of fracturing to create a permeable zone in natural geothermal systems inadequate in either total fluids, permeability, or extent of natural fractures. The idea was further developed by Kennedy (1964), who examined the economics of explosive stimulation for producing steam from a 500°C (930°F) geothermal formation at a depth of 10,000 ft. From these early speculations

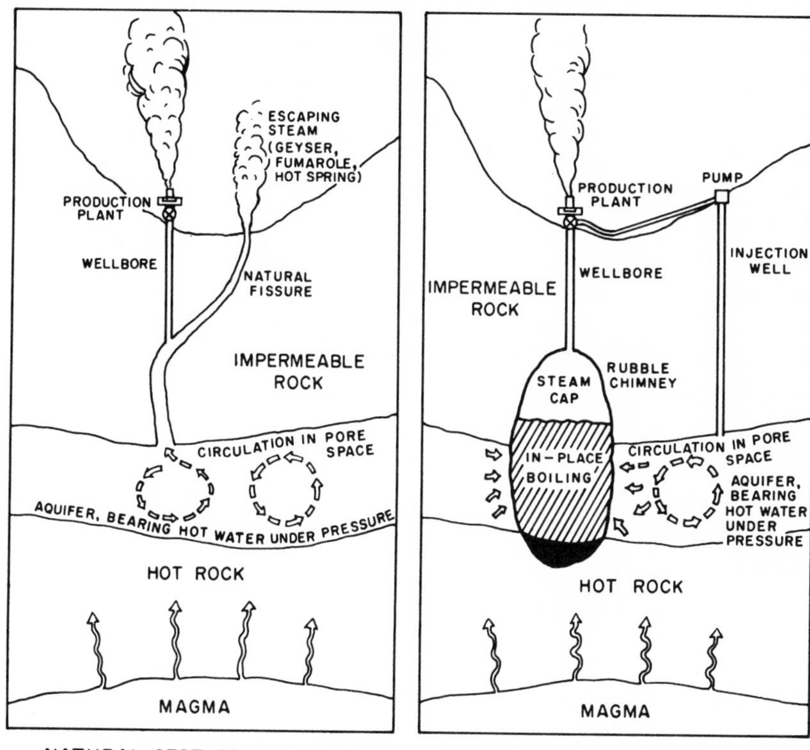

Fig. 1. Schematic comparison of a natural and an explosion-stimulated geothermal field.

came the concept of using surface waters in a closed cycle to produce steam from explosively fractured dry geothermal formations (see Burnham and Stewart, this volume).

The use of explosive stimulation of geothermal aquifers in a manner similar to that of the stimulation of hydrocarbon production was proposed by Raghavan, Ramey, and Kruger (1971). The concept considers the use of low-yield nuclear explosives to create a large-diameter wellbore (rubble chimney) in a geothermal aquifer presumed capable of replenishing its own produced fluids. Figure 1 contrasts, in schematic cross section, an explosion-stimulated geothermal aquifer and a hypothetical natural geothermal system. By appropriately controlling production rate and steam quality in the explosion-fractured wellbore, it may be possible to optimize the production of energy from the reservoir.

Environmental Effects

Although the effects of ground motion from an underground nuclear explosion are discussed elsewhere (Sandquist and Whan, this volume), a few comments in passing may be useful. It is of course unlikely in the extreme that nuclear stimulation for geothermal-steam production would be considered for a populated area. And for explosives of yield smaller than 100 kt, the effects can be predicted; calculations of expected damage to structures as a function of structure type and distance from the detonation can be made by the methods developed by Blume (1969). For feasible development of geothermal resources in remote areas, the costs incurred in comprehensively repairing structural damage would be a small or negligible fraction of the total costs of the program and would be specifically allocated in the cost of developing the geothermal reservoir.

The impact of geothermal steam production on the land, the air, and surface water and ground water is generally limited to the immediate area surrounding the zone, as discussed by Bowen (this volume).

The chemical quality of geothermal waters is variable, ranging from the few parts per thousand of dissolved materials in reasonably clear waters to the extensive and complex burden of concentrated brines. It is believed that the Imperial Valley of southern California, where the dissolved-solids concentration of some brines is in excess of 300,000 ppm, may indeed offer economic potential in the recovery of minerals from the produced brine. Whatever the quality of the geothermal waters, in normal operation they would be wholly contained within an extraction/reinjection cycle, and would have no effect on areal surface or ground waters. The effects of released geothermal fluids are also discussed by Bowen.

The effluent of explosion products must also be considered. The radioactivity history of a fully contained nuclear explosion entails the types and quantities of radionuclides produced, their initial distribution during the hydrodynamic phases of the detonation, and their long-range redistribution. The production of radionuclides depends upon the type of explosive used; the mode of energy release can range from all fission of transuranium elements to almost all thermonuclear reactions of light elements. Radionuclides are produced as fission products by nuclear fission, as tritium (^{3}H) from thermonuclear reactions, or as activation products from the capture of released neutrons by the ambient atomic environment. A review of these aspects is given by Sandquist and Whan (this volume). Nuclear explosives such as those of the Diamond series, which are being developed for commercial use, not only have been reduced in diameter and improved in temperature tolerance and other physical properties, but also are improved with respect to the reduction of radioactivity, especially that of tritium. And other techniques, such as neutron-absorbing shields, are being developed to reduce the total production of radioactivity.

During the expansion of the initial cavity produced by the detonation, a considerable thickness of rock is vaporized and molten. Upon cooling of the cavity, a puddle of molten rock vitrifies and a substantial fraction of the nonvolatile radionuclides becomes fixed in the vitrified layer at the bottom of the cavity. The volatile radionuclides, such as ^{85}Kr and ^{3}H, are distributed throughout the void volume of the resulting configuration. The ^{85}Kr as an inert gas can be flushed from the chimney upon reentry and production; the ^{3}H may associate as HTO, HT, CH_3T, etc. Studies on the chemical distribution and production history of the tritium are part of the objectives of the Gasbuggy and Rulison gas-stimulation experiments. The results of these studies now appear likely to show that the environmental radioactivity resulting from nuclear-stimulated geothermal-energy production will constitute a negligible public risk, and one well worth taking when balanced against the increased power production from geothermal resources.

Production Methods

The generation of electricity from geothermal heat requires the production of about 1.7×10^4 lb/hr of saturated steam per megawatt of electricity. And tolerable amortization of a power plant and production system may require steam-production rates in excess of 1 million lb/hr for periods of 20 to 40 years. The energy-content calculations given above indicate that to sustain steam-production on such a massive scale, it may be necessary to recover energy from the hot, porous rock as well as from the thermal fluids contained in the aquifer. Thus the economic feasibility of producing electric power from low-grade geothermal resources appears to depend on engineering design that can deal with (1) limited deliverability per well and (2) limited volumetric energy content in the aquifer or limited reserves of the geothermal fluid.

Several methods of production can be considered with respect to these two goals. Three of these methods have been described conceptually by Raghavan, Ramey, and Kruger (1971). In one method the hot, compressed water is produced to the surface and flashed to steam in surface equipment; the cooler residual water is returned to the aquifer. In a second method, a gradual, controlled reduction of pressure, throughout the production period, causes a boiling front to recede into the fractured rubble chimney and eventually into the aquifer formation. In the third method, one of particular utility where the geothermal waters are high in salinity or valuable mineral content, two rubble chimneys are operated cyclically; steam is flashed in one chimney until the residual brine reaches a maximal concentration, whereupon flashing is transferred to the second chimney, while the first is pumped free of its saturation brine, chiefly to allow it to self-recharge but perhaps also to extract its mineral content.

Production of hot liquid for surface flashing. Figure 2 illustrates the concept of surface flashing of geothermal waters. Pressure regulation at the wellhead maintains the hydrostatic pressure of the fluid, and the water is pumped at pressure to surface-flashing equipment. As we have seen, about 32 weight percent of fluid at 500°F can be flashed initially, declining as the pressure and temperature decline. With this method of production, energy for vaporizing is available only from the compressed liquid. The separator removes the cooler residual water as well as chemical contaminants; this water, as well as the condensate from the subsequent generator-condenser operations, can be either ponded on the surface or returned as recharge to the geothermal aquifer.

Production of varying steam ratio by receding flashing. Figure 3 illustrates the concept of flashing the hot water underground with a continuously receding flashing front. Without pumping, a geothermal-aquifer system at hydrostatic pressure might not flow to the surface. In such cases, steam production within the wellbore can be initiated by a controlled reduction in pressure. Initially, water rises in the wellbore to a level where its surface reaches the saturated vapor pressure corresponding to its temperature, and flash-evaporates continuously into steam at that point. As production continues and the temperature drops, the

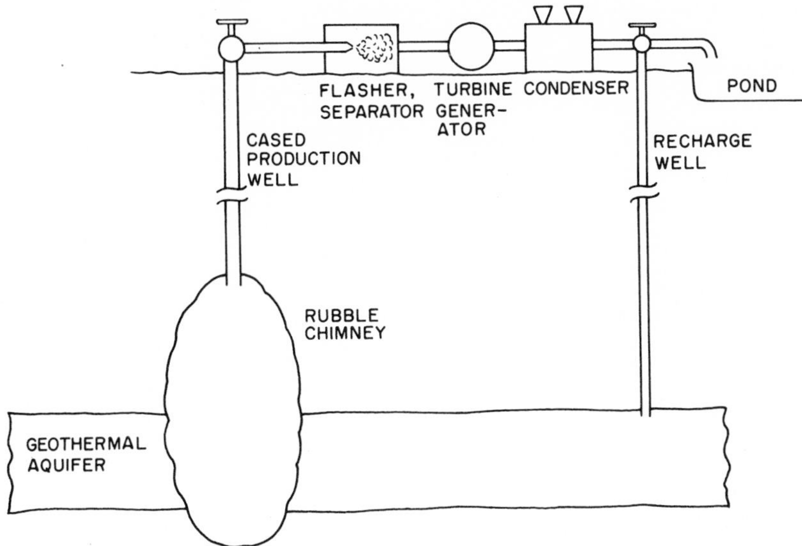

Fig. 2. Surface flashing of geothermal water.

flashing front recedes continuously down the chimney and into the aquifer. By controlling the production rate, the operator can time the process so as to deplete the reservoir over the desired life of the resource. This mode of production requires handling of both liquid and steam, and the production of both separated liquid and turbine condensate could be recharged to the aquifer.

The control of steam production by controlling boiling underground within the geothermal system proper allows a considerable fraction of the heat content of the chimney rubble and aquifer rock to be removed. Under these conditions, a good deal of the in-place liquid can be vaporized to steam. The time scale of such a production method has not yet been determined. There may not be sufficient heat content in the rubble chimney and aquifer to sustain a sufficient production rate over the lifetime of the reservoir, for energy is produced from the rubble and host rock at the expense of a decrease in temperature. Thus, steam quality changes with time, owing to two effects: the change in steam temperature and the change in gradient with a receding flashing front in the wellbore. Though this production method appears to be more efficient than surface flashing, its evaluation requires the development of con-

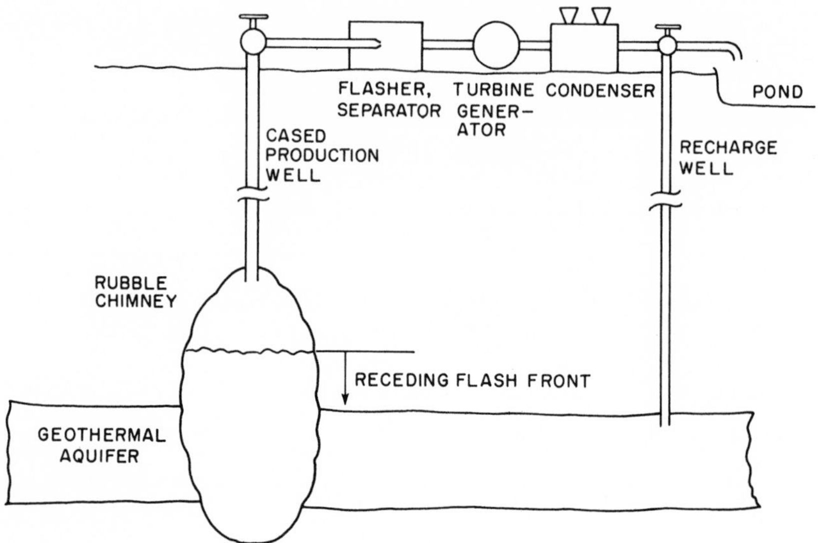

Fig. 3. Receding flash front into the geothermal aquifer.

siderably more information on two-phase boiling in porous media, especially at elevated temperatures; on transient flow characteristics in the nuclear-fractured aquifer; and on the effect of increasing salt concentration in the system.

Cyclic production wth flashing in paired chimneys. Figure 4 shows the third method of production that might achieve the increased thermal efficiency of in-place boiling while maintaining a somewhat constant steam quality and production. The production cycle begins in one of two rubble chimneys, with flashing at the top of the chimney under aquifer conditions of temperature, pressure, and salt concentration. Steam production continues with the flash front receding to the aquifer level in the chimney or until the salt concentration of the cooled water in the aquifer reaches saturation. At that time, the colder, saturated water in the chimney is pumped to evaporation ponds to recover its mineral content; pumping continues until near-normal aquifer conditions are reestablished in the rubble chimney. At the same time, steam production is initiated from the second rubble chimney to maintain the steam-production rate.

The evaluation of this method of production would require information beyond that needed for the continuously receding flashing front. In

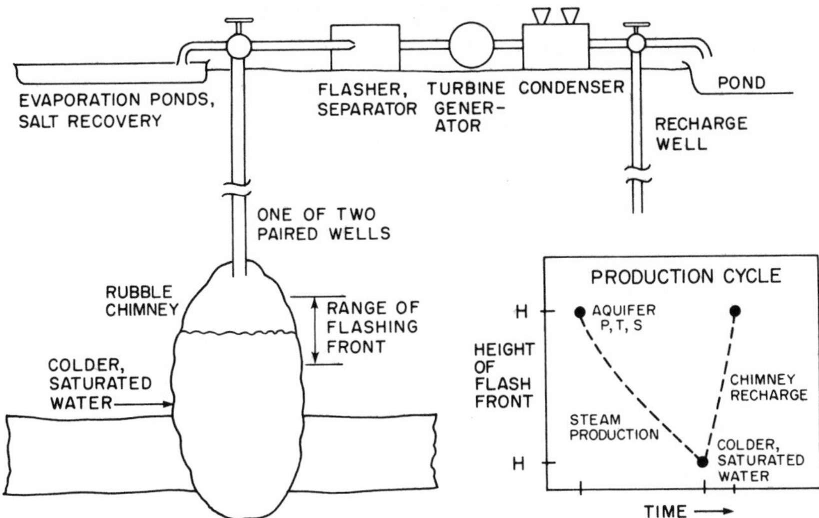

Fig. 4. Cyclic production with flashing front in paired wells.

particular, it would be necessary to develop experimental data on the properties of the dissolved solids in the geothermal system and on the kinetics of their precipitation from the cooled aquifer waters.

Performance Forecasting

We have noted that the amortization of a power plant and steam-reservoir development may require steam-production rates in excess of 10^6 lb/hr for periods of 20 to 40 years. Thus the discovery and stimulation of a geothermal aquifer containing water at temperatures sufficient to produce usable steam rates is a necessary but insufficient condition for its development as a geothermal-energy resource. Before capital investment can be justified, an evaluation must be made of the aquifer's total reserves and the optimum rates of producing steam over the lifetime of the surface equipment.

It is generally considered that geothermal reservoirs are subject to recharge by surface waters at some natural, but unknown, rate. Thus development at production rates equal to, or less than, the natural-recharge rate could lead to essentially limitless production of steam. Development at higher rates could lead to rapid exhaustion of fluid production. Unfortunately, determining the natural-recharge rate is usually not a simple matter. Much geochemical, hydrological, and meteorological research has examined natural recharge in existing producing areas. And although we may with some confidence doubt that natural recharge equals the extraordinary localized fluid-production rates in geothermal areas, we do not yet have the ability to forecast the optimum extraction rate for a particular geothermal field, an ability considered crucial to the proper development of geothermal resources. Fortunately, much useful information can be drawn from related studies of groundwater hydrology and oil- and gas-reservoir engineering. Both fields are concerned with the production of a valuable fluid from the pore space of subsurface formations.

One approach to the forecasting problem may be called the "volumetric estimate." The bulk size of the reservoir is estimated from geological and geophysical data. Knowledge of the porosity and volumetric fluid saturations (for both hot liquid and steam) within the pore space permits calculation of the initial mass of hot water and/or steam. Estimating deliverability of the wells and the unit recovery of fluid (from calculation or experience) then permits completion of the "volumetric" forecast. Unfortunately, it appears that a volumetric estimate has not been possible for any of the currently producing geothermal-fluid fields. Development of these fields has proceeded on a trial-and-error basis—power plants of small size were installed and operated for a time to gain confidence before expansion to larger power plants was undertaken.

A second approach to performance forecasting is "performance matching." Where some development already exists and production is in progress, a mathematical model of the fluid reservoir is postulated. The system size and productivity are then established by matching measured production data (mass produced, enthalpy produced, reservoir and producing pressures and temperatures, etc.) with the like parameters of the mathematical model. When the full range of model parameters and their relationships have been determined, the model can be used to forecast the production performance of other development schemes. This approach is used extensively in oil and gas production. The first such model for geothermal-fluid production has been described by Whiting and Ramey (1969) and applied to production at Wairakei, New Zealand.

To develop a model of geothermal-reservoir behavior, both the mass and the enthalpy of the produced fluid must be specified. Moreover, if

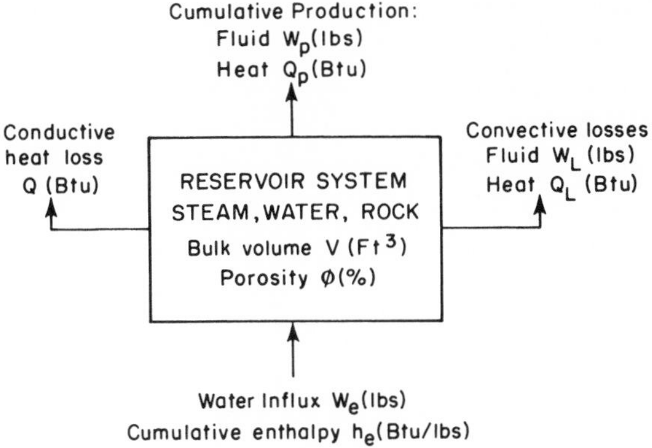

Fig. 5. Schematic diagram of reservoir model (after Whiting and Ramey, 1969).

liquid water and steam coexist in the reservoir, the method of production plays an important role. The fluid produced could be saturated liquid, saturated steam from the steam cap (see Fig. 1), or any combination of the two. Also, the specific enthalpy of the produced fluids can vary significantly. Depending on the initial state of the reservoir, the fluid depletion could be isothermal or isenthalpic. Finally, the rock matrix contains large amounts of sensible heat, which can affect the production process considerably.

A model of a reservoir system developed by Whiting and Ramey (1969) is shown in Fig. 5. The system contains rock, water, and steam. The bulk volume of the reservoir system, V, is given in cubic feet; its porosity, ϕ, is given in percent. The cumulative production at any given instant is given by W_p, lbs of fluid. Associated with this mass of fluid is a cumulative heat production, Q_p, in Btu. The model accounts for heat loss, Q_L, and for mass (fluid) loss, W_L, due to convection in, for example, fumaroles, hot springs, or wild wells. Whiting and Ramey have shown that conductive heat loss, Q, at the margins of the reservoir should be negligible and should not affect reservoir performance over time periods for which normal forecasting is made. The influx of water, W_e, together with its associated energy (cumulative enthalpy, h_e), are considered separately. This influx, from a contiguous aquifer, could be steady or varying. The specific enthalpy of the liquid recharge is assumed to be constant.

The analysis is based on the zero-dimensional material-balance method commonly used in petroleum-reservoir engineering. The following assumptions are inherent:

1. Complete thermodynamic equilibrium.
2. Uniform pressures and saturations of fluids.
3. Uniform withdrawals, which implies that fluids are produced from all parts of the reservoir and that the effects of well sinks, as such, are ignored.

Simultaneous solution of mass, energy, and volumetric balances leads to the expression

$$W_p(h_p - E_c) + W_L(h_L - E_c) + Q = W\{E_i - E_c + [(1-\phi)/\phi]\,[x_i \nu_{gi} + (1-x_i)\,\nu_{fi}]\,\rho_r C_{vr}\,(T_i - T_c)\} + (h_e - E_c)\,(B/\nu_{fe}) \sum Q_D(t_D)\Delta p_n \quad (1)$$

where

h_p = enthalpy of produced fluids
W = initial mass of hot water and steam in reservoir bulk volume, V
E = internal energy
Q = conductive heat loss
ϕ = porosity of rock matrix
x = steam quality
ν = specific volume
ρ_r = density of rock and contained fluids
C_{vr} = specific heat at constant volume of reservoir rock and contained fluids
T = temperature
B = water-influx constant
$Q_D(t_D)$ = dimensionless cumulative influx, corresponding to dimensionless time, t_D
Δp_n = pressure drop at any time n

The subscripts p, L, c, i, r, g, f, and e indicate produced, loss, current, initial, reservoir, steam (gas), liquid, and influx values, respectively. For compressed liquid only, Eq. (1) reduces to

$$(W_p + W_L)\nu_f = W(\nu_f - \nu_{fi}) + B \sum Q_D(t_D)\,\Delta p_n \quad (2)$$

Most underground reservoirs are associated with contiguous aquifers. As the pressure drops within the reservoir, the aquifer reacts to the pressure decline by yielding up further water. If our estimate of reservoir performance is to be dependable over time, we must evaluate the quantity of aquifer water that encroaches into the reservoir. A convenient method of calculating water influx is that of van Everdingen and Hurst (1949). The dimensionless influx, $Q_D(t_D)$, is generally a function of aquifer flow geometry and rock and fluid properties, as well as time.

Whiting and Ramey have provided values of $Q_D(t_D)$ for an infinite aquifer for three hypothetical geometries—linear, radial, and hemispherical. The values given are for unit pressure drop at the reservoir/aquifer boundary. In general, the pressure at the boundary will vary with time. To account for the actual variable-pressure history at the field boundary, the production period is divided into a series of small time intervals and the pressure is considered to be constant during each interval. Thus the actual pressure history is replaced by a step function (linear superposition). The water influx, $\tilde{W}_e$, at any given instance is given by

$$\tilde{W}_c = [B\Delta p_n Q_D(t_D)]/\nu_{fc} \tag{3}$$

where Δp_n is the pressure drop and $Q_D(t_D)$ is the appropriate influx function. The cumulative water influx, W_e, is then given by

$$W_e = B/\nu_{fe} \sum Q_D(t_D)\Delta p_n \tag{4}$$

For Eq. (1) to be used to predict reservoir performance, a number of parameters pertaining to the reservoir must be determined. These include aquifer size and type, initial reservoir size, aquifer porosity and permeability, initial pressure, and temperature. A large number of production data sets is generally available and can be used to estimate the desired constants. Usually a least-squares analysis of the material-balance equation is required. In this method, best values of fluid in place, aquifer constants, reservoir size, etc., are determined from values yielding a minimum sum of the squares of differences between observed and calculated pressures.

Having determined the necessary system constants, we can analyze the future performance of the reservoir for any assumed production-time relationship, as well as the potential profitability of various hypothetical schedules of production.

Consideration of the reservoir mechanics discussed above suggests that some of the principal assumptions inherent in the Whiting and Ramey material-energy balance equation must be relaxed for performance matching of nonideal real systems. It has already been pointed out that more information is required regarding thermodynamic behavior. It should be noted that the equation does not account for pressure and temperature variations within the reservoir, or for segregation of fluids. However, the inclusion of these effects is not an intractable problem. Cady (1969) and Bilhartz (1971) have applied the material-energy balance equation with success to laboratory experiments. But both found that it was necessary to apply the equation to the steam and boiling-liquid zones separately to account for the temperature differences between the two zones.

Conclusions

The stimulation of geothermal-fluid production from hydrothermal aquifers for producing steam to generate electricity appears to be technically feasible. Several potential methods of production are possible to optimize the recovery of the heat stored in the geothermal aquifer. It is apparent that research is needed not only to study these methods, but also to develop a method for evaluating the changing potential resources across the fluid-production lifespan needed to depreciate capital investment.

The ability to evaluate the economic feasibility of explosion stimulation for development of geothermal resources and to determine optimum modes of production depends on the development of improved methods of geothermal reservoir engineering. Methods for the optimum removal of geothermal fluids must be based on a thorough knowledge of the recovery mechanisms operating within the reservoir, including fluid expansion, liquid vaporization, and the segregation of liquid and vapor (gravity drainage), as well as the influx of fluids from adjacent or underlying aquifers. The deposition and production of minerals must be considered, as well as the environmental effects due to the explosion and subsequent geothermal production. Thus a rigorous evaluation of geothermal-aquifer stimulation requires study of a number of thermophysical, hydrodynamic, and chemical parameters. These parameters may be summarized as follows:

Thermophysical:

- Water and rock temperatures
- Enthalpy of discharged steam/water mixture
- Steam pressure and temperature
- Thermal effects of discharged fluids
- Reservoir response to recycled water

Hydrodynamic:

Saturation vapor-pressure level (SVPL) in wellbore
Deliverability as a function of wellhead pressure
Deliverability as a function of time
Deliverability as a function of SVPL
Reservoir total reserves

Chemical:

Composition of aquifer fluids
Composition of produced fluid
Steam quality at wellhead
Salt buildup in wellbore
Chemical effects of discharged fluids

Although much basic information is available, it is clear that additional analytical, laboratory, and field studies should be made to assess the ultimate importance of explosive stimulation of geothermal aquifers.

REFERENCES

Bilhartz, H. L., Jr. 1971. Fluid production from geothermal steam reservoirs. MS report, Stanford University, Stanford, Calif.

Blume, J. A. 1969. Ground motion effects. *In* Proc. Symp. Public Health Aspects of Peaceful Uses of Nuclear Explosives, SWRHL-82.

Cady, G. V. 1969. Model studies of geothermal fluid production. Ph.D. dissertation, Stanford University, Stanford, Calif.

Calhoun, J. C., M. Lewis, and R. C. Newman. 1949. Experiments on the capillary properties of porous solids. Trans. AIME, v. 186, pp. 189–96.

Carlson, R. H. 1959. Utilizing nuclear explosives in the construction of geothermal power plants. Proc. Second Plowshare Symposium, UCRL-5677.

Donaldson, I. G. 1968. The flow of steam/water mixtures through permeable beds: A simple simulation of a natural undisturbed hydrothermal region. New Zealand J. Science, v. 11, no. 1, pp. 3–23.

Kennedy, G. C. 1964. A proposal for a nuclear power program. Proc. Third Plowshare Symposium, TID-7695.

Kruger, P. 1972. Nuclear explosion engineering. *In* McGraw-Hill yearbook of science and technology (New York: McGraw-Hill).

Raghavan, R., H. J. Ramey, Jr., and P. Kruger. 1971. Calculation of steam extraction from nuclear-explosion fractured geothermal aquifers. Trans. Am. Nucl. Soc., v. 14, p. 695.

van Everdingen, A. F., and W. Hurst. 1949. The application of the Laplace transformation to flow problems in reservoirs. Trans. AIME, v. 186, pp. 305–24.

Whiting, R. L., and H. J. Ramey, Jr. 1969. Application of material and energy balances to geothermal steam production. J. Petrol. Techn., v. 21, pp. 893–900.

14. Induction and Growth of Fractures in Hot Rock

MORTON SMITH, R. POTTER, D. BROWN & R. L. AAMODT

While the consumption of energy in the United States has been doubling every 28 years during this century, the output of electric power, other than hydroelectric, has been doubling every 8 years. The environmental consequences of energy production, especially in this rapidly expanding sector, are now causing much concern. Modern electric power plants are large, and there is a growing trend toward building them away from cities. Transmission lines, moreover, are necessary to take the power to the consumer, and these lines now occupy an area larger than the State of Connecticut (Legislative Reference Service, 1970). An equally large area has been devastated by strip mining, much of which was for coal (*ibid.*). Obviously, the doubling rate must soon be stopped, and better methods of generating and distributing power are needed.

Geothermal energy offers much promise as a source of clean power. If we divide its development into the three stages defined by Banwell and Meidav (1971), the United States is now in the first stage, using geothermal energy only where it is expressed in surface manifestations. The next stage will be the development of geothermal resources in various large regions where the thermal gradient in the Earth is several times normal; such regions have been estimated to include about one-tenth of the land surface of the Earth (Banwell and Meidav, 1971), and are exemplified in the United States by hundreds of thousands of square miles around the Gulf of Mexico and in the West. In the third stage, deeper wells will exploit the normal geothermal gradient of about 16°F per thousand feet. Each succeeding step in the development will

The authors are at the Los Alamos Scientific Laboratory, University of California, Los Alamos, New Mexico.

reduce the need for transmission lines by placing the geothermal installation closer to the consumer. As this distance becomes shorter, the utilization efficiency of geothermal power will be increased as it becomes economical to pipe the hot water directly to consumers for space heating, cooling, or industrial processing. Experience in Iceland indicates that this can be done efficiently for over 40 miles (Bodvarsson, 1961). Cooling systems using water as the refrigerant, a lithium bromide solution as the absorbent, and hot water as the power source have been used successfully in New Zealand (Barnea, 1972), and are reportedly being mass produced in the U.S.S.R., where they are run in reverse as heat pumps in cold weather (Tikhonov and Dvorov, 1971). These direct uses of low-temperature geothermal fluids greatly increase the efficiency of geothermal-energy systems.

Widespread use of geothermal energy will be facilitated if a way can be found to produce it from hot, relatively dry rock by circulating a fluid introduced from the surface through the rock and bringing the heated fluid (probably water) back to the surface. Some concepts of how this might be done are discussed in this paper.

Hydraulic Fracturing: Concepts and Options

The idea of using the heat from slowly cooling lava beds is an old one. The difficulty involved is made clear by Ingersoll, Zobel, and Ingersoll (1954). They consider a spherical cavity with a radius of 4 ft, located deep in an old bed, where the temperature is 500°F. Water is poured down a pipe extending from the surface to the cavity and steam is withdrawn through another pipe at a rate such that the water in the cavity is maintained at 300°F. The power withdrawn decreases with time; after one week it is 7.09 kw. In order to produce 100 Mw under similar conditions, the radius of the cavity would have to be 670 ft. Though this system is not particularly interesting, with a change in geometry the same surface area can probably be produced quite easily by the method of hydraulic fracturing (Gatlin, 1960).

Hydraulic fracturing is begun by pumping water from the surface into a borehole. When the borehole becomes sufficiently pressurized, a tensile crack will form somewhere along its margin. The fracture will grow as the pumping continues, and will continue to grow as long as water can be pumped in fast enough to exceed the leakage taken off by the permeability of the rock. This procedure has not, to our knowl-

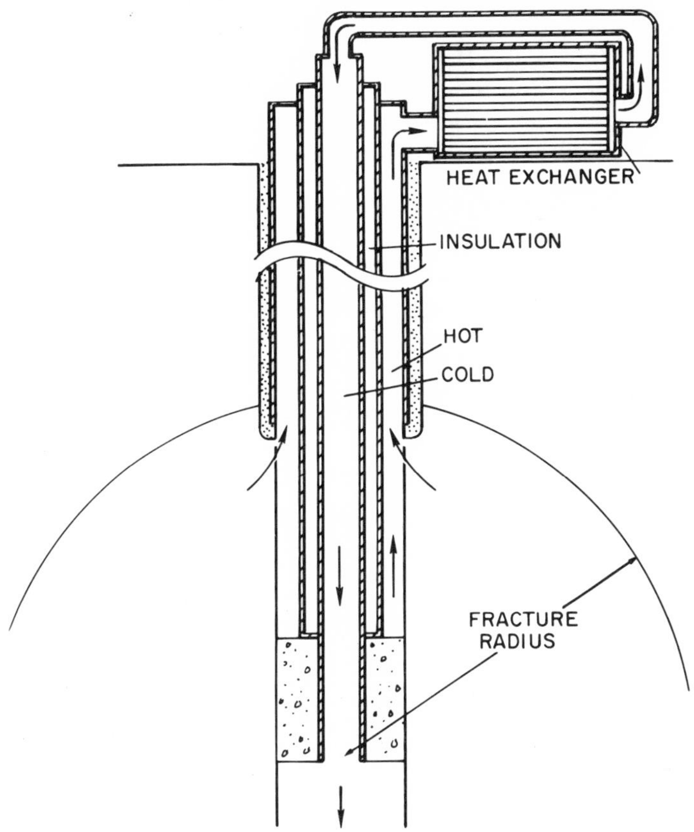

Fig. 1. Circulation loop with concentric pipes.

edge, been attempted in hot granite, but our discussions with experts in the field warrant confidence that it will prove effective. Computer modeling of the hydraulic-fracture process at the Los Alamos Scientific Laboratory (Harlow and Pracht, 1972) indicates that thermal stresses, set up by the differential shrinkage of the rock as it is cooled by the inflowing water, will induce additional cracks, which may themselves propagate. The crack pattern is thus expected to be more complex than it would be in cold rock.

The working fluid used to extract the geothermal energy may be circulated in several ways. One way is to install a concentric-flow system of the configuration shown in Fig. 1 in the same hole—drilled oversize for the purpose—that has been used to create the hydrofrac-

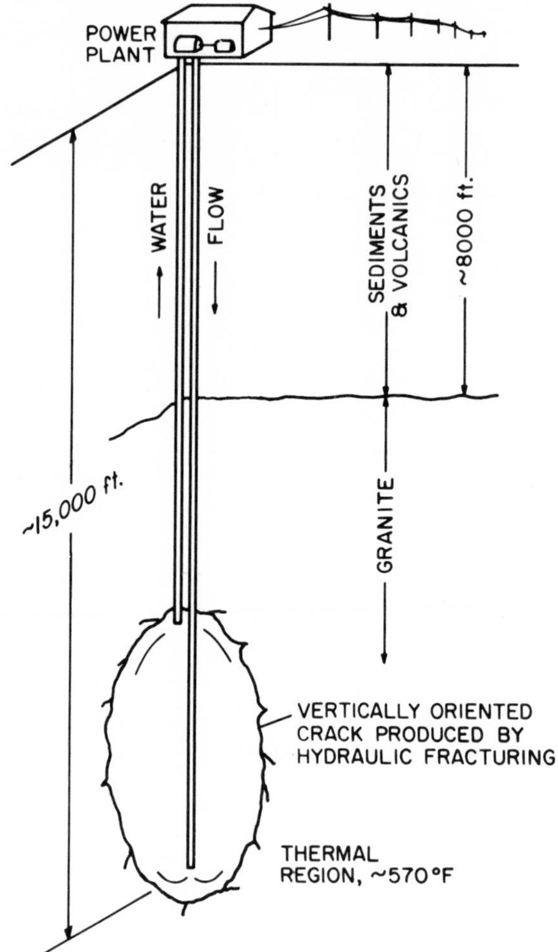

Fig. 2. Circulation loop with separate pipes.

ture. This approach has the disadvantage of complexity in the pipe string, but has the advantage that we know where the bottom of the return pipe is, and can ensure that it is connected to the top of the reservoir by inducing a series of small, overlapping fractures from the bottom to the top of the uncased hole before the center pipes are emplaced. However, where it does not prove too difficult to drill a second hole and intersect the top of the fracture from a position perhaps 200 feet to one side, the two-hole circulation system shown in Fig. 2 will

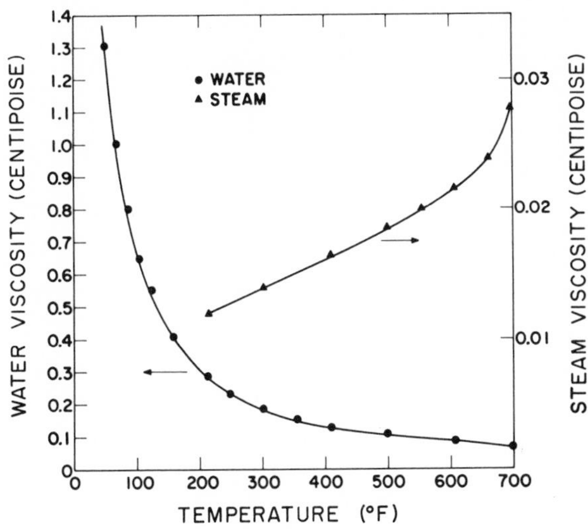

Fig. 3. Viscosity of water and steam vs. temperature.

probably be the one most used. (Concentric pipe is useful in one more case, as discussed below.)

If the crack system is capable of supporting the weight of a column of hot water to the surface, plus the additional pressure needed to prevent boiling, a water cycle in which the hot water rises, passes through a heat exchanger at the surface, and is returned to the reservoir, may be used. Appendix A to this paper shows that as much as 10 times more thermal energy may be transmitted up a given pipe if the working fluid is water instead of superheated steam. Though a steam system brings less dissolved solids to the surface than an all-water system, dissolved gases such as sulfur dioxide are a more serious problem in a steam system, since in a hot-water system the gas remains dissolved as the water passes unflashed through the heat exchanger and down the return pipe. Another major advantage of the all-water system is that the viscosity of water decreases as its temperature increases (Fig. 3), whereas, for superheated steam, the opposite is true. Thus, high-temperature circulating water tends to follow the hottest path, all else being equal, whereas steam flows best through the coolest path. This becomes especially important if the cavity grows by thermal-stress cracking, as we shall see.

The advantages of an all-water system need not be lost even if the crack system should lose too much water at full pressure. Deep-well

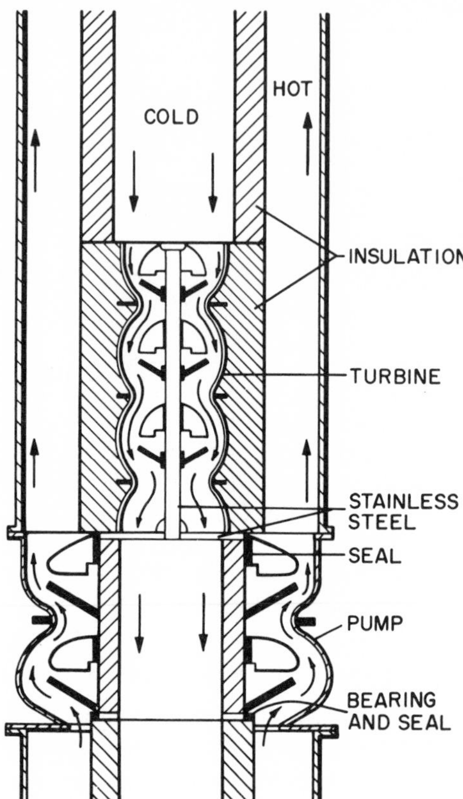

Fig. 4. Turbine/pump for use in concentric circulation loop.

submersible pumps, hardened to withstand the high temperature and pressure of the environment, are probably within the state of the art of pump design (Stepanoff, 1948). If such a pump is placed in the hot-water line, the pressure on the reservoir may be reduced while that at the surface is maintained. The water in the cold-water return pipe can be held at a level compatible with the desired reservoir pressure.

Returning to the concentric-flow geometry, we see now the additional advantage alluded to above; the pump in the hot-water line may be driven by a turbine in the cold-water line, as shown in Fig. 4, with no additional power from the surface. Pressure in the hot-water line differs with depth from that in the cold-water line, because of the difference in density of hot and cold water. The pressure at depth in the hot and cold lines is shown in Fig. 5; at the surface the pressure in the two lines is the same, 2,900 psi. For the temperatures shown, the pressure

Fig. 5. Pressure in hot and cold water vs. depth, with 2,900 psi surface pressure.

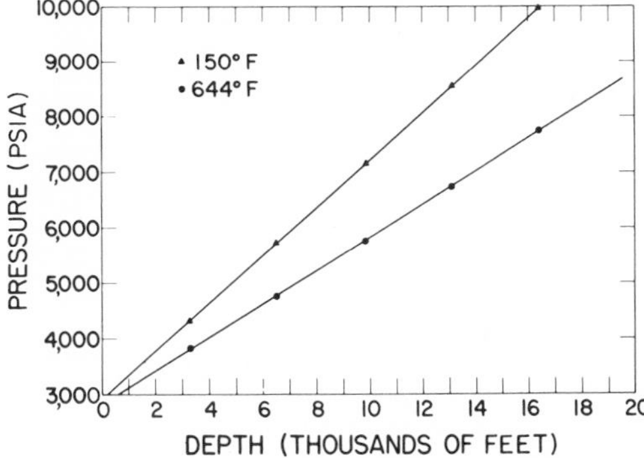

difference is 2,050 psi at a depth of 15,000 ft. This pressure difference is available to drive the circulation of the water, and the system will indeed, according to the computer-simulation calculations, have to be throttled down in most cases by a valve in the cold-water pipe at the surface. The natural excess pressure may be used to drive the turbine-pump, or, if this pressure is too low, an additional surface pump may be placed in the cold-water line in order to increase the turbine driving-pressure to the desired value.

Hydraulic Fracturing: Practice

The first step in the development of a dry geothermal reservoir is the selection of an appropriate site. This would be in one of the known regions in which the geothermal gradient is abnormally high and hot igneous rocks are present not far below the Earth's surface. There are many locations in the western United States that satisfy these requirements, including one near Los Alamos, New Mexico.

If, for example, a rock temperature high enough to be useful for power generation (say, near 600°F) is reached at a depth of 15,000 ft—as might be expected over much of the western third of the United States—then the geothermal reservoir is entered by drilling to that depth. The hole is cased with steel pipe, which is cemented in place by conventional methods. Since the lower part of the hole is in hot igneous rock, these operations are more difficult than the drilling and casing of conventional oil wells. But with a fluid circulation sufficient both to cool the drill bit and to remove the cuttings, the drilling can be done with a conventional oil-field rotary-drilling rig. Suitable casing and cementing materials are also available commercially.

To assist in understanding the subsequent fracturing behavior of the underground system, the character of the rock and of the hole are investigated as fully as possible before the casing is inserted.

Using methods common in the drilling industry, the well casing is perforated at a point several hundred feet above the bottom of the hole, and a temporary seal is inserted just above that level. By the use of a high-pressure pump at the surface and a high-pressure line extending through this seal, a fluid pressure sufficient to produce hydraulic fracturing is developed in the sealed-off section of the hole. Through the perforations and in the open section of the hole, this pressure (of the order of 7,000 psi above hydrostatic pressure at that point) is exerted against the rock adjacent to the hole, developing tensile stresses in the rock eventually sufficient to cause it to crack. The stress required to extend a crack is in general much less than that needed to form it. Therefore, once breakdown (cracking) has begun, pumping is continued at reduced pressure until the principal cracks have been extended to the desired radius. Mathematical analysis agrees with field experience in indicating that the resulting crack will be in the form of a thin, vertically oriented disk of elliptical cross section,* as is suggested by Fig. 2. Since the crack is filled with water, fluid pressure in the crack can be expected to hold it open without other support.

At a distance of about 200 ft from the first hole, a second hole of the same size is drilled, to intersect the upper part of the crack system created by hydraulic fracturing in the first hole. The orientation of the crack may have been determined by listening with strategically placed geophones to noise generated at points on the rim, as the crack grows intermittently at various places (McClain, 1971). If not, the orientation at the start of the hydraulic-fracturing operation can be determined. If for some reason the crack system is not intersected by directional drilling from the second hole, then—after this hole has been cased and cemented—hydraulic fracturing can be repeated from the bottom of the second hole, to produce a second crack system that intersects with the first. Several variations of the hydraulic-fracturing procedure have been suggested, including: hydraulic fracturing from the bottom of the first hole, to ensure communication between this point and the crack system established by previous fracturing at a high level; hydraulic fracturing simultaneously from two holes; and hydraulic frac-

* In hot rock this may not be the case, as we shall see below in connection with computer simulation.

turing from an open (uncased) section at the bottom of the first hole. Selection among these and other possibilities must await more detailed analysis, and agreement among drilling and hydraulic-fracturing specialists. However, the relatively straightforward procedure outlined above has a high probability of producing the interconnected crack system required.

After the second drilling, a pump located at the surface is used to force cold water down the deeper hole, through the completed underground fracture system, and up the shallower hole. Once a moderate temperature difference has been established between cold water descending in the first hole and hot water rising in the second hole, circulation through the system is maintained by natural convection, and pumping is discontinued except for the injection of such additional water as is required to keep the system full. In Appendix B of this paper it is shown that one-fourth to one-half cubic foot of void space is formed for each million Btu withdrawn from the rock. The void space formed each day when 100 Mw of thermal energy is being drawn from the rock requires 25,000 gallons of make-up water even if there is no underground leakage from the reservoir.

As mentioned previously, an area equal to the surface area of a sphere 670 ft in radius is required in order to recover 100 Mw of thermal power from a relatively cool (500°F) rock mass. A crack with a 1,000-ft radius is thus slightly larger than necessary. But if a crack of 3,000-ft radius can be formed in rock that is at 570°F, it has been calculated that an average power of 89 Mw (thermal) could be extracted for 20 years through the initial crack faces (Robinson et al., 1971). Such calculations are probably too conservative, however, because of the effects of thermal-stress cracking, which we shall take up next.

Thermal-Stress Cracking

In Appendix B the void volume resulting from the removal of 1 million Btu of heat from rock is calculated. This volume is, at least to a first approximation, independent of the magnitude of the decrease in temperature. In Appendix C, it is shown that if the temperature increase is linear with depth the rock must be cooled to a particular fraction of its initial temperature if it is to be fractured by thermal contraction, and that the fraction is independent of depth, again at least to first order. The fraction is close to 1 if the thermal gradient is large, but may become so small, if the thermal gradient is small, that thermal fracture will not take place, or will proceed very slowly. Some geologists have indeed expressed doubt that granite rock will fracture when cooled; the alternative suggested was an increase in porosity, which, however, would not necessarily increase permeability.

In any event, when a cube of granite about one foot on a side was slowly heated to about 600°F over several days, the oven removed, and water applied to one face, large tensile fractures appeared in about 20 minutes. Similar results were obtained with basalt. It is expected, therefore, that the void formed by cooling will expose fresh hot rock by cracking, and, if the new fractures are sufficiently permeable, the life of the reservoir will be greatly extended. A given amount of empty space, and therefore surface area, can be distributed through the rock in many ways. If it occurs as parallel cracks, the permeability is proportional to the square of the distance between cracks (Harlow and Pracht, 1972). The decrease of viscosity with temperature should be helpful in inducing circulation through newly created cracks. Finally, the growth of the reservoir should be fastest at the bottom, where the coldest water overlies the hottest rock.

Computer simulation of reservoir behavior. Although the processes taking place when water circulates throughout a mass of broken rock are individually simple, their interaction renders the situation complex. Such problems are well suited for numerical solution with a computer, and a program that includes the flow of heat through the rock, the cooling, shrinkage, and fracture of the rock, and the growth of the reservoir has been devised by F. Harlow and W. Pracht at Los Alamos. The problem also includes heat transfer to the water, and the temperature, density, viscosity, and velocity of the water. Unknown factors such as crack spacing were absorbed into parameters that could be varied, and the sensitivity of cavity behavior to these unknown factors could be determined. As expected, the calculations show the reservoir growing preferentially downward and sideways, unless the crack spacing is made too small (about 1/2 inch or less). Figure 6 shows for a typical model the contours of equal porosity at various times, using a crack spacing of about 2 inches. Figure 7 shows the variation of power output with time for this model: an initial rise as the top of the cavity is warmed from below; a period of decreasing power output, which is finally overcome

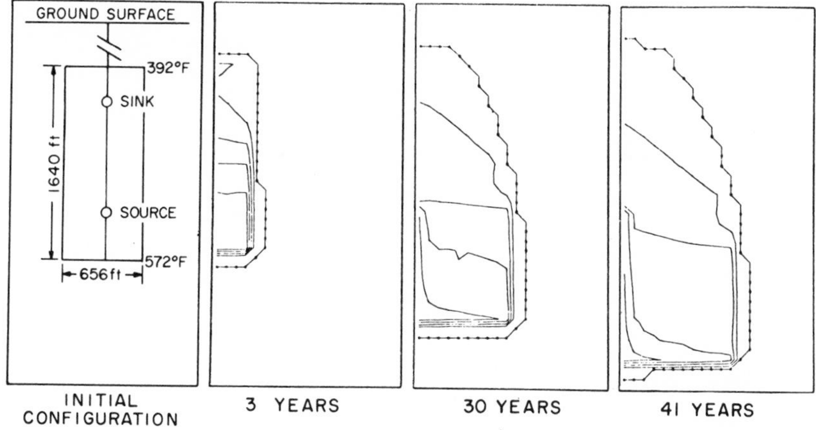

Fig. 6. Contours of equal porosity from a typical computer calculation. The outer line shows the growth of the initial configuration. The sink in the initial configuration is the outlet of heated water; the source, the inlet of cold water.

by cavity growth; and a subsequent rise to more than the initial power level. In this calculation, the flow rate was held constant at 26.5 gallons per second, or about 2.3 million gallons per day.

Calculations for the case of the single thin crack, i.e. where no thermal-stress fracturing has taken place, verify that such cracks could indeed supply many tens of megawatts of thermal power for several decades. However, the rock is undergoing cooling while the initial fracture is being formed. Thermal stresses will create new cracks, preferentially near the source of the first one, where the water is coldest, and these cracks may also propagate under the high pressure exerted at that point. Thus, the reservoir may resemble a cylinder rather than a thin flat crack when the initial fracture is complete.

Geochemical complications. Most natural minerals have significant solubility in hot water, and their solubilities increase with temperature and pressure. This fact may have both favorable and unfavorable consequences. If the reservoir does in fact grow downward and the temperature of the circulating water increases above the initial temperature of the rock in the initial reservoir, expansion would cause the thermal fractures to close again if it were not for the solution and perhaps reprecipitation of minerals. The water rising to the surface will contain minerals that may be expected to corrode and plug the heat exchanger. However, some of these minerals may have commercial value, and their

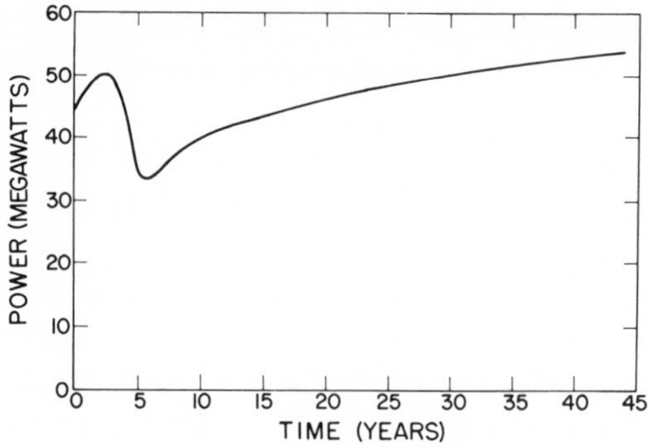

Fig. 7. Power vs. time at constant circulation rate.

recovery from the water may be profitable. Certainly, intensive study must be made, both in the laboratory and in the field, of the solubility and precipitation of minerals in solution under realistic conditions of temperature and pressure, if we are to understand and exploit the thermal-cracking process.

Economics

If, from any point on the Earth's surface, a hole were drilled downward to a sufficient depth, it would eventually reach rock hot enough to constitute a potentially useful energy source. The thermal reservoir represented by the Earth's crust and mantle is large, and it exists everywhere beneath the Earth's surface. A general method of extracting energy from this source has been outlined above, a method drawing only upon existing—if newly combined—engineering techniques. The chief question remaining is whether this method, or some variant of it, can be used to extract energy from hot, dry rock in a usable form, at a usefully high rate, at a cost that is comparable with that of energy from other sources.

Water at 200° to 300°F has many important uses, including space-heating, water-distillation, and chemical processing. In many cases, such uses could be served economically by relatively low-grade geothermal energy (as, indeed, they are already served in several parts of the world). However, the major contemporary interest in geothermal energy lies in its promise as a new, nonpolluting source of electric power. In considering the economics of man-made geothermal-energy systems, therefore, it has been assumed that thermal energy will be useful only if it is brought to the surface at a temperature high enough to operate generating equipment. The minimum temperature required for a steam power plant has here been assumed to be about 500°F, although lower-temperature systems are operating. And isobutane cycles operating at temperatures of 340°F or lower may be available soon (see Anderson, this volume).

It has also been assumed, on the basis of the present state of the drilling art, that it would probably not be practical to drill holes deeper than about 20,000 ft to reach sufficiently hot rock (this is probably a temporary limitation).

Finally, it has been assumed that waste heat from the power plant would be dissipated to the atmosphere by forced-draft, air-cooled heat exchangers. These are more expensive than wet cooling towers, but save a great deal of water and avoid thermal pollution of lakes and streams. (The only pollution from the contained power system proposed would be a rising plume of warm air, which, it is believed, would have no significant environmental effect.)

On this basis, preliminary cost estimates have been made for a typical 100-Mw steam/electric power plant, operating in the western United States. Typically, to provide the necessary thermal energy from the geothermal source, this would require two pairs of 15,000-ft holes into rock at about 570°F. The plant cost, including development of the underground circulation system, is estimated to be $186 per kw of generating capacity, and the generating cost is estimated to be 4.7 mill per kwh at the busbar. Even if these estimates are low by a factor of 2, which is unlikely, such a system would be competitive with nuclear or fossil-fuel power plants.

Although deep-rock temperatures are in general significantly lower in the eastern United States than in the West, there are large regions in the East (Van Orstrand, 1951) in which rock temperatures above 340°F can be reached at depths less than 20,000 ft. Accordingly, cost estimates have been made for a typical case in which five pairs of 18,000-foot holes would be required to supply energy in the form of water at 340° to a 100-Mw isobutane-cycle electric power plant (see Anderson, this volume). Because of the high cost of drilling ten deep holes, plant cost in this case is high—an estimated $316 per kw, which is higher than that of a coal-fired plant and approaching that of a nuclear power station. But because the fuel cost is essentially just the cost of amortizing the underground system, the generating cost is estimated to be only 8.0 mill/kwh. If these estimates are in error by even 50 percent, it still appears that this type of geothermal-energy development may be competitive in large sections of the eastern United States.

Proposed Experiments

The engineering and economic aspects of the development of dry-rock geothermal-energy systems appear to be sufficiently favorable to justify a field experiment. The intent of such an experiment would be to demonstrate the engineering feasibility and operation of such a system, and to investigate its problems and economics. An experiment of this sort has in fact been proposed to the U.S. Atomic Energy Commission, and exploration for a suitable site has been in progress for several months.

On the basis of geological information so far collected and heat-flow measurements made in a series of holes drilled for the purpose, it now appears that an area in the Jemez Mountains (Smith, Bailey, and Ross, 1970)—about 2 to 5 miles west of the Valles Caldera, and about 30 miles west of Los Alamos Scientific Laboratory—is particularly suitable for a field experiment. Here the geology is relatively uncomplicated, the depth to the basement granite is only about 2,500 ft, and the geothermal gradient is about 6 to 10 times normal. An exploratory hole in this area has recently been drilled and cased to 2,430 ft depth, to examine heat-flow, lithology, hydrology, and drilling problems. It is intended to extend this hole a few hundred feet into the granite, with continuous coring, to investigate the character, probable competence, temperatures, and heat-flow properties of the granite, and to permit preliminary hydraulic-fracturing and pressurization experiments. If the basement rock is found to be capable of containing a pressurized system, then a specific site will be selected for development of an experimental underground circulation system.

Preliminary heat-flow measurements indicate that in this area a rock temperature of about 500° to 600°F can be expected at a depth of only 7,500 ft—which is not unusual in regions of geologically recent volcanism. It is proposed to create, at approximately this depth, a circulation loop of the general type sketched in Fig. 2 but with a relatively small hydraulic fracture—perhaps 1,500 ft in radius—and with the extracted heat wasted to the atmosphere by an air-cooled heat exchanger.

Simply creating such a system and establishing convective circulation in it will represent an important feasibility demonstration. Operating the loop for approximately 10 months should—with relatively small initial heat-transfer surface—indicate whether the system simply decays, or whether instead it perpetuates itself by some combination of thermal-stress cracking and conduction from the magmatic rock below. Continued circulation for another 2 or 3 years will then permit detailed studies of the thermal, chemical, and mechanical behavior of the system, and of its economics. Problems are, of course, expected to develop in such areas as fluid loss through the natural void system in the rock, and plugging of circulation channels by mineral deposition. However, it is believed that engineering solutions to such problems can be found.

Scientific investigations will form an important part of this experiment. A long list has been developed of parallel and follow-on programs that should be undertaken, and of systems modifications and alternative operating modes that should be tried. However, the principal objective of the first major experiment will be a very practical one: to demonstrate a new, nonpolluting, geothermal-energy system that might be producing commercial power within about 10 years, and that could within about 20 years contribute significantly to the solution of the energy and pollution problems of the world.

Appendix A: Comparison of Steam- and Water-Dominated Energy Transport from Reservoir

Steam cycle. Superheated steam from a reservoir at a temperature of 640°F and a depth of 9,842 feet flows to the surface through a pipe of inside diameter d (inches), arriving as dry saturated steam at pressure P. From saturated-steam tables we find the entropy at pressure P. Since the flow is isentropic, superheated-steam tables will yield the pressure corresponding to the same entropy at the reservoir temperature. From pressure and entropy, the specific volume $\widetilde{V}$ may be found by interpolation. The empirical Babcock formula for steam flow (Crane Co., 1957) is

$$\delta P\,(\text{psi}) = 0.47\,W^2 L\widetilde{V}[(d+3.6)/d^6] \qquad \text{(A1)}$$

where L is length in feet and W is flow rate in lb/sec. The results of calculations using (A1) under the assumptions and conditions given above are shown in Table A1.

In these calculations the length L was divided into 10 equal increments and $\widetilde{V}$ was recalculated for each increment. To calculate the flow rate for 100 Mw (94,826 Btu/sec), we calculate the difference in enthalpy between satu-

TABLE A1

Stem-Cycle Calculations for Given Assumptions and Conditions

Parameter	Assumed surface pressure (dry saturated steam) 1,000 psi	800 psi
Entropy [Btu/(lb°F)]	1.3897	1.4153
Enthalpy	1,191.8	1,198.6
Enthalpy after isentropic expansion to 150°F	834.1	849.8
Available energy (Btu/lb)	357.7	348.8
Flow (lb/sec)	265.1	271.8
Initial specific volume $\tilde{V}$	.3232	.3767

rated steam at the surface pressure P and the same steam after it has been isentropically expanded to a temperature of 150°F (the turbine-outlet temperature), and divide the result into 94,826 Btu/sec. It is seen (Fig. 8) that a 12-inch pipe is not quite large enough to transport 100 Mw under the assumed conditions, but that a 13-inch pipe will manage.

Water cycle. Water at the reservoir temperature of 640°F flows to the surface through a 12-inch pipe. After passing through a heat exchanger, the water flows down a similar pipe at a temperature of 150°F. The pressure difference at the bottom of the hot- and cold-water pipes is 1,400 psi (Fig. 5). Part of this differential will appear as pressure drop across the heat exchanger and across the reservoir. We may assume that 1,000 psi is available to overcome the impedance of the pipe, by pumping if necessary. The viscosity of water at 640°F is about one-twelfth that at 60°F, as may be seen from Fig. 3; but at 150°F, viscosity is approximately one-half that at 60°F. We may take this disparity into account, within the accuracy of this water/steam-flow comparison, by considering the impedance of only one of the pipes and assuming that the water is at 60°F. Hazen and Williams's formula (Crane Co., 1957) for the flow of water in a smooth pipe

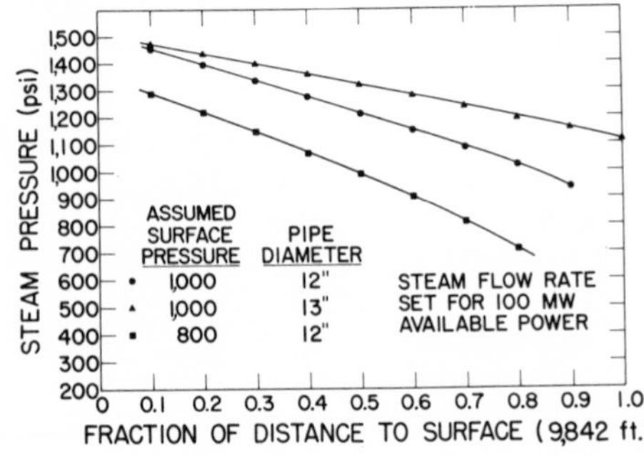

Fig. 8. Variation in steam pressure with distance from surface.

of length L (feet), diameter d (inches), and pressure difference $P_1 - P_2$ (psi) is

$$Q\ (\text{lb/sec}) = 8.59d^{2.63}[(P_1 - P_2)/L]^{0.54} \qquad \text{(A2)}$$

If we assume that $L = 9{,}842$, $P_1 - P_2 = 1{,}000$ psi, and $d = 12$ inches, then $Q = 1{,}721$ lb/sec. Since the enthalpy difference between 640°F water and 150°F water is 560.7 Btu/lb, the available energy is 964,982 Btu/sec, or 1,000 Mw, as contrasted with 100 Mw in the case of steam.

Appendix B: Void Generated per Million Btu of Heat Energy Withdrawn from Rock

If the specific heat of rock is c [Btu/(lb°F)], its density is ρ (lb/ft³), and its volume coefficient of expansion is B (per °F), then the withdrawal of $Q \times 10^6$ Btu of heat energy from a volume V produces a temperature drop $\delta T = Q \times 10^6/(\rho c V)$. Since this change of temperature causes a thermal contraction $\delta V/V = B\delta T = 10^6 BQ/(\rho c V)$,

$$\delta V = 10^6 BQ/(\rho c) \qquad \text{(B1)}$$

Typical values of the constants (Ingersoll, Zobel, and Ingersoll, 1954; Geological Society of America, 1950) are $B = 13.8 \times 10^{-6}$, $c = 0.19$, and $\rho = 168$. Inserting these values in (B1), we have $\delta V/Q = 0.43$. If 100 Mw of power is withdrawn for 24 hours, $Q = 8{,}191$, and the void is 3,541 ft³, requiring 26,559 gallons of make-up water.

Appendix C: Temperature at Which Thermal Fracture Occurs, vs. Depth

Consider a mass of rock in which the temperature gradient, directed downward, is G. At a depth h (ft), the temperature before cooling is

$$T(°\text{F}) = Gh + T_s \qquad \text{(C1)}$$

where T_s is the surface temperature, averaged over the entire year.

From hydraulic-fracturing experience, we note that fracture occurs at some multiple of hydrostatic pressure, say at $b \times 0.433h$. An average value of b is about 1.5 (Perkins and Kern, 1961). Since we assume that at least full hydrostatic pressure is operative, an additional tension, arising from thermal contraction, equal to $(b - 1) \times 0.433h$, is required to induce cracking. Let the thermal-expansion coefficient be a (per °F), and let Young's modulus be E. We consider a process in which the shrinkage, $\delta L/L = a(T_0 - T)$ (T_0 is the temperature to which the rock has been cooled), is followed by a stretching to bring the rock back to its original length, which requires a tension

$$E|\delta L/L| = Ea(T - T_0) \qquad \text{(C2)}$$

We equate this to the extra tension required for cracking, using (C1) for h, and arrive at

$$T_0 = [1 - (b-1) \times 0.433/(GEa)]T + [(b-1) \times 0.433/(GEa)]T_s \qquad \text{(C3)}$$

Some typical values (Geological Society of America, 1950) are $E = 8 \times 10^6$ psi, $a = 4.6 \times 10^{-6}$, $G = 0.1$ °F/ft, $T_s = 40$ °F, and $b = 1.5$. Inserting these values in (C3), we have

$$T_0 = 0.94T + 2.35, \text{ or } T_0 \simeq 0.94T \tag{C4}$$

In this example, cooling by only 36°F would crack 600°F rock. The coefficient of T in (C4) is designated K in the computer-simulation calculations (Harlow and Pracht, 1972). It becomes smaller if the thermal gradient is low, since rock of temperature T will lie at greater depths.

REFERENCES

Banwell, J., and T. Meidav. 1971. Geothermal energy for the future. Paper presented at 138th Annual Meeting of the American Association for the Advancement of Science, Philadelphia.

Barnea, J. 1972. Geothermal power. Scientific American, v. 226, no. 1, pp. 70–77.

Bodvarsson, G. 1961. Utilization of geothermal energy for heating purposes and combined schemes involving power generation, heating and/or by-products. Proc. United Nations Conf. on New Sources of Energy, Rome, 1961.

Crane Co. 1957. Flow of fluids through valves, fittings, and pipe. Technical paper no. 410, p. 3–3.

Gatlin, C. 1960. Petroleum Engineering. Englewood Cliffs, N.J.: Prentice-Hall, pp. 323–24.

Geological Society of America. 1950. Handbook of physical constants (published by the Society, reprinted at Ann Arbor, Mich.).

Harlow, F., and W. Pracht. 1972. A theoretical study of geothermal energy extraction, J. Geo. Res., v. 77, p. 7038.

Ingersoll, L. R., O. J. Zobel, and A. C. Ingersoll. 1954. Heat conduction with engineering, geological, and other applications. Madison: Univ. of Wisconsin Press, pp. 42, 142–43.

Legislative Reference Service, Library of Congress. 1970. The economy, energy, and the environment. Washington, D.C.: U.S. Govt. Printing Office, p. 116/45.

McClain, W. C. 1971. Seismic mapping of hydraulic fractures. Oak Ridge National Laboratory Report TM3502.

Perkins, T. K., and L. R. Kern. 1961. Widths of hydraulic fractures. Trans. AIME, v. 222, p. 944.

Robinson, E., R. Potter, B. McInteer, J. Rowley, D. Armstrong, R. Mills, and M. Smith. 1971. A preliminary study of the nuclear subterrene. Los Alamos Scientific Laboratory Report LA-4547, App. E.

Smith, R. L., R. A. Bailey, and C. S. Ross. 1970. Geological map of the Jemez Mountains, New Mexico (U.S. Geological Survey Map I-571).

Stepanoff, A. J. 1948. Centrifugal and axial flow pumps. New York: Wiley, pp. 381–83.

Tikhonov, A., and I. Dvorov. 1971. Development of research and utilization of geothermal resources in the USSR. Proc. Symp. on Dev. and Util. of Geothermal Resources, Pisa, Italy, 1970. New York: United Nations.

Van Orstrand, C. 1951. Observed temperatures in the Earth's crust. *In* B. Gutenberg, ed., Internal constitution of the Earth. New York: Dover, chap. VI.

15. Chemical Explosive Stimulation of Geothermal Wells

CARL F. AUSTIN AND GUY WILLIAM LEONARD

Just as there are more dry holes for oil, gas, or, now, geothermal fluids than there are productive holes, so also are there more marginal and submarginal producers drilled than initially prolific producers. Thus the operator, unless blessed with uncanny luck, will find himself after a time saddled with wells that fail to produce in commercial quantities. For the geothermal well this may be the result of insufficient fluid recharge to the reservoir, loss of porosity or permeability of the host rock through mechanical alteration, loss of permeability owing to mineral deposition, etc. The operator's problem then becomes how to stimulate a formerly economical producer to an acceptable rate of production. And indeed he will do well to consider also how to further stimulate an already commercial well, to help cover the losses from his marginal wells.

How best to establish and maintain a reinjection well is a third consideration. Initially, the fluid-acceptance rate might need improving; later, problems with rock alteration and precipitation may arise. These problems are amenable to explosive-stimulation techniques.

The Geothermal Environment

A discussion of the efficacy of explosive-stimulation methods must make certain assumptions about the nature of the environment within geothermal deposits. To illustrate the environmental diversity of the geothermal well, we quote Austin's (1966) definition: "Geothermal

Carl F. Austin is Research Geologist, Research Department, and Guy William Leonard is Assistant Technical Director for Development (Propulsion and Explosives) and Head, Propulsion Development Department, at the Naval Weapons Center, China Lake, California.

prospects are areas of steam emission, fumarolic- and volcanic-type gas emissions (other than from active volcanoes or contemporary lava flows), mineral springs of any temperature, mineral deposition indicating young-to-recent liquid and gas leakage of the preceding types, and young intrusions at modest depths, with emphasis on domes and caldera structures." Although the definition emphasizes igneous geology, the host geology per se is not defined or restricted. In a recent publication, Austin, Austin, and Leonard (1971) list five fundamental models of geothermal deposits: granite-stock heat sources, metamorphic-zone heat sources, basaltic-magma heat sources, wet geothermal-gradient heat sources, dry geothermal-gradient heat sources. Examples of the diversity of host rock can easily be shown for the granite-stock deposit type, for these have generally been the target for exploration drilling to date. Thus Larderello, in Italy, produces from anhydritic dolomite and cavernous limestones; Wairakei, in New Zealand, from volcanic debris, including pumice and crystal and vitric tuffs; and The Geysers, in California, from graywacke.

A more specific example of the diversity of what at first glance might seem a simple geothermal system would be a granitic stock area of eastern California. Host rocks for these deposits include granitic rocks ranging in composition from granite to diorite, crystalline metamorphic rocks (equivalent to gabbro in composition), fine-grained basic dikes, quartz veins and simple pegmatites, volcanic necks of rhyolitic or andesitic composition, olivine basalt, and areas of valley fill composed of debris from all these rock types. Production of geothermal fluids from a single deposit could be expected, for example, from fracture zones, collapse-caused breccia zones, or fractured volcanics. Fluid reinjection could be undertaken in any of these zones or, more likely, in fracture zones along the margins of the main geothermal-structural features. Our examination of ways in which explosive stimulation might be employed will use an eastern California deposit as its example.

Bore-Hole Enlargement

Chemical explosives can be used to "spring" a drill hole for the purpose of creating a larger-diameter hole at a selected horizon. Proper springing can result in a cavity of several times the original bore-hole diameter. Rather than resorting to under-reaming, the completion program could utilize explosives to create a larger-diameter bore hole in a producing zone. The advantage of explosives is that only a light ser-

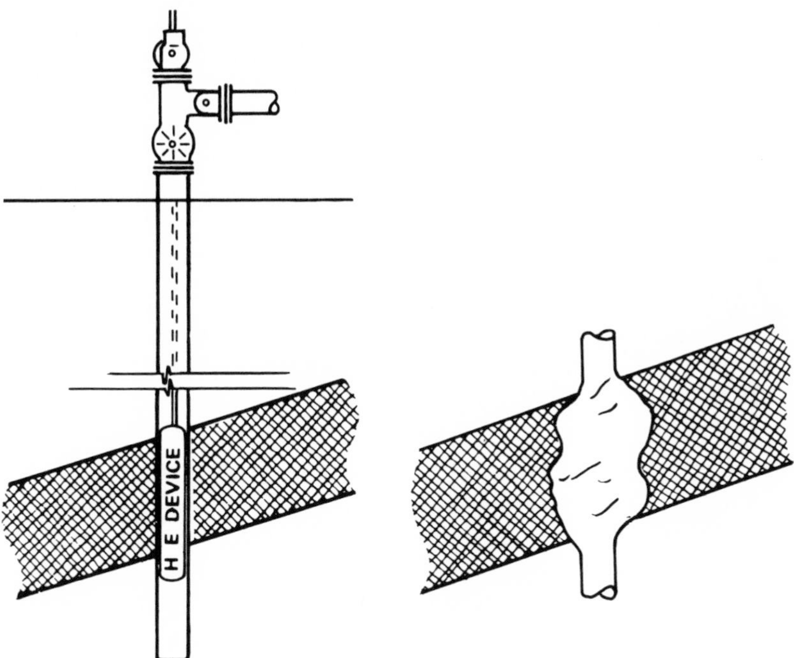

Fig. 1. Use of explosives to enlarge a bore hole in a producing zone: *left*, device lowered to producing zone; *right*, bore-hole radius increased in producing zone. A 12-inch bore hole is assumed in Figs. 1 through 5.

vice rig is needed and the probability of entrapping major "junk," debris, or equipment in the hole is minimal. Furthermore, the shock and vibration damage to the well perimeter due to springing the hole can of itself significantly enhance production rates. An impressive example of this phenomenon was examined by one of the authors recently when a tunnel initially excavated by drilling and blasting in a wet carbonate host rock underneath St. Louis was extended by boring; the bored portion was virtually dry, whereas the drill-and-blast section produced significant quantities of water.

In the strong granitic or metamorphic host an attempt at under-reaming to enlarge a bore hole could very easily result in the loss of the tools once the hole approximated the size of either individual rubble in a fracture zone or individual joint blocks. The occurrence of a fine-grained dike in the zone to be enlarged would present a further complication for the under-reaming process. Fine-grained dikes are quite strong, and though often associated with post-dike open fractures would

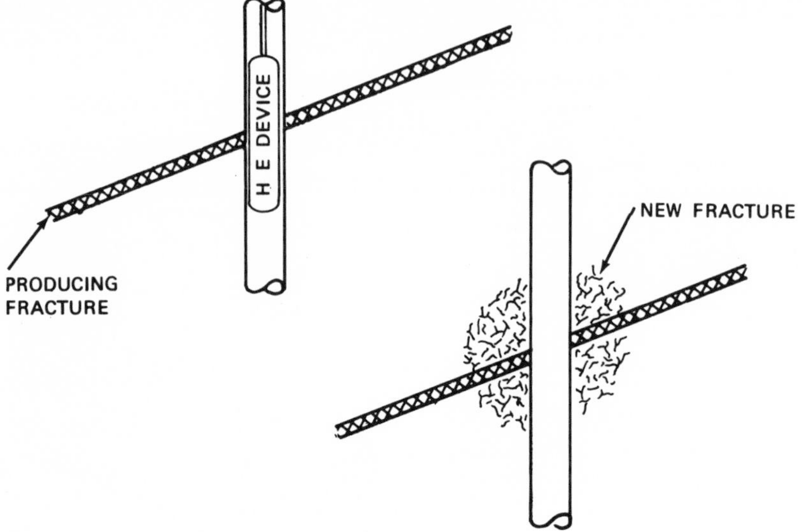

Fig. 2. Use of explosives to fracture the host rock adjacent to a crosscutting fracture: *left*, as drilled; *right*, after explosion.

present serious eccentric loads on any rotating tools. The use of explosives to enlarge the bore hole would avoid these problems. Figure 1 illustrates this approach. Springing is best accomplished with brissant explosives. The method should be used with caution in thin-bedded, steeply dipping horizons.

Permeability increase from stress waves. As already mentioned, blasting in a bore hole can increase the flow rates into the hole. The increase is accomplished by destruction of grain-to-grain bonding, by destruction of physical blockages of fractures and pores, or by the creation of entirely new fractures. The use of explosive charges for this purpose—i.e., to increase the effective diameter of a well through fracture propagation or to enhance the nearby porosity—has the advantages that the equipment required is minimal, little weight need be handled, and little or no junk is likely to remain in the hole. An example of the application of a simple blasting program in the eastern California environment is the fracturing of the host rocks adjacent to a crosscutting fracture (Fig. 2). A second example (Fig. 3) is the breakup of a mineral-cemented breccia zone. A third (Fig. 4) is the destruction of grain-to-grain bonding in a granitic host. These methods are equally appli-

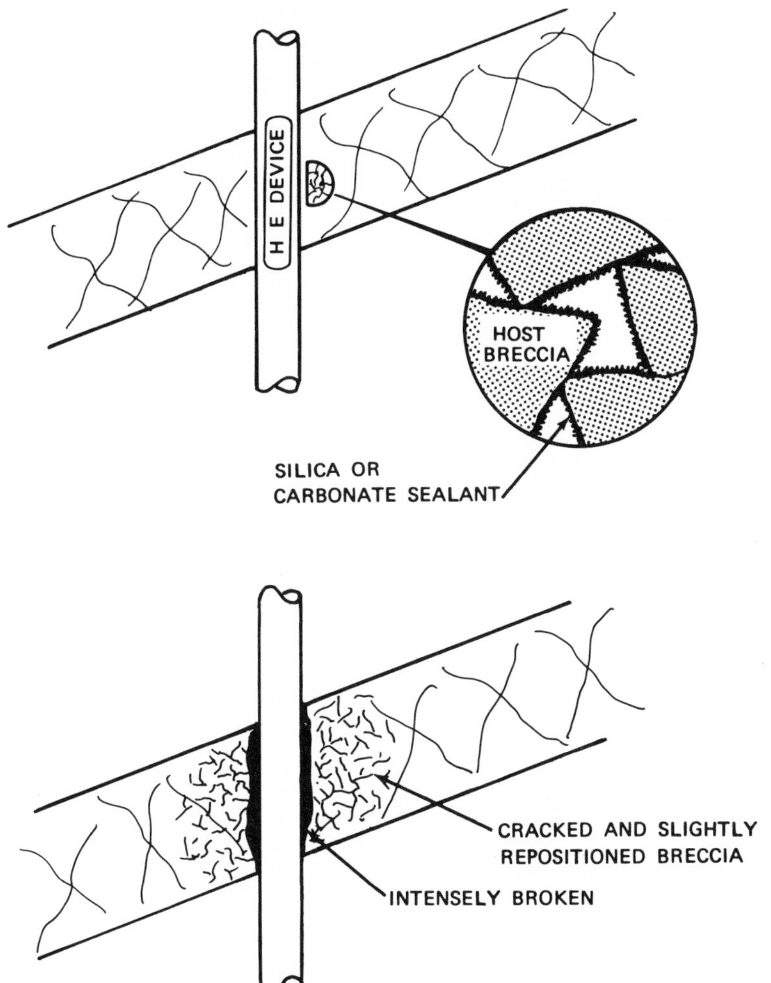

Fig. 3. Use of explosives to shatter brittle minerals plugging a breccia zone: *above*, as drilled; *below*, after explosion.

cable to the stimulation of a new well, the reactivation of an older well (i.e., the removal of plugging owing to reservoir alteration or precipitation), and the modification of a reinjection well.

Permeability increase from perforation (shaped charge). The conical shaped charge has a long history of use in the field of well completions. It can be used in lieu of bulk-explosive charges to increase the effective

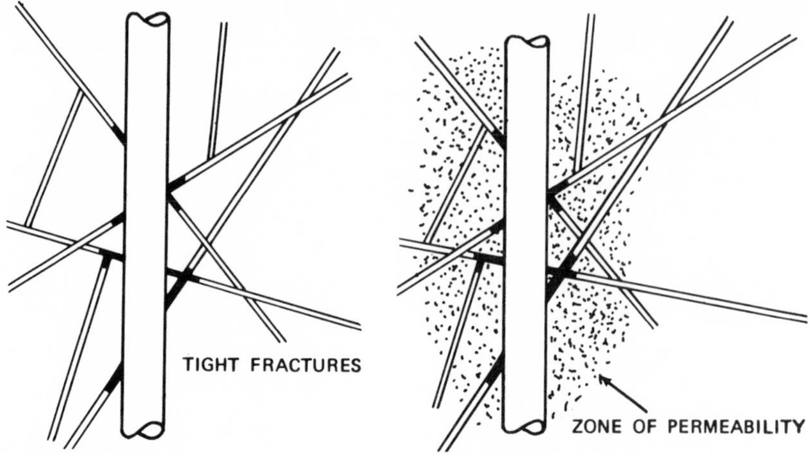

Fig. 4. Use of explosives to destroy grain-to-grain bonding to enhance production from adjacent tight fractures: *left*, low production due to plugging; *right*, stimulation due to bond failures. Dark areas are mineral plugging. The case shown illustrates stimulation to remove fracture plugging that has resulted from deposition during production.

diameter of a well by means of perforations at right angles to the borehole axis. Although often considered to be useful chiefly as a means of perforating casing and cement in order to provide access from the reservoir to the production tubing, the conical shaped charge can, under some conditions, increase the effective diameter of a well by a factor of 2 or 3. This is a familiar oil-field technique that is directly applicable to geothermal well-completion and well-stimulation programs (see Fig. 5).

Permeability increase from perforation (gun-fired projectiles). Bullets or "gun perforators" have been used to perforate rock in lieu of the jet from a conical shaped charge. Indeed, this method of perforation antedates the conical shaped-charge techniques in use today. The lower two parts of Fig. 5 are equally illustrative of both conical shaped-charge and bullet-perforation programs.

Rock-Mechanics Considerations

To be successful, the explosive stimulation of a geothermal well requires the careful understanding and matching of three major factors: (1) the method of explosive-energy delivery (liquid, solid, encased,

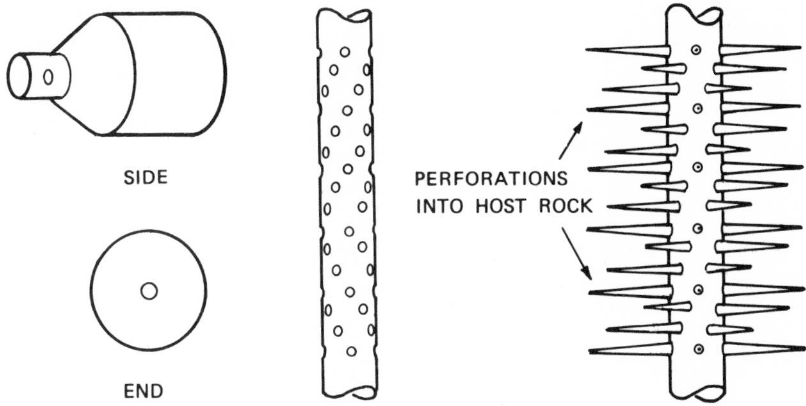

Fig. 5. Use of conical lined-cavity shaped charges to increase the effective diameter of a bore hole: *left*, a typical charge; *middle*, a string of charges ready to shoot; *right*, the end results.

bare, etc.), (2) the type of host rock (mineralogy, grain structure, degree of anelasticity, etc.), and (3) the fluid phase in the formation (steam or gas, liquid).

Stress-wave passage. The detonation of an explosive charge in a bore hole results in two distinct and recognizable zones in the host rocks: a *close-in* zone, where the rock fabric is destroyed by crushing, compaction, and phase changes; and a *distant* zone, where fracturing and grain-bond failure occur. We shall give due consideration to both zones, since the fluids produced by a bore hole must traverse both.

Stress-wave passage is the principal phenomenon of the distant zone. The effects of stress-wave passage on grain structure in igneous rocks have been investigated by Goldsmith and Austin (1964) and Goldsmith et al. (1968). The rocks studied in these two reports are typical of the rocks that form the bulk of the host and reservoir at a number of potential geothermal deposits in eastern California. These rocks, or at least rocks comparable in structure and stress-wave response, can be expected at a great many geothermal sites throughout the world. Let us examine how the results of these laboratory studies can be applied to the design of a stimulation program. Goldsmith et al. concluded that "for diorite, the attenuation is high and the dispersion is low. The ultimate strength of diorite and the static Young's modulus of diorite both decrease in direct proportion to the number of impacts previously experienced by the specimen." Since these tests were run on samples in air, the assumption is made that the zone to be stimulated either is gas-filled naturally or has been gas-filled as part of the stimulation program in order to control the rock-failure mechanisms. In contrast to diorite, these authors noted, "for leucogranite the attenuation is large, and pulses in it fluctuate drastically at the pulse front. The compressive strength of leucogranite appeared to be somewhat affected by the reflected tensile wave but did not appear to be influenced by the incident compressive wave." Further, "spessartite (dark, fine-grained dike rock) was found to be highly elastic, to exhibit virtually no attenuation or dispersion, and to be essentially unaffected by the repeated passage of stress waves."

These results, then, indicate that, given a bore hole in mixed granite rocks and dikes, the explosive fracturing of dikes that might be serious barriers to flow would be difficult if not impossible beyond the close-in zone. But given potential production in a zone of mixed leucogranite and diorite, extensive fracturing and grain damage could be accomplished by repeated explosions in the diorite zones. Indeed, repeated shocks were shown to have a general tendency to decrease both the value of Young's modulus and ultimate strength in the coarse-grained igneous rocks. Experimental data have shown that when a transient pulse with a stress level above a critical value traverses a rock, energy is absorbed by fractures, grain cleavages, or damage to grain-to-grain bonds. Thus, repeated explosions can be looked upon as a stimulation method per se or as a preliminary to hydraulic fracturing.

The high probability that geothermal deposits of the "basaltic-magma heat-source type," such as the basalts of Hawaii, will receive exploration interest in the near future warrants a close look at how this fine-grained rock behaves when subjected to stress waves. (See Austin et al. (1971) for the suggested testing of the potential geothermal deposit at and adjacent to the Kaneohe Marine Air Station on Oahu.) Tests by Goldsmith et al. (1968) have shown that "basalt shows slight attenuation and virtually no dispersion. Both the coefficient of attenuation and Young's modulus for basalt tend to decrease as the number of shocks experienced by the specimen, at a given initial shock level, is increased." The strength of basalt did not appear to be affected by the pulse intensity used in these tests, and basalt was shown to be an excellent example of the capacity of a brittle material to reach damage saturation with a given severity of repeated impacts. This implies strongly that attempts to produce extensive fracturing in a reservoir of basalt or comparable material may not be successful beyond the first or first few detonations.

Also worth noting is the fact that stress-wave-induced damage is highly localized in porous media (which we assume to be gas-filled). Tests by Austin, Cosner, and Pringle (1966) on scoria and comparable material (bulk density of 1 g/cm^3 vs. material density of 3 g/cm^3) have shown it to be a superb energy-absorbing medium.

Available literature on the subject of stress-wave-passage effects in rocks is distinctly limited with respect to types of rocks actually tested, and the conditions of testing seldom, if ever, resemble the conditions to be expected in a down-hole stimulation program. But broad concepts do exist. Some rocks are highly elastic, and damage saturation for a given charge size can be expected. By contrast, other rocks fatigue rapidly upon repeated stress-wave passage. Cores from potential producing zones should be obtained and tested under down-hole conditions, or the down-hole conditions should be modified in the zone of stimulation to ensure reasonable prediction accuracies for distant-zone effects. Figure 6 shows an explosively induced fracture in a diorite. Although the rock fabric is extensively damaged, only the megascopic fracture can be seen in the thin section shown. The ability of the damaged fabric to transmit fluids has not been tested to date under controlled conditions, although this is a vitally important consideration.

When anticipating or predicting the fracture behavior of rocks and their mineral constituents in the design of a stimulation program, the investigator must bear carefully in mind the fact that rock materials follow their own fundamental behavior. Thus in studies with diorite, Austin and Pringle (1964a) concluded that

1. In terms of specific experimental results, a medium- to coarse-grained intrusive rock such as diorite will yield abundant fractures with the appearance of multiple scabs when explosively loaded, but will fail to yield corner fractures of the type believed caused by stress-wave reinforcement.
2. In more general terms, the behavior of medium- to coarse-grained rocks subjected to impulsive loading cannot be simulated or duplicated by the use of metallic samples or by the use of nongranular plastic samples such as Plexiglas. These experiments show that attempts to model rock-material behavior with other than rock materials must be approached with extreme caution.

Fig. 6. Explosively induced fractures and grain damage in diorite. Average grain size in the photomicrograph is 3/16 inch, as indicated by the bar.

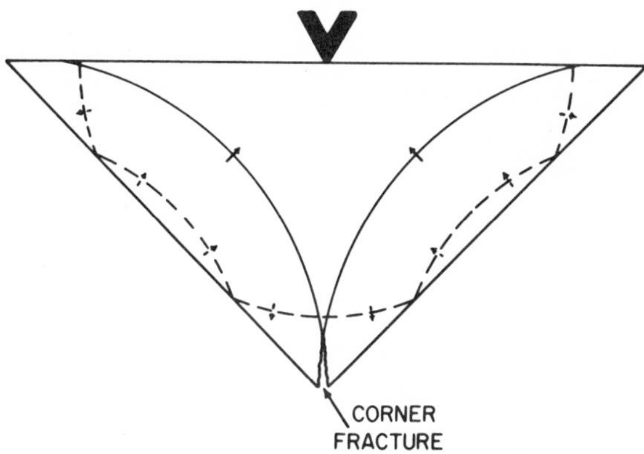

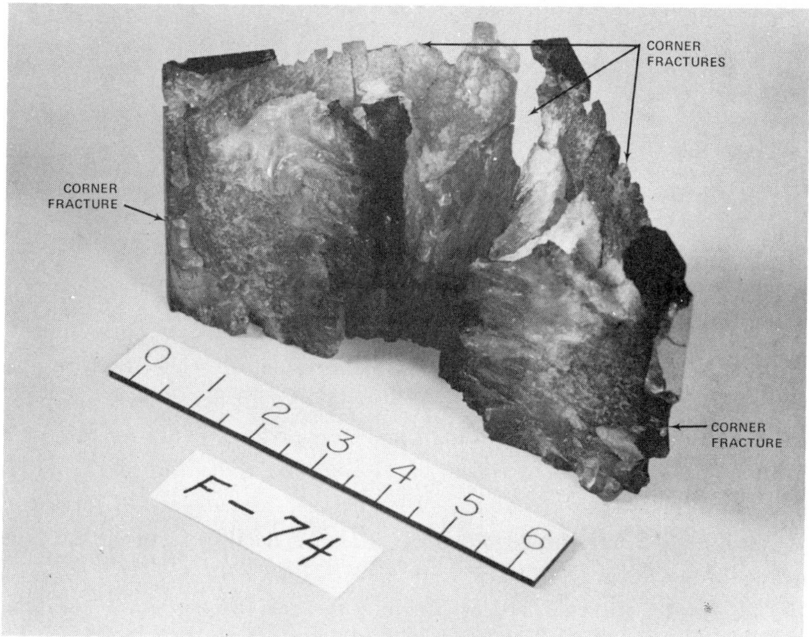

Fig. 7. Typical corner fractures due to stress-wave reinforcement, as seen in metals and in Plexiglas: *above*, predicted pattern; *below*, actual pattern in Plexiglas (the scale is in inches).

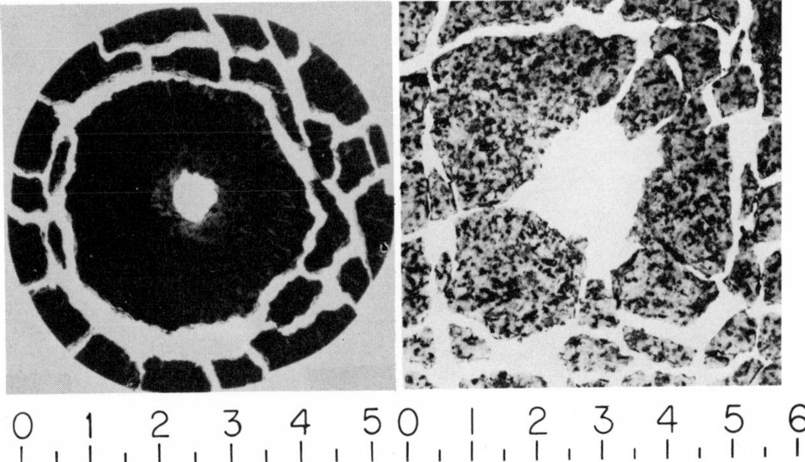

Fig. 8. Typical fracture patterns seen in explosively loaded diorite and limestone sheets. The scales are in inches.

Figure 7 shows the typical corner fracture of metals and of Plexiglas. Though widely cited in texts as a fundamental failure mechanism of rocks subjected to impulsive loading, this type of fracturing does not occur in the rocks tested to date. Figures 8 and 9 show the types of fracture patterns that should be anticipated in explosively loaded rocks, whether coarse-grained igneous rock, limestone, sandstone, or conglomerate, all of which have been studied experimentally. The fracture patterns shown are considered typical of what should occur in explosively loaded joint blocks with bedding planes or parting planes (Fig. 8), in tabular dikes or silicified zones, or in more massive joint blocks (Fig. 9). The fracture pattern of Fig. 9 is "expanding-conical" in form and is combined with fractures caused by tensile reflection from the flat face opposite the point of initiation.

The reason for the lack of tensile-wave reinforcement in common rock materials has been discussed on theoretical grounds by Rinehart (1964), and is further illustrated in work by Rinehart and Pearson (1963). Briefly, the value of Poisson's ratio for rocks is in the range of 0.2–0.25, and is such that, at large angles of incidence, shear waves are reflected rather than tensile waves. Some minerals, such as quartz, *do* undergo corner fracturing in single crystals but *do not* in crystalline aggregates.

Fig. 9. Fracture pattern in a limestone block that has been explosively loaded. Detonation has proceeded upward from the bottom. The block initially measured 6 inches by 6 inches at the base by 5 inches tall.

Close-in effects in rock have received relatively little attention, with the exception of studies in the field of nuclear blasting. Key features that are of importance in explosive stimulation can be recognized and illustrated, however. Close-in failure in rocks with gas-filled pores is generally the result of pore-space collapse. If this collapse is extensive and the resulting damaged zone is symmetric and confined, i.e. not cut by radial fractures, the collapse zone will remain intact in a metastable and highly stressed state. This zone can be highly impermeable, so that the net result of a symmetric blast may well be a loss of production, despite extensive formation fracturing and disaggregation. This problem can be overcome by the use of shaped explosive charges, which cause asymmetries, by the use of repeated blasting to produce a hole large enough to result in spontaneous collapse of this zone, or by exploiting natural asymmetry in the bore-hole-wall rock properties to produce spontaneous collapse. Figure 9 illustrates these phenomena. Close-in effects are difficult to duplicate in the laboratory, since once fracturing extends to the sample edge, confinement is lost and the potentially impermeable damaged zone collapses spontaneously. This zone is in fact not present in Figs. 8 or 9. Other close-in effects, including melting, are also possible in certain rock types, as are crystalline phase changes. The major problem close in, however, appears to be inducing collapse of the potentially impermeable zone of "crushed" rock.

Conical shaped-charge jets. The use of conical shaped-charge jets in rock has been described in detail by Austin (1959) and by Austin and Pringle (1964b). In these studies, rock was found to respond to the shaped-charge jet in three fundamental modes: the crushed-zone mode, the compaction mode, and the stable mode.

The effect on the rock of a penetration that follows the crushed-zone mode is comparable to the effect of detonating an explosive charge in a small bore hole in a semi-infinite target block. The resulting hole is lined with a crushed, metastable, dense zone whose permeability in the down-hole environment is apt to be very low. Rocks in which this mode of penetration occurs are those that are coarsely crystalline, e.g. diorite, adamellite, granite, and nonvesicular forms of basalt and rhyolite. If the symmetry of the crushed zone can be destroyed, the zone will collapse spontaneously, yielding a hole larger than the original penetration and a hole whose walls would appear far more permeable than the crushed-zone portion. Symmetry destruction is often spontaneous, as in the case of the holes illustrated in Fig. 10, where the dynamic strength of the rock is markedly different in different directions along the penetration path. If extensive stimulation is to be attempted in rocks following the crushed-zone mode, a design should be developed for an asymmetric jet that would encourage spontaneous collapse of the crushed zone. In the case of Fig. 10, the original hole was both large and elliptic, so that in the course of 5 minutes the hole enlarged

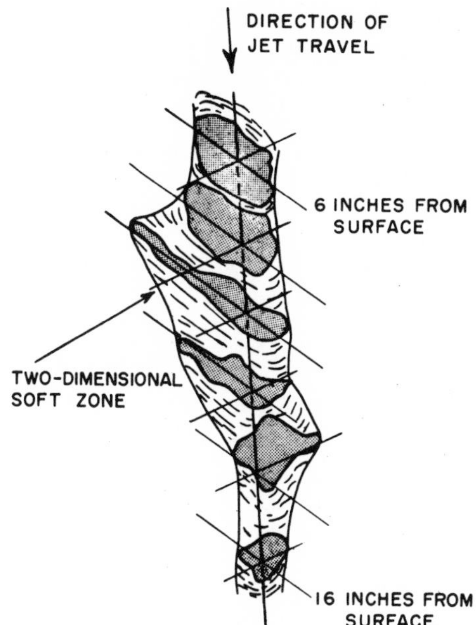

Fig. 10. A conical shaped-charge hole-shot into a hydrothermally altered granite rock having variations in dynamic strength in various directions.

from some 2 inches in diameter to nearly 3 inches as the crushed zone collapsed inward.

Compaction-zone penetrations would be found typically in such rock as sandstone or pumiceous tuff. In this type of rock target, the hole is formed largely by pore-space collapse, but although a large, dense collapse zone is formed, this zone stores little or no energy and spontaneous collapse is unlikely. A stimulation program using conical charge jets in rocks following the compaction mode must rely on differential pressure and flow to wash the compaction zone out of the penetration.

Stable-mode holes have been found only in limestones to date, but should be anticipated in other massive carbonates, including dolomite and such hydrothermal carbonates as ankevite. The formation of a stable-mode hole in a massive or semi-infinite target does not result in damage to the host rock beyond a thin film (1 mm or less) of calcined rock, lightly plated with liner metal and liner metal oxides. This type of penetration will provide an enlarged well perimeter but will not significantly modify the host-rock properties.

Lined-cavity shaped-charge jets do at times cause fracturing in rock targets. If the target is a breccia, extensive fracturing may occur in the first solid piece encountered, but if a jet enters a fracture at a shallow angle of incidence or encounters a slanting rock face, the jet will be

Fig. 11. Crushed zones in rhyolite, with fractures filled with liner metal and liner metal oxide.

deflected and no significant penetration will occur. The critical angle for deflection depends on both the rock type and the jet design; tests against diorite in air showed extensive jet deflection at 20° and 30° angles of incidence.

Two interesting groups of fracture are seen in conical shaped-charge penetrations: one occurs in the crushed zone, the other in limestones that have minor freedom of motion (usually on joint or bedding planes). Figure 11 shows a rhyolite crushed-zone sample obtained by overcoring. Here, the crushed zone failed as a solid and the resulting fractures were then filled with injections of liner metal and liner metal oxides. The resulting zone would appear to be of a lower permeability than that desired for stimulation. Figures 12 and 13 show fracture patterns in limestone, with the fractures again filled with injections of both liner metal and liner metal oxides. In the fractured limestones, the apparent prerequisite for the formation of this pattern was the presence of a nearby open joint. Figure 14 shows idealized cross sections of three principal types of conical shaped-charge penetrations into rock.

Interestingly enough, experiments have shown that gas-filled, poorly consolidated rocks tend to fail by means of pore-space collapse, whereas

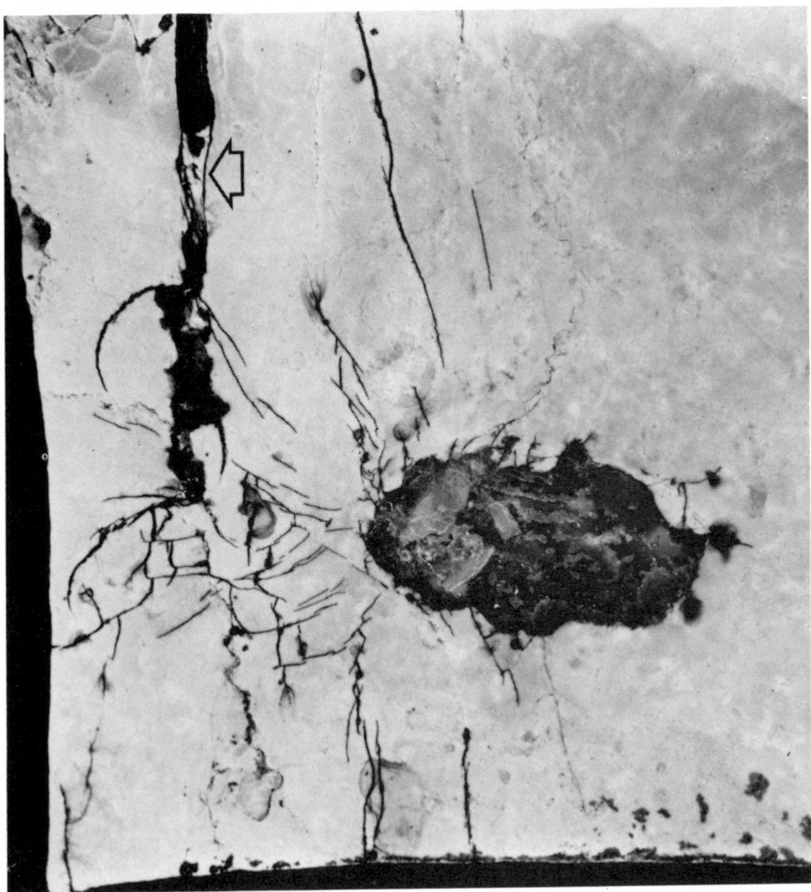

Fig. 12. Fractures developing asymmetrically in limestone as the shaped-charge hole approaches an open joint. Fractures are filled with liner metal and liner metal oxides. The arrow indicates a natural fracture; the area shown is about 3½ inches wide.

fluid- (liquid-) filled rocks tend to fail by shear. The implication is that wet, hydrothermally altered rocks may behave very differently from the dry rocks studied and reported on in the literature. Extensive testing needs to be done in this area.

Gun-fired projectiles. Gun-fired projectiles (gun perforators) offer two distinct advantages over jets from conical charges: established penetration formulae permit reasonably accurate prediction of penetration depth; and bullets can be shaped so as to favor the collapse of the compaction zones. Crushed-zone collapse appears to be an unlikely

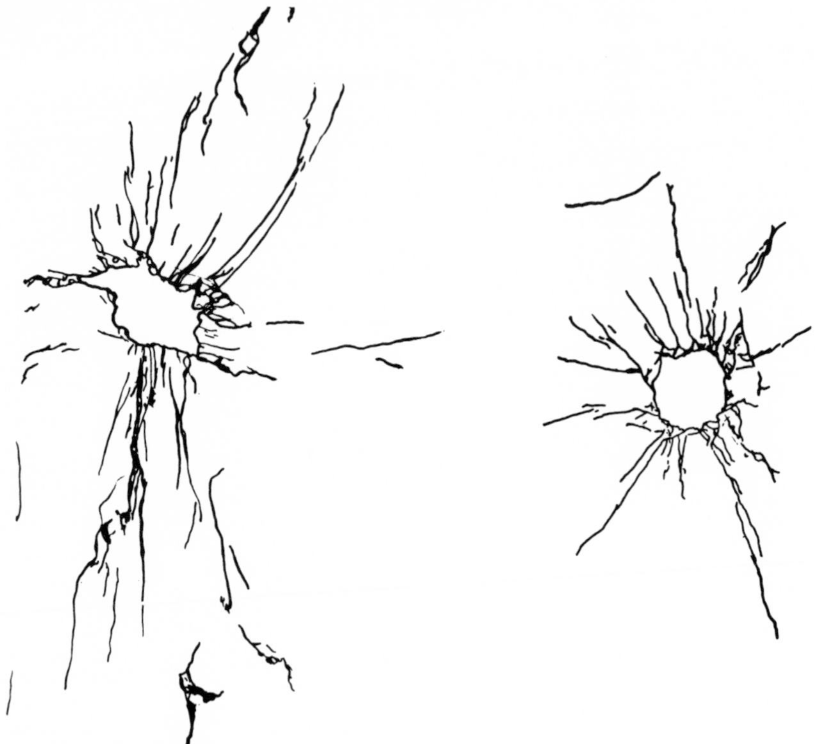

Fig. 13. Fracture pattern developed in jointed and relatively thin-bedded limestone. Fractures are filled with liner metal and liner metal oxides. Orientations: *left*, from near hole bottom, looking up the hole (i.e. looking toward the source of the jet); *right*, 2.15 inches farther up the hole, looking down-hole. The fracture at the left is about 1 inch broad.

outcome, owing to the limited penetrability of bullets into the stronger igneous and metamorphic rocks that follow this mode. Square bullets appear especially attractive for the control of compaction-zone collapse. Although jet penetrations are often cited as being junk- or metal-free, this condition is difficult to achieve in most rock types, and there is no question that a bullet perforation will generally plug its hole unless the bullet is soluble, meltable, or corrodible, or has some other suitable property.

An excellent study on rock penetration by small-caliber bullets is that of Tolch and Bushkovitch (1947). They use the equation

$$X = K_p(W/d^2)d^{1/6}(V/1{,}000)$$

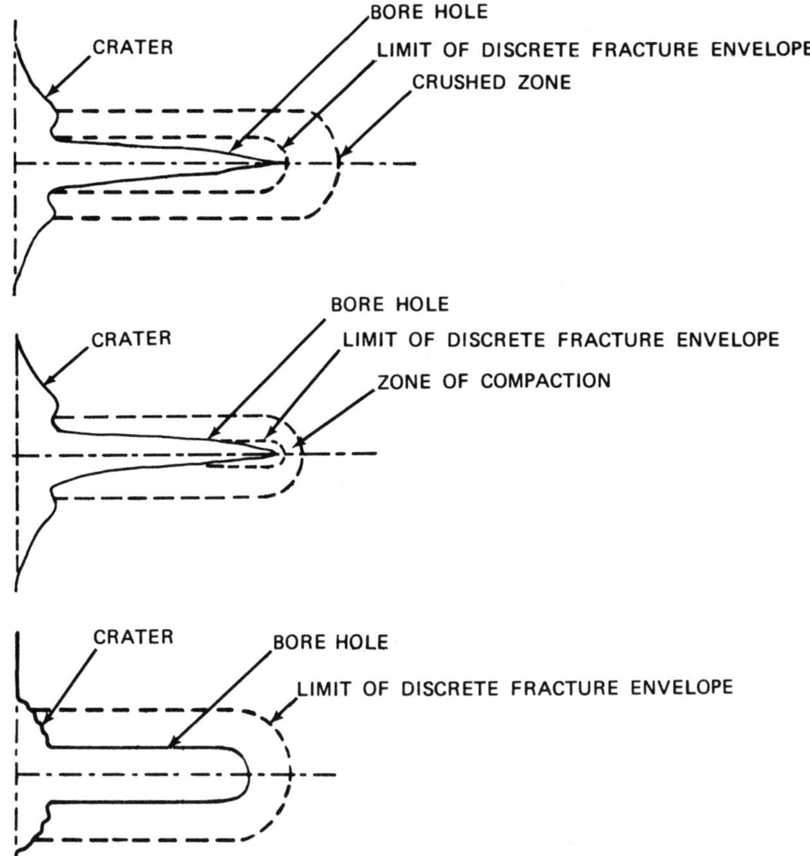

Fig. 14. Idealized cross sections of the three principal types of conical shaped-charge penetrations: *above*, crushed-zone mode; *middle*, compaction mode; *below*, stable mode.

where

X = penetration, in inches
K_p = constant dependent on rock type
W = weight of projectile, in lbs
d = projectile diameter, in inches
V = striking velocity, in ft/sec

A second equation, shown here in a widely used form, is that given by Robertson (1941):

$$X = (P/2gbi) \ln [1 + (b/a) V^2]$$

where

X = maximum penetration depth, in ft
P = sectional pressure, in psi ($= W/A$)
W = weight of projectile, in lbs
A = maximum cross-sectional area of projectile, in in^2
g = acceleration due to gravity, in ft/sec^2
b = inertial coefficient, in $lb\text{-}sec^2/in^2\text{-}ft^2$
i = dimensionless shape factor for penetrator
a = shatter strength, in psi
V = striking velocity, in ft/sec

The Explosive

Current explosive technology permits the construction of explosive-stimulation charges for temperatures up through 400°C (752°F); at higher temperatures, such devices become extremely expensive. Whether a stimulation is to be by stress-wave effects, jet penetration, or bullet penetration, the energy source is presumed to be an explosive whose stability and performance are compatible with the conditions of intended utilization. Geothermal-well conditions that can significantly affect explosive choice include such temperature effects as melting, which can totally alter the anticipated explosive geometry, and explosive dispersal and cookoff, which is the spontaneous burning, explosion, or detonation of an explosive heated above some specific temperature and pressure prevailing at the location of the explosives. Moreover, the well gases or fluids can interact with the explosive, resulting in sensitization or desensitization of the explosive, if it is not completely isolated from the fluids. It is also important to consider the potential for sensitizing between the explosive and the materials used in well boring, including joint compounds and lubricants.

The possibility of post-shot toxicity or contamination must also be considered; thus charges with high outputs of mercury or other poisons (whether biologic or metallurgic) should be carefully controlled with respect to quantity. And the disposal of unused charges or duds must be duly considered. Matching the pulse shape to the type of rock was taken up in earlier sections of this paper; thermal stability, beyond the considerations just mentioned, will be taken up in detail below, in connection with the use of encased explosives.

Uncased explosives. These explosives, especially in the form of slurries and liquids, offer the very real advantages of intimate bore-hole wall contact and, in some instances, injectability as well into rock pores; but predicting the behavior of an unconfined explosive is difficult. An excellent example of the problem is the recent experience of the metal-mining industry with the use of certain types of slurried explosive in drill holes containing pyrite in the hole walls, resulting in an apparent instability. The stabilities of explosive melts, slurries, and liquids with dense metal brines, with such sulfide minerals as pyrite, and with the corrosion products of drilling equipment are unknown or at best poorly understood, though the hazardous and sensitizing nature of a number of heavy metal oxides and salts is well known, as noted by Von Egidy et al. (1960) and by Davis (1956).

Another major problem with uncased explosives is the high cost of these materials in a form that will survive at temperatures as high as 400°C. Such explosives have been made in laboratory quantities, but their costs vary from $150 to $1,800/lb. At the same time, given a commercial application and reasonable market potential, costs could be reduced to the range of $5/lb in 100-lb lots for some of the very stable materials, and in some instances could approach the $0.16/lb cost of TNT. One of the more heat-resistant explosives is now commercially available from DuPont under the name TACOT.

The density and solubility of an uncased explosive can also lead to problems: the explosive may tend to float or sink in the localized down-hole environment; and, with changing conditions up the well, it is quite conceivable that the explosive could precipitate and thus accumulate in unwanted quantities at some unexpected location.

Encased explosives. The fully encased explosive, offering as it does freedom from many down-hole uncertainties, is always attractive as a stimulation tool. It is predesigned, and because it can be supplied with operating instructions a minimum of expertise is needed in the field.

For fully encased charges today, TNT is readily available, inexpensive, and relatively stable. Though TNT melts above 80°C (176°F), its properties as a liquid explosive are fairly well known. Liquid TNT can be detonated with a lead-azide detonator and an HMX booster. Since most geothermal wells being drilled and produced today are located in areas of natural surface-fluid leakage and are intimately involved with near-surface ground water, they are in the epithermal-temperature range of 50°–200°C (120°–390°F). For this range, TNT needs only a light case to prevent intermixing of the well fluids and the explosive.

A few wells today have tapped mesothermal fluids with temperatures of 200° to 300°C; and in the future, test wells may encounter hypothermal fluids with temperatures in the range of 300° to 550°C. (In the mid-1960s, for example, the Union Oil Company Well, River Ranch No. 1, at over 8,000 ft depth, encountered temperatures of approximately 370°C.) Although TNT is stable for long-term storage at 185°, decomposition begins above 190° and becomes rather rapid above 200°. At a heating rate of 20°C per second, the decomposition becomes very rapid above 240°, with the charge either burning or exploding before reaching 300°. Because of the limited thermal stability of TNT at temperatures above 200°, where wells encounter temperatures beyond this point, an insulated case can be used, or the well can be quenched prior to explosive emplacement, or the two methods can be employed together. Quenching, however, may induce both pore-space precipitation and significant grain damage in the host rocks near the bore hole, thus modifying the energy transfer into the host. The presence of the case will also modify the energy transfer into the bore-hole walls. In all instances case design must be such as to minimize the junk left in the hole.

To demonstrate the feasibility of using TNT for well stimulation with down-hole temperatures of 400°C, tests were conducted at the Naval Weapons Center in 1972 using samples of TNT encased in Teflon. Teflon was chosen because of its low thermal diffusivity, its compatibility with TNT, and its capacity to resist rapid decomposition at temperatures up to 440°. Also, the use of Teflon avoids introducing metallic junk into the well. For these tests, the Teflon containers were machined out of 1.75-inch-diameter solid stock, producing selected inside diameters from wall thicknesses of 0.125, 0.25, 0.375, and 0.50 inch. The Teflon cylinders were filled with TNT to within 0.5 inch of the top, and the top was then closed with a 1-inch-thick Teflon cap. Thermocouples were placed in the center of the explosive and also at the interface between the explosive and the Teflon wall. Each container was then placed in an aluminum tube furnace and heated. From the experimental results obtained for the 1.75-inch-diameter containers, the predicted cookoff times were calculated for various larger-diameter containers. These preliminary findings indicate that a 0.25-inch Teflon wall is sufficient to assure protection in excess of 30 minutes for cylinders of 7 inches or larger diameter. Even longer times before cookoff may be obtained by increasing the wall thickness. Thus the selection of an appropriate thickness for the Teflon wall assures reasonable handling times and enables the emplacement and detonation of TNT charges in unquenched wells at up to 400°C (750°F).

Extremes of environmental temperature will be encountered when attempts are made to put current speculation over the extraction of heat from magma chambers into actual practice, as for example in Hawaii. Here, temperatures can be expected in the range of 650° to 1,200°C in the zone of heat exchange. For temperatures above 400°, explosive charges of TNT could be used if adequately protected. Such protection, however, is good for only a limited time and both reduces the explosive volume available and shields the bore-hole walls. The best approach with current technology would appear to be insulation combined with ablation.

If longer times are required than can be achieved with encased TNT in the mesothermal-temperature range of 200° to 300°C, then a main charge of TACOT could be used, with the TACOT serving as a heat sink and insulator for a lead-azide detonator with a NONA booster. In the hypothermal-temperature range of 300° to 500°, main charges of NONA could be used at temperatures of up to 400°. A vented ablative system would be needed with a NONA charge, however, to protect the lead-azide detonator.

Conclusions

Only three broad conclusions are given:

1. Explosive stimulation of geothermal wells appears to be feasible with today's technology.

2. The study of the behavior of rocks subjected to impulsive loading is in its infancy, especially at elevated temperatures and with fluid saturation.

3. The technology of high-temperature explosives is likewise in its infancy; perhaps the stimulation of geothermal wells will provide the impetus for advances in the state of the art as well as reductions in cost.

REFERENCES

Austin, C. F. 1959. Lined cavity shaped charges and their use in rock and earth materials. Bulletin 69, New Mexico Bureau of Mines, Socorro, N.M.

Austin, C. F. 1966. Selection criteria for geothermal prospects. Papers presented at AIME Pacific Southwest Mineral Industry Conference, Sparks, Nev., May 5–7, 1965, pt. C, report 13.

Austin, C. F., W. H. Austin, Jr., and G. W. Leonard. 1971. Geothermal science and technology: A national program. Technical series 45-029-72, U.S. Naval Weapons Center, China Lake, Calif.

Austin, C. F., L. N. Cosner, and J. K. Pringle. 1966. Shock wave attenuation in elastic and anelastic media. Trans. Soc. Mining Engineers.

Austin, C. F., and J. K. Pringle. 1964a. Comments on explosively formed fractures in rock. Trans. Soc. Mining Engineers.

——— 1964b. Detailed response of some rock targets to jets from lined-cavity shaped charges. J. Petrol. Techn.

Davis, T. L. 1956. The chemistry of powder and explosives (New York: Wiley).

Goldsmith, W., and C. F. Austin. 1964. Some dynamic characteristics of rock. *In* H. Kolsky, ed., Stress waves in alesastic solids (Berlin: Springer Verlag).

Goldsmith, W., C. F. Austin, H.-C. Wang, and S. Finnegan. 1968. Stress waves in igneous rocks. J. Geophys. Res., v. 71, no. 8.

Rinehart, J. S. 1964. Influence of Poisson's ratio on the fracture patterns generated by impulsive loads. NOTS TP 3624. Naval Ordnance Test Station, China Lake, Calif.

Rinehart, J. S., and J. Pearson. 1963. Explosive working of metals (New York: Pergamon-MacMillan).

Robertson, H. P. 1941. Terminal ballistics. Committee on Passive Protection Against Bombing, National Research Council, Washington, D.C.

Tolch, N. A., and A. V. Bushkovitch. 1947. Penetration and crater volume in various kinds of rocks as dependent on caliber, mass, and striking velocity of projectile. Report 641, Ballistic Research Laboratories, Aberdeen Proving Ground, Md.

Von Egidy, A., M. Finger, M. Hill, D. Ornellas, E. Ellison, and J. Kury. 1960. A new liquid explosive, NTN. UCRL-5861, Lawrence Radiation Laboratory, University of California, Livermore, Calif.

16. Environmental Aspects of Nuclear Stimulation

GARY M. SANDQUIST AND GLENN A. WHAN

The United States is experiencing an energy crisis that is particularly acute in the electric-power sector of the nation's energy economy. The 1970 National Power Survey, assembled by the Federal Power Commission (FPC), predicts that the nation's electric-energy requirements in 1990 will nearly quadruple to 1.26 million megawatts of generating capacity from the 0.34 million megawatts consumed in 1970. The increase is predicated upon both a growing population and an increasing per capita consumption.

The FPC further predicts that recurring and spreading power shortages may occur, owing to delays in expanding electrical-generating capacity caused by labor disputes and low labor productivity, inadequate forecasting of power needs, changing regulatory standards, litigation and licensing delays, and strong vocal concern over environmental pollution and degradation resulting from the production and utilization of energy. In particular, the consumption of fossil fuels by electrical utilities accounts for over 50 percent of the oxides of sulfur in the country's air-pollution load.

Public concern over environmental deterioration was awakened in the early 1960s and today amounts to almost a crusade for a national commitment to improve and preserve environmental quality. This commitment has resulted in extensive government legislation at both the Federal and local levels. The national will to force man's energy technology to conform with his environmental conscience is reflected in

---

Gary M. Sandquist is Professor of Mechanical Engineering, University of Utah, Salt Lake City; and Glenn A. Whan is Professor of Nuclear Engineering, University of New Mexico, Albuquerque.

the following statement of the 1970 FPC National Power Survey Task Force on the Environment: "In our time, we have seen commanding new social values arise, and among the most important of these is a new respect for the conservation of the environment, and the need to adapt our energy resources and supply to the restrictions this imposes upon us."

Currently, fossil fuel and hydroelectric resources provide the bulk of our electric power, but electric power generated from geothermal-energy resources is claimed by many to be less injurious to the environment than any other currently feasible method of thermal power generation. Operating experience with the few, modestly scaled geothermal plants in use around the world seems to justify this claim.

The number and magnitude of naturally occurring geothermal sites, such as The Geysers, 80 miles north of San Francisco, that can feasibly be developed appear to be limited (Hubbert, 1969; White, 1965) and incapable of supplying a significant portion of the nation's electric-power needs. However, if under the Plowshare Program established by the U.S. Atomic Energy Commission (AEC) in 1957, the stimulation of potential geothermal sites by nuclear explosives can be shown to be safe and environmentally tolerable, then the impact of geothermally generated electricity upon the nation's energy economy could be greatly amplified. But the environmental effects of nuclear stimulation are incompletely understood, and the issue remains broad, vital, and charged with emotion. Inquiry into the environmental risks involved must be both extensive and intensive. Bowen (this volume) discusses the environmental effects of producing electric power from natural geothermal sources; the additional impact of nuclear stimulation, which has not been experimentally investigated, will be considered here.

Although there are at present no operational nuclear-stimulated geothermal sites, the AEC has accumulated extensive theoretical and experimental data related to underground nuclear tests. Since the signing of the Limited Test Ban Treaty in August 1963, the United States has announced about 250 underground nuclear detonations with yields ranging from less than 1 kiloton to more than 1 megaton of TNT equivalent. Most of these tests have been conducted at the Nevada Test Site. Three Plowshare tests were conducted outside the Nevada Test Site: the "Gnome" event near Carlsbad, New Mexico, in 1969; and the two gas-stimulation experiments—"Gasbuggy," near Farmington, New

Mexico, in 1967, and "Rulison," near Grand Valley, Colorado, in 1969. Other gas-stimulation experiments, such as "Rio Blanco" and "Wagon Wheel," are scheduled for the future. These experiments provide some foundation upon which to develop and predict the environmental consequences of using nuclear explosives underground.

For nuclear stimulation of geothermal resources, we shall assume here that the primary use of geothermal heat is to generate electric power. The manifold environmental aspects of geothermal-energy extraction stimulated by nuclear explosives may then be conveniently classified into three distinct time periods associated with the life cycle of a geothermal site: pre-plant operations, plant operations, and post-plant operations. Each of these will be taken up in turn.

Pre-Plant Operations

The first period in the development of a nuclear-stimulated geothermal site, though of relatively short duration compared to the potential plant lifetime, is vitally important; it is during this phase that the safety, feasibility, and environmental acceptability of the entire site development must be established. The greatest environmental concerns are the consequences of employing nuclear explosives to stimulate the potential geothermal site, thus permitting the release of adequate heat energy to operate the power plant.

Geologic considerations. The need for an extensive and exhaustive geologic study of the site is apparent. First, it must be established that a potentially exploitable geothermal-energy source exists at the site and that the use of nuclear explosives will yield access to the best source. The geologic studies and tests must also provide basic data on the nature, structure, and significant features of the rock and soil formations at the site. These data are necessary to evaluate and predict ground-motion effects and the potential for venting of radioactive materials.

Hydrologic and meteorologic considerations. Hydrologic studies must be conducted in the vicinity of the site, chiefly to identify existing ground and surface waters and to determine where and how fast this water flows, both before and after the nuclear event. It is incumbent upon these studies to verify that unacceptable contamination of usable water supplies will not occur.

Federal and local government approval. The National Environmental Policy Act of 1969, passed by the 91st Congress, was signed into law by President Nixon on January 1, 1970. The act requires that for major Federal actions significantly affecting the quality of the human environment, a detailed statement be prepared on (1) the overall environmental impact of the proposed action, (2) all adverse environmental effects that cannot be avoided, (3) alternatives to the proposed action, (4) the relationship between local short-term uses of man's environment and the maintenance and enhancement of long-term productivity, and (5) any irreversible and irretrievable commitments of resources that would be involved in the proposed action. It is apparent from these Federal actions that the total environmental consequences must be determined and evaluated before the utilization of nuclear explosives is approved for geothermal stimulation. An environmental-impact statement is prepared for each nuclear-stimulation experiment; this was done, for example, for the Rio Blanco and Wagon Wheel natural-gas projects.

In September 1971 the U.S. Department of the Interior, in implementing the Geothermal Steam Act of 1970, prepared a Draft Statement entitled "Environmental Impact Statement for the Geothermal Leasing Program" that is intended to comply with the requirements of the National Environmental Policy Act. Besides satisfying the Environmental Protection Agency (which is an outgrowth of the Act) the U.S. Atomic Energy Commission, under the Atomic Energy Act of 1946 and as amended in 1954, controls and is responsible for public health and safety in connection with nuclear detonations under the Plowshare Program. Furthermore, Presidential approval is required for all nuclear tests outside the Nevada Test Site. And in addition to these required Federal approvals, it is evident that the environmental consequences of the program must be well established at extensive state and local government hearings before local approval is obtained.

Nuclear-Explosion Effects

Ground motion. At the time of detonation of an underground nuclear explosion a tremendous amount of energy is liberated, vaporizing the rock and creating an expanding spherical cavity. The great pressure of the expanding cavity drives a shock wave outward in all directions from the point of the detonation. When the shock wave reaches the surface of the earth, it causes the surface to vibrate, much in the manner of a natural earthquake. Geological formations and man-made structures, even at considerable distances from the explosion point, can suffer damage from the resulting ground motion.

As is the case with a natural earthquake, humans, animals, and marine life can be killed or injured by falling debris and other consequences of the ground motion. Vegetation such as trees, plants, and ground cover can be destroyed by breakage, soil changes, rock slides, or root damage. The indirect effects of ground motion, however, such as changes in the drainage patterns of the area or in the quality of local water, must also be considered. Building damage can result directly from ground motion, but the mechanism for industrial damage is primarily indirect—a loss of production resulting from work stoppages and the evacuation of personnel for safety reasons within a radius of many miles of ground zero at the time of detonation. Other effects constitute a public inconvenience: automobile and train traffic must be halted and recreational areas evacuated.

An extensive inventory of man-made structures must be made within a radius of 20 to 50 miles from the point of detonation, depending upon the depth of burial, the yield, and the number of nuclear devices employed. Buildings, bridges, tunnels, power and telephone lines, gas and oil wells, water wells, pipelines, mines, airports, dams, roads, and railroads within the potential-damage radius must all be catalogued and assessed for potential seismic damage.

In most instances, direct damage to buildings will be the major result of seismic motion. Structural damage will depend upon distance from the detonation point, geological characteristics of the terrain, depth of burial and the number and yield of the explosive devices, and the natural vibration-response frequency of the individual structures. Methods for considering these factors are available (Blume, 1969), and relatively accurate predictions of anticipated structural damage can be made once an accurate survey within the proposed potential-damage radius has been made. Damage predictions for a given structure are derived by relating the predicted ground-motion intensity to the structure's age, location, condition, and capacity to withstand ground motion. Experience with underground nuclear detonations at the Nevada Test Site can be used to estimate potential structural damage from a particular detonation at a particular location. Fig. 1 shows ground motion as a function of distance from detonation point for detonations of 5, 100, and 1,000 kt and illustrates the various zones of structural damage.

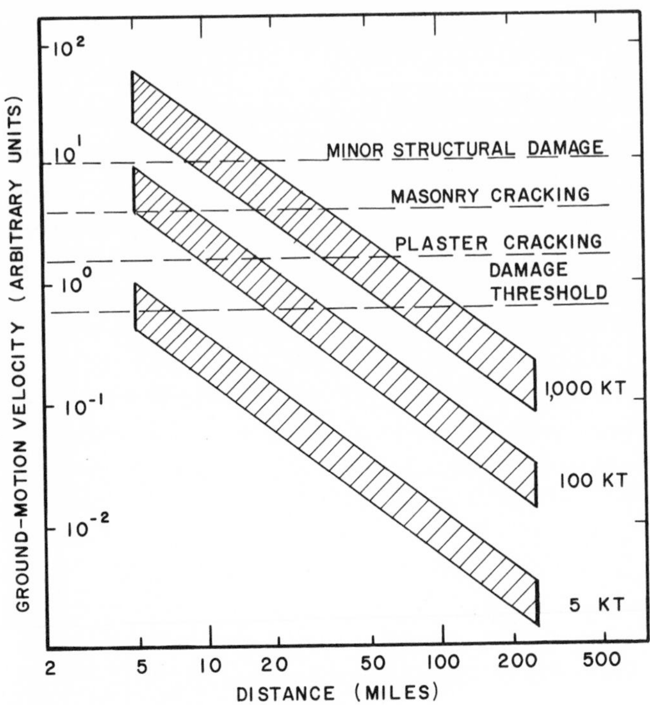

Fig. 1. Typical seismic damage associated with underground nuclear explosions.

Building damage can be significant at very large distances from a 1,000-kt explosion and indeed even from a 100-kt explosion. Thus the use of multimegaton nuclear devices under the Plowshare Program in most of the continental United States appears impractical. The reported cost of building damage for the 40-kt Rulison experiment, which was conducted in a relatively isolated area of Colorado, was roughly \$90,000. It is not difficult to anticipate that a 1-Mt explosion 20 or 30 miles from a town of 10,000 population could produce damage claims of \$500,000 to \$1 million.

Underground nuclear-detonation experiments at the Nevada Test Site have created small aftershocks. These shocks result from small movements on preexisting fault planes and constitute the release of natural strain energy. In an area of high natural seismic activity, which might well be the situation at a potential geothermal site, a seismic event might occur sooner than it otherwise would as the result of a nuclear detonation. It could be argued, however, that this could only help to reduce the environmental impact of a natural earthquake, since it would promote the release of the natural strain energy before it builds up to a magnitude that could produce a major earthquake. It could of course also be argued that the nuclear detonation could hasten the occurrence of the major earthquake, or that in the absence of the nuclear detonation the inevitable major quake might be generations or centuries in coming. But the experience of the Plowshare Program appears likely to provide satisfactory answers to all these questions.

Radiation releases. Currently, all nuclear explosives, even those relatively "clean" devices that derive the bulk of their energy from fusion, require a fission trigger. This trigger, composed of 3 kt or more of fission energy (from U-235 or Pu-239), provides the high temperature necessary to initiate fusion. The fissioning of this trigger produces about 200 different isotopes of about 36 elements ranging in atomic mass from about 75 to 160. Most of these isotopes are radioactive, but fortunately, most of them have short half-lives. A rule-of-thumb estimate is that the radioactivity of mixed fission products decays by a factor of 10 for every sevenfold increase in time. Thus, the fission products decay by a factor of 10 below the first-hour level of 7 hours, by another factor of 10 below the first-hour level in 7 hours, by another factor of 10 after 2 days, by another factor of 10 after 2 weeks, and so forth. Thus most of the short-lived fission products would decay to negligible levels after a suitable waiting period.

Induced radioactive materials are produced by neutron capture in stable elements within the nuclear device and in elements in the surrounding rock material, elements such as sodium, silicon, aluminum, iron, and manganese. Induced radioactivity can be reduced by surrounding the nuclear device with neutron-absorbing materials, such as boron. It would appear now that the radioactivity induced from devices designed specifically for geothermal stimulation could probably be maintained at a minimum level. Additional sources of radioactivity are the nonreacted nuclear fuels, such as uranium, plutonium, and tritium. Tritium produced from thermonuclear reactions accounts for the largest source of initial radioactivity when a fusion device is used and is probably the most significant radiological problem in geothermal-resource stimulation. It will be taken up below.

Experimental evidence obtained from radiation logs of drill holes in-

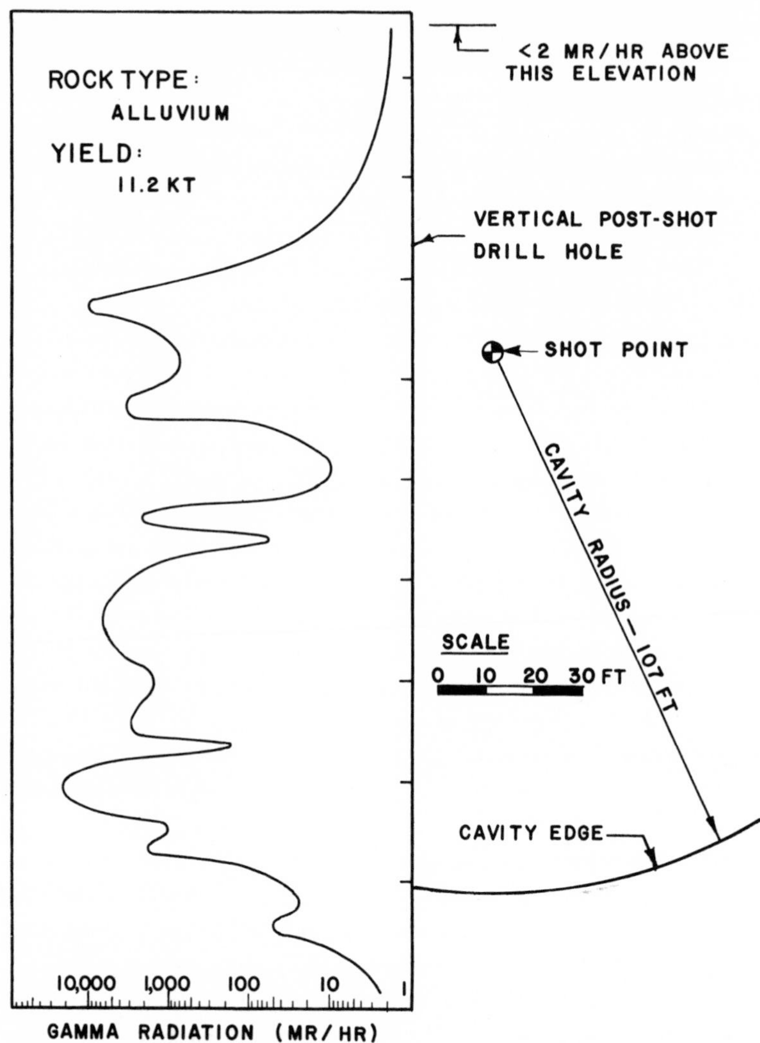

Fig. 2. Approximate gamma-radiation profile 1 year after detonation, Cimarron event (source: UCRL-7350, Rev. 1).

dicates that the bulk of the radioactivity yielded by underground nuclear explosions is condensed within a fused glassy material that amasses primarily at the base of the cavity. Figure 2 is a gamma-radiation profile of the Cimarron event made 1 year after detonation. In this figure, the radiation profile was made in a reentry hole drilled some 15 ft from the original shot-point emplacement hole. The gamma-radiation dose rate, in mr/hr, is given in relation to the depth of the shot point. The cavity radius was 107 ft before collapse of the overburden. As is indicated, most of the radiation is located between the shot point and the bottom edge of the cavity. Above the shot point, the radiation falls rapidly to a value of about 2 mr/hr. The level of radiation dose rate can be compared to the average dose rate of 125 mr/yr to man due to the natural radioactivity in the world and incoming cosmic radiation.

The glassy material condensed at the bottom of the cavity contains most of the produced radioactivity, and experience indicates that less than 10 percent of the radioactivity content is released by leaching with water over a long period of time. The problem of hot-water leaching of a nuclear-stimulated geothermal reservoir, however, will require considerable investigation to determine the long-term radioactivity behavior during production.

Potential geothermal-stimulation applications fall basically into two categories: the stimulation of geothermal aquifers, at depths of perhaps 3,000 to 8,000 ft, with low-yield 5- to 100-kt devices; and the use of nuclear explosives with yields of 100 to 1,000 kt to fracture hot, dry rock at depths of 8,000 to 12,000 ft. The probability of prompt venting, i.e. the immediate release of radioactivity to the atmosphere at the time of detonation, differs in these two applications. For the geothermal-aquifer stimulation, the probability of venting and the fraction of initial radioactivity released would be higher than for very deep dry-rock applications. However, because the nuclear explosive used for aquifer stimulation would probably be smaller than that used to break up deep, dry rock, the resulting initial quantity of radioactivity released would be correspondingly smaller. It is not unreasonable to suppose, then, that the product of the probability of release times the quantity released might be of a similar order of magnitude for the two cases.

Though leakage is theoretically possible, there has been no recorded leakage into the atmosphere from any underground nuclear explosion detonated at a depth greater than 4,000 ft. Seepage of radioactive material into ground-water aquifers is equally unlikely, but could occur, either as a result of improper cementing or stemming of the drill hole or through unexpected faults or fissures in the area surrounding the implacement hole. Because cementing techniques at high temperatures are not yet an established art, verifying that radioactivity cannot escape could prove difficult.

Table 1 lists the initial radioactivity from nuclides of greatest envi-

TABLE 1

Initial Radioactivity from Some Radionuclides of Greatest Environmental Concern, in Kilocuries, for Various Sizes and Types of Detonations: A, Fission-Product Activity; B, Induced Activity in Igneous Rock

	Fission explosion				Fusion explosion					
	5 kt (3.5×10^6 kCi)		100 kt (7.0×10^7 kCi)		3-kt trigger (2.1×10^6 kCi)		100 kt (97-kt fusion)		1,000 kt (997-kt fusion)	
Radionuclide	A	B	A	B	A	B	A	B	A	B
H-3	0.005	1.10	0.1	22.0	0.003	0.66	0.0	[a]	0.0	[a]
C-14	0.0	0.0	0.0	0.002	0.0	0.0	0.0	0.002	0.0	0.023
Ar-37	0.0	12.5	0.0	250	0.0	7.5	0.0	700	0.0	7,000
Kr-85	0.10	0.0	2.0	0.0	0.06	0.0				
Sr-90	0.75	0.0	15.0	0.0	0.45	0.0				
I-131	725	0.0	14,500	0.0	435	0.0				
Xe-131	5.0	0.0	100	0.0	3.0	0.0				
Xe-133	1,650	0.0	33,000	0.0	1,000	0.0				
Cs-137	0.90	0.0	18.0	0.0	0.54	0.0				

SOURCE: Ref. UCRL-50230, Rev. 1; UCRL-50656.

[a] 100 kt, 28.0 + 1,940; 1,000 kt, 280.0 + 19,400 (latter figure in both cases indicates residual tritium at rate of 20,000 Ci/kt).

TABLE 2
Potentially Volatile Radioactive Nuclides Present 180 Days After a 1,000-kt Nuclear Explosion in Igneous Rock

Nuclide	Half-life	Fission (kCi)	Fusion[a] (kCi)	RCG[b] (pCi/ml) Air	Water	Volatility in steam (80 atm, 623°K)
Kr-85	10.8 yrs	20	0.06	0.3	—	Permanent gas
Sr-90	28.8 yrs	150	0.45	3×10^{-5}	0.3	Gaseous precursor
Ru-103	40 days	1,150	3.45	0.003	80	Oxidizing, high; reducing, low
Ru-106	1 yr	1,000	3.0	2×10^{-4}	10	Oxidizing, high; reducing, low
Sb-125	2.7 yrs	60	0.18	9×10^{-4}	100	Apparently high
Te-127m	109 days	90	0.27	0.001	50	High
Cs-137	30 yrs	180	0.54	5×10^{-4}	20	With CO_2, low; without CO_2, high (30%)
			Radioisotopes Induced in Soil			
H-3	12.3 yrs	220	20,290[c]	0.2	3,000	As permanent gas and tritiated vapors
Na-22	2.6 yrs	—	0.6	3×10^{-4}	30	—
P-32	14.3 days	2	2.5	0.002	20	—
S-35	88 days	29	40.0	0.009	60	—
Ar-37	35 days	70	200	100	—	Permanent gas
Cs-134	2 yrs	14	18.3	4×10^{-4}	9	With CO_2, low; without CO_2, high (30%)

SOURCE: American Oil Shale Corporation et al., 1971.

NOTE: Other radioisotopes of interest that generally exist as nonvolatile compounds include Fe-59, Co-60, Ce-144, Pu-239, Pu-241, and many transuranic isotopes.

[a] These amounts assume linear scaling of 50-kt fission and fusion devices. Assume a nominal 3-kt fission trigger for fusion device.

[b] These radiation-concentration guides, taken from CFR Part 20 for unrestricted areas, are maximum values ignoring solubility of compound.

[c] Assumes 2 grams of residual tritium per kt of fusion yield.

ronmental concern for four typical devices that could be considered for nuclear geothermal stimulation. The radioactive nuclides most likely to be involved in prompt venting or seepage would be primarily the gaseous elements xenon, argon, krypton, iodine, and tritium. But even these gases would have to follow a tortuous path through cracks and fissures to reach the surface. On the basis of experience with cratering experiments, it would seem very unlikely that more than 5 percent of the total radioactivity would escape even in the event of a maximum-credible prompt-venting accident. And with delayed venting or seepage the maximum percentage escaping should be even lower (see Table 2).

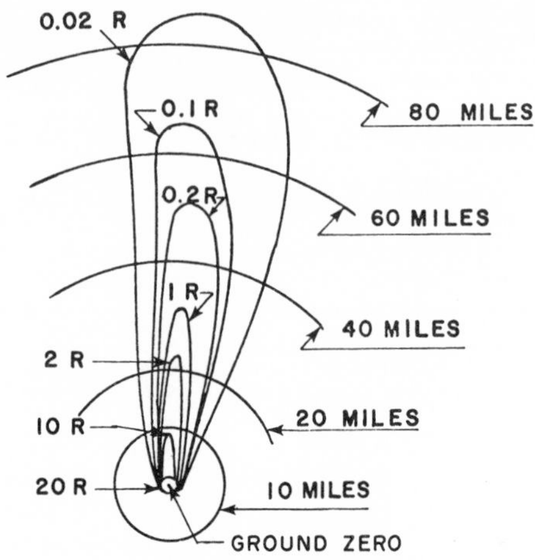

Fig. 3. Example infinite-isodose curves for a 100-kt explosion, 2–5 percent prompt-venting release (source: TID-7695).

In a prompt-venting accident, the initial spread of radioactive material would be dependent upon meteorological conditions and the local geography. Generally, the radioactive-fallout area is restricted to a very narrow corridor in the prevailing-wind direction. Estimated infinite-isodose lines for a 100-kt explosion followed by a postulated 2- to 5-percent release of radioactivity are shown in Fig. 3. It is obvious from this graph that a major venting accident of this magnitude could have a serious environmental impact out to distances of 30 to 60 miles along a relatively narrow front.

In assessing the environmental impact of prompt releases of radioactivity, the effects can be categorized in the manner of those for seismic effects. Although the direct exposure of persons to radioactivity released from a prompt-venting accident is by far the greatest environmental concern, the analysis of a new site should also consider the deposition of radioactivity on plants and in water aquifers and streams and the subsequent secondary effects on humans through food pathways. Fig. 4 presents an idealized network of potential food pathways leading to man. An environmental program that monitors and controls the transport and concentration of radioactive materials in the biosphere would be essential for an acceptable and safe application of nuclear explosives for geothermal-heat stimulation. And again, the accumulated experience of the Plowshare Program would be drawn upon in assessing the risks of particular planned detonations.

Other effects of nuclear explosions. The geologic analysis of a potential geothermal site must assess carefully the prospects for two other

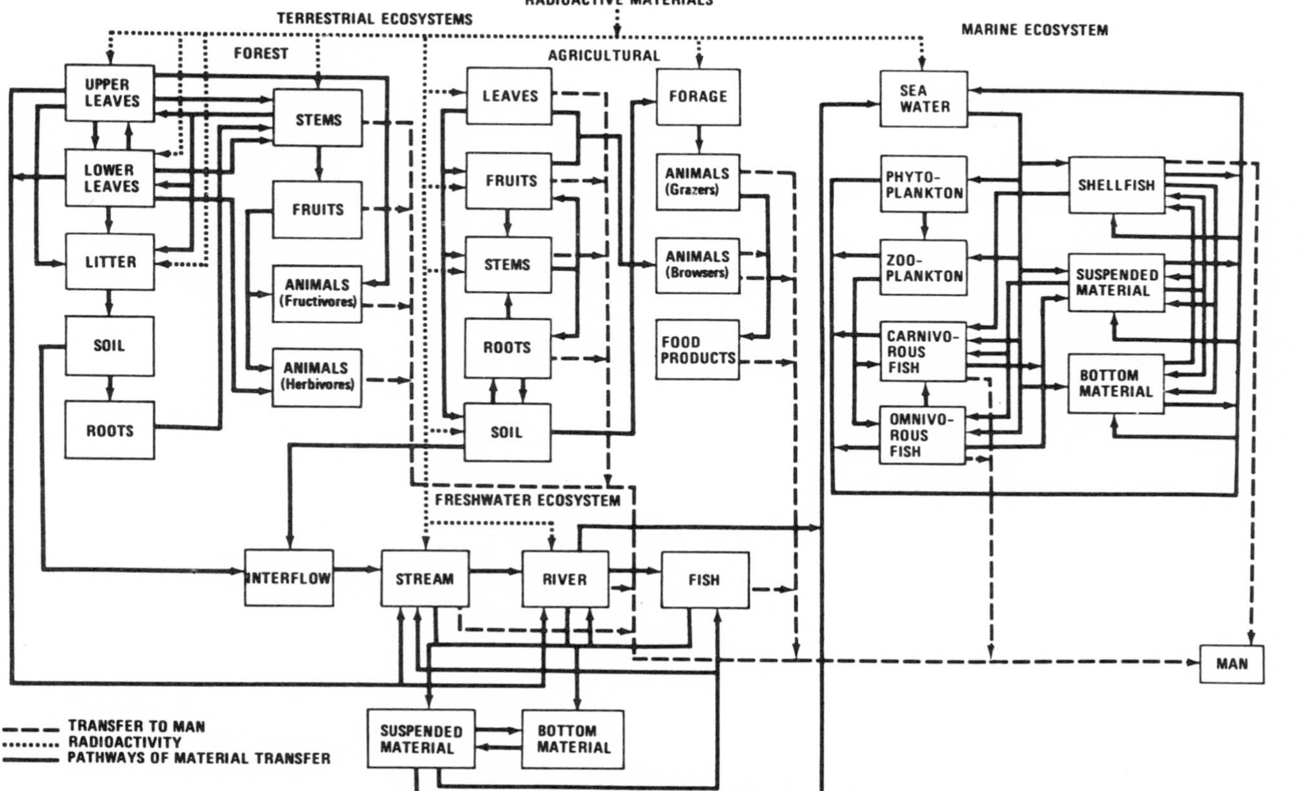

Fig. 4. Idealized network diagram of food pathways leading to man (source: TID-25708).

phenomena, though the probability that either might occur is remote.

Surface extrusion of molten rock resulting from the underground nuclear explosion is a possibility that must be evaluated for a given geothermal site. The likelihood of triggering a volcanic eruption depends upon the proximity of the preexisting magma and the availability of open fractures or older vents through which the magma could rise to the surface.

Water near the surface may be very close to the temperature at which it will flash to steam. An instability such as a pressure release at the surface caused by the nuclear detonation might cause the water to flash to steam and produce a hydrothermal explosion.

Plant and Site Construction

We turn now to the preparation of the implacement well and the physical construction of the plant and auxiliary facilities. Though the environmental consequences of drilling, excavation, and construction are considered well established, experience has shown that generally more ecological damage results from these activities than from the actual underground nuclear detonation. The actual operations that would be of environmental significance would depend upon the particular geothermal site under development; but some of the more common construction activities and effects that must be examined for their environmental impact are (1) drilling for geologic and hydrologic studies, nuclear-explosive deposition, and radiation-monitoring wells; (2) land-excavation and -filling operations for access roads, railroad tracks, pipelines, transmission lines, canals, cooling ponds, and building sites; (3) movement and storage of building materials, temporary housing for construction workers, and the like; (4) redistribution and loss of materials and effluents; and (5) plant design and layout, with respect to compatibility with the surrounding environment.

The quantity of heat energy required to operate even a moderate-size thermal-electric plant over a three-decade lifetime is large. A typical 250-Mw electric plant utilizes about 250 billion kwhr of energy over 30 years. Furthermore, since the permissible yield of a nuclear device peacefully employed in the United States would be limited by potential structural damage, radiological hazards, possible seismic coupling, public opinion, and other ecological considerations, it seems evident that in most cases many moderate- to low-yield devices would be required to successfully and safely develop a nuclear-stimulated geothermal site, particularly where the source is hot, dry rock.

A geothermal-development program requiring multiple explosions could be undertaken in either of two ways. Either all of the devices required to supply the plant's lifetime energy needs could be fired before plant construction begins, or the devices could be fired as needed during the plant's lifetime and the plant constructed to withstand ground shock or made mobile to permit withdrawal to a safe position during firing. The plant could be mobilized by employing a waterway and barge system (American Oil Shale Corp. et al., 1971).

From an environmental viewpoint, extended firings through the plant's lifetime offer certain advantages, such as improved radiological measurement and control and other environmental studies requiring long time duration before significant effects are discernible.

Plant Operations

The second period in the life cycle of a geothermal site sees the plant generating useful power. This is a period of long duration, perhaps decades, depending upon the magnitude of the available geothermal-energy resource at the site and the prevailing economics of power generation.

During this period the environmental aspects of greatest concern are the transport of radioactive materials and their distribution through the biosphere owing either to slow operational losses or to rapid releases resulting from accidents or containment failures. And there are other environmental factors to be examined.

Release of radioactive materials to the biosphere. The distribution and percentage contribution of fission-product activities at 50 days and 1 and 2 years after detonation are graphically portrayed in Fig. 5. It is to be expected that small quantities of these materials would exist in the working fluid (e.g. water) used to transport heat energy from the geothermal source to the plant. The quantity of each isotope released would depend upon its chemical and physical makeup, the temperatures and pressures in the cavity, and the nature of the working fluid.

Of greater concern, perhaps, are the volatile radioactive products that are not condensed within the cavity and are easily transported to the ground surface via the working fluid or fissures and cracks. Although careful geologic studies and deep nuclear-device emplacement should prevent inadvertent loss of these gases through fissures or cracks, trans-

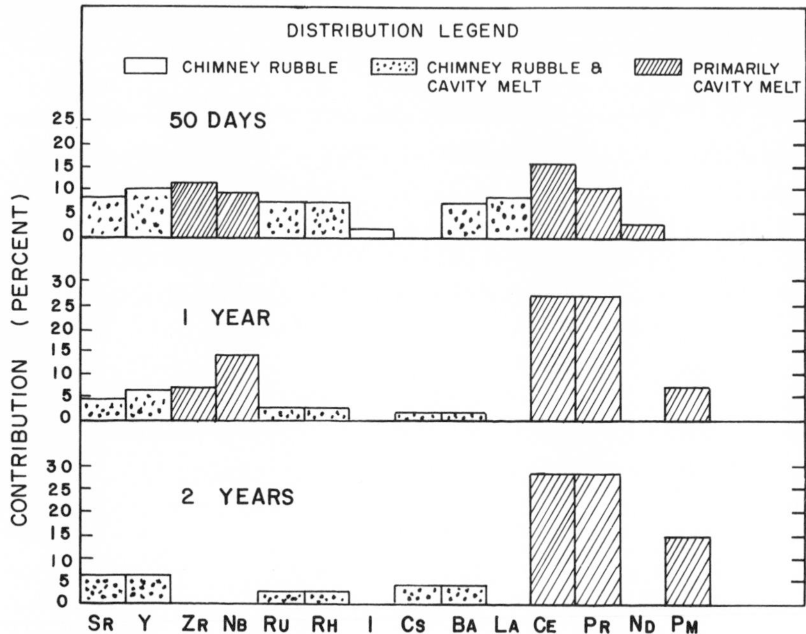

Fig. 5. Distribution and percentage contribution of principal radionuclides relative to total fission-product activity (source: UCRL-7350, Rev. 1).

port in the working fluid seems unavoidable. Table 2 lists those potentially volatile radioactive nuclides present 6 months after a 1,000-kt nuclear explosion. In a geothermal system, tritium would primarily appear molecularly bound in water, hydrogen gas, H_2S, CH_4, NH_3, or similar compounds.

Noncondensable gases present in the steam of a thermal-electric plant tend to collect in the condenser. In natural fields these gases are generally eliminated by venting to the atmosphere. In a nuclear-stimulated geothermal system the operational venting of large quantities of gases with radioactive components might be unacceptable, and it would be necessary to monitor the effluent; and, if the radioactivity exceeds the established maximum concentration, either to compress and store the radioactive components or to recycle them into the cavity.

If a direct power cycle using steam from the cavity to operate the turbines is to be employed, then contamination by deposition of radioactive materials throughout the entire working-fluid system becomes an important consideration, and the risk of radiation release to the atmosphere is increased. The use of a noncontaminated secondary working fluid would be more acceptable from an environmental-safety standpoint, but plant efficiency would be correspondingly reduced. Regardless of the energy-transport system employed, adequate monitoring and control of all system components (piping, tanks, turbines, condensers, valves, etc.), transporting or storing of radioactive materials, and compliance with appropriate Government radiation standards would be required.

Another mechanism for the potential transport of radioactive materials contained within the nuclear cavity is the flow of ground water in the vicinity of the cavity. Typical flow rates of ground water in aquifers range from 1 to 1,000 ft per year, although fluid flow within the neighborhood of the cavity may be much greater, owing to fissures and cracks produced by the detonation. Most radioactive materials, however, actually move much more slowly than the transporting water, because of sorption and dispersion processes. Distribution by ground-water flow thus represents the maximum rate at which radioisotopes can be transported in the water system.

Tritium as tritiated water is an exception, however—traveling, as it does, at the same rate as ground water. A time lapse of 10 half-lives for a specific radioactive isotope produces a reduction in activity by a factor of about 1,000, and the lapse of 20 half-lives reduces the activity by 1 million. Thus even in ground water flowing at a rate of 1,000 ft/yr, tritium activity, with a half-life of about 12 years, would be reduced a thousandfold within 23 miles of the detonation site. Actually, tritium does not appear to concentrate in any part of the human body and has a biological half-life in the body of about 12 days, owing to natural elimination processes. It is thus one of the less hazardous radioactive materials produced by nuclear explosives, as reflected in the Federal Radiation Council concentration guide of 3,000 pCi/ml in unrestricted water. Nevertheless, the large quantity of unreacted tritium (typically about 2 grams or 20,000 Ci/kt of fusion yield) and the problems associated with monitoring (tritium is a low-energy beta emitter with no photon emission except for beta-produced bremsstrahlung) and controlling the various phases and compounds that can incorporate tritium make management of tritiated materials perhaps the most significant radiological problem associated with nuclear-stimulated geothermal development.

Another area of concern is the possible release of radioactive materials resulting from the rupture of a containment system or from a piping failure. The steam-transport and deposition problems associated with nuclear-stimulated geothermal sites are described by Krikorian (this volume).

Consequences of plant accidents and natural catastrophes. An operating geothermal plant, like any other major industrial installation, supports a diverse spectrum of activities and facilities. The environmental consequences of natural disasters or of severe accidents involving plant facilities must be evaluated in determining whether developing a geothermal site will be acceptable. The potential for such environmentally hazardous accidents at a geothermal site would be similar to that of conventional nuclear and fossil-fuel power plants. Specific environmental aspects would be determined by the character and location of the geothermal plant and site.

Nonradioactive plant emissions and effluents. Although radioactive contamination of the biosphere is the greatest potential environmental hazard in a nuclear-stimulated geothermal site, there are other potential toxic and hazardous emissions and effluents to be expected. The release and disposition of gases (CO_2, H_2S, H_2, CH_4, NH_3, halogens), fluids (e.g. brines), solids (e.g. salts), and energy (thermal effects) from the operating system would require study. This problem is treated in detail by Bowen (this volume).

Biological and societal environmental effects. To ensure compliance with the National Environmental Policy Act of 1969 it is necessary that the environmental statement pursuant to a nuclear-stimulated geothermal site contain a bioenvironmental safety study that (1) predicts the potential effects and aftereffects of the nuclear detonations upon the ecological systems of the surrounding area; (2) prescribes actions to be taken to avoid or minimize harmful consequences that may occur; and (3) recommends corrective measures should there be any undesirable consequences. The primary bioenvironmental goal would be generally to minimize ecological changes resulting from the development and utilization of the geothermal site. Thus it would be necessary to study the ecological dynamics of the area and to characterize the terrestrial and aquatic ecosystems. It would be necessary also to evaluate possible disturbances of these ecosystems, including threats to endangered species, and to take preventive or remedial actions.

A significant environmental aspect that has often not been adequately considered in the development of large industrial facilities is the societal impact that may result. Changes in housing, business patterns, population distribution, transportation flow, employment, and local government are often produced by major industrial activities. Although the prediction and evaluation of these effects may be difficult, public sentiment and support will be heavily dependent upon these factors and a complete environmental-impact study would accord them careful attention.

Post-Plant Operations

After the useful production of energy has ceased, certain controls and safeguards would have to be maintained to ensure continued public safety and ecological compatibility. Because of the longevity of many of the radioactive isotopes produced within the nuclear cavity (e.g. Kr-85, Cs-137, tritium, Sr-90, plutonium, Ce-144, and Ru-106), a quasi-perpetual program of monitoring and control of radiation losses from the cavity and of other contaminated facilities should continue. Other site emissions and effluents should be controlled, if necessary, and site cleanup and land restoration considered. Finally, an important consideration in the post-plant period would be the assignment of perpetual responsibility and liability for the spent geothermal site.

Conclusions

Because the development of geothermal-energy resources appears to be attractive from an environmental and perhaps economic standpoint, the utilization of nuclear explosives to stimulate latent geothermal sites merits further examination. However, the contained underground detonation of one or more nuclear devices raises considerable questions and problems that demand critical evaluation. Deep dry-rock stimulation seems at this time to favor the use of very large (200- to 1,000-kt) explosives, which would produce severe environmental impact from seismic effects. Geothermal-aquifer stimulation would make use of smaller nuclear devices, but because of the probable affinity to tectonic-plate boundaries of the geologic formations addressed by this application, the potential environmental impact of radioactive contamination would be a serious concern, as would possible induced seismic fault movement.

Nuclear stimulation of deep, tight, natural-gas reservoirs was initiated in 1967, and to date only two field experiments have been conducted, with two more scheduled within the next two years. Environmental impact data obtained in these experiments could provide invaluable information for the design of a potential nuclear geothermal-stimulation project. It would seem sensible to demonstrate the environmental acceptability of natural-gas stimulation before proceeding with nuclear-geothermal stimulation.

The excellent safety record of the Plowshare Program is reassuring—no personal injuries and only minor damage to structures and minimal impact on the environment attributable specifically to the detonations. Furthermore, the official position of the AEC is that a nuclear device will be detonated under the Plowshare Program only when it is ascertained that the operation can be accomplished without direct or indirect injury to people or animals and without unacceptable risk to ecological systems or to natural or man-made structures. Nonetheless, when underground nuclear-explosive stimulation is undertaken, the primary concern will remain the large quantity of radioactive materials deposited within the ground and the potential for unacceptable release of these materials in the future.

REFERENCES

[Atomic Energy Commission technical reports are available from The National Technical Information Service, Springfield, Virginia 22151, at a price of $3.00 each.]

American Oil Shale Corporation, Battelle-Northwest, Westinghouse Electric Corporation, U.S. Atomic Energy Commission, Lawrence Livermore Laboratory, and Nevada Operations Office–AEC. 1971. A feasibility study of a geothermal power plant. USAEC Technical Report No. PNE-1550.

Barge, D. C., and W. C. King. 1969. Fission gas release curves. University of California Lawrence Livermore Laboratory Technical Report UCRL-50656.

Battelle Memorial Institute. 1968. Bioenvironmental safety studies, Amchitka Island. Progress report for fiscal year. USAEC Technical Report No. BMI-171-116.

Blume, J. A. 1969. Ground motion effects. *In* Proc. Symp. Public Health Aspects of Peaceful Uses of Nuclear Explosives. USAEC Technical Report No. SWRHL-82.

Eisenbud, M. 1969. Environmental radioactivity. New York: McGraw-Hill.

El Paso Natural Gas Company. 1971. Technical studies report, Project Wagon Wheel. USAEC Technical Report No. PNE-WW-1.

Fuller, R. G. 1969. Amchitka biological information summary. Battelle Memorial Institute.

Glasstone, S. 1971. Public safety and underground nuclear detonations. USAEC Technical Report No. TID-25708.

Green, J. B., and R. M. Lessler. 1971. Reduction of tritium from underground nuclear explosives. University of California Lawrence Livermore Laboratory Technical Report UCRL-73258.

Holzer, A. 1965. Calculation of seismic source mechanisms. University of California Lawrence Livermore Laboratory Technical Report UCRL-12219.

——— 1970. Plowshore and the environment. University of California Lawrence Livermore Laboratory Technical Report UCRL-72830.

Hubbert, M. K. 1969. Resources and man, "energy resources." Report to National Academy of Sciences. San Francisco: Freeman.

Knox, J. B. 1968. Water quality in flood nuclear craters. University of California Lawrence Livermore Laboratory Technical Report UCRL-50531.

Knox, J. B., et al. 1970. Radioactivity released from underground nuclear detonations: Source, transport, diffusion, and deposition. University of California Lawrence Livermore Laboratory Technical Report UCRL-50230 (Rev. 1).

Larson, J. D., and W. A. Beetem. 1970. Chemical and radiochemical analysis of water from streams, reservoirs, wells and springs in the Rulison Project area. U.S. Geological Survey Report No. USGS-474-67.

Lawman, F. G., et al. 1967. Effects of the marine biosphere, hydrosphere, and geosphere upon the specific activity of contaminant radionuclides. USAEC Technical Report No. CONF-690406-10.

Merritt, M. L. 1970. Physical and biological effects: Milrow event, U.S. Atomic Energy Commission Nevada Operations Office. USAEC Technical Report No. NVO-79.

Robison, W. L., and L. R. Anspaugh. 1969. Assessment of potential biological hazards from Project Rulison. University of California Lawrence Livermore Laboratory Technical Report UCRL-50791.

Schultz, V. 1966. References on Nevada Test Site biological research. Great Basin Naturalist, v. 26, nos. 3–4, p. 79.

Smith, C. F. 1971. Gas analysis results for Project Rulison calibration flaring samples. University of California Lawrence Livermore Laboratory Technical Report UCRL-50986.

Teller, E., et al. 1968. The constructive uses of nuclear explosives. New York: McGraw-Hill.

U.S. Atomic Energy Commission. 1970. Environmental statement: Underground nuclear test programs, Nevada Test Site, fiscal year.

——— 1971. Environmental statement: Cannikin.

——— 1972. Environmental statement, Rio Blanco Gas Stimulation Project. USAEC Technical Report No. WASH-1519.

U.S. Atomic Energy Commission Nevada Operations Office. 1969a. Technical discussion of offsite safety programs for underground nuclear detonations. USAEC Technical Report No. NVO-40 (Rev. 2).

——— 1969b. Project Rulison: Post-shot plans and evaluations. USAEC Technical Report No. NVO-61.

U.S. Bureau of Mines. 1971. Blasting vibrations and their effects on structures. Bull. 656.

U.S. Dept. of Health, Education, and Welfare. 1969. Proc. Symposium on Public Health Aspects of Peaceful Uses of Nuclear Explosives, Las Vegas, Nevada. USAEC Technical Report No. SWRHL-81.

White, D. E. 1965. Geothermal energy. U.S. Geological Survey Circular 519.

Williamson, M. M. 1964. Fallout calculations and measurement. Proc. Third Plowshare Symposium. USAEC Technical Report No. TID-7695.

17. Corrosion and Scaling in Nuclear-Stimulated Geothermal Power Plants

OSCAR H. KRIKORIAN

The purity of the steam produced in a nuclear chimney or surrounding fracture zone will have a major effect on the design, cost, and operation of a nuclear-stimulated geothermal power plant. Both the quantity and the type of contaminants carried by the steam may affect the well life, the turbine-condenser design, the operating efficiency of the plant, and the safety and environmental-protection regulations that must be established.

For calculation purposes, the geothermal plant is assumed to be stimulated by a total nuclear-explosive yield of 1 Mt comprising one or more fission or thermonuclear explosives, including 3 kt of fission for each thermonuclear explosive. The production well is assumed to be tested 180 days after detonation and to produce steam at approximately 662°F (623°K) and 80 atm pressure in the chimney. The corresponding wellhead pressure would be about 50 atm after well testing, and an additional shut-in period of 1½ to 2 years may be anticipated before the well is put into production. We shall consider the effect of this shut-in period on the amounts and types of radionuclides carried by the steam.

The assumptions made in this paper limit the problem areas to steam effects in hot, dry rock systems or dry-steam geothermal fields. Corrosion problems in a mixed water/steam field or in a hot-water field can be assumed to be, in general, far more severe than those in dry-steam systems. Hot-water problems should of course also be analyzed, espe-

---

Oscar H. Krikorian is with the Lawrence Livermore Laboratory, University of California, Livermore. This work was performed under the auspices of the U.S. Atomic Energy Commission.

cially since the reservoirs of hot water that have been located at moderate depths are more extensive than the steam fields at the same depths, but such an analysis has not been the goal of this investigation.

Problems with steam quality are associated with mineral content, radioactive-materials content, and noncondensable-gas content, and will be taken up in that order.

Mineral Content of Steam

A number of minerals are known to form volatile gaseous compounds by reacting with steam and may, thereby, be transported by the produced steam. The main concern of previous work on mineral transport by steam has been to obtain a better understanding of hydrothermal phenomena, though considerable interest has also been shown in the processes of scaling on steam-turbine blades. Experimental information in these two areas, especially on the scaling of steam-turbine blades, is useful in interpreting and predicting the steam-transport behavior of minerals in geothermal applications, both in the underground well and through the power cycle. The transport of radioactive species and their relationship to mineral transport are discussed in the next section.

The problems that can result from mineral deposition on the piping, in the steam turbine, and in the condenser of a nuclear-stimulated geothermal power plant can be extremely serious. Straub (1964) has shown that the deposition of salts and silica (and hence of associated radioactivity) on turbine blades is dependent on the impurities in the steam and on the steam pressure. Salt deposits on steam-turbine blades are water-soluble and, with proper operating procedures, can be washed off with water. The more important of these salts are NaCl, NaOH, sodium silicates, Na_2CO_3, and sometimes Na_2SO_4. Deposits that are insoluble in water are most commonly silicic acid, amorphous silica, and the various crystalline forms of silica. Water-insoluble deposits of iron oxides and $CaCO_3$ may also occur.

The water-insoluble deposits, *in particular the various forms of silica*, present the most serious problems. These deposits can lead to the distortion of turbine-blade configuration and hence lower turbine efficiency. More seriously, they can lead to turbine imbalance and ultimately to vibrations that can materially damage the turbine. It may be possible to wash off the water-insoluble deposits with NaOH solutions if the rate of deposition is not high, but the operation is time-consuming and requires considerable care if damage to the turbine is to be avoided.

To reduce silica deposits to a level tolerable in the operation of a nuclear-stimulated geothermal plant, the maximum permissible silica concentration in steam with 80 atm of steam pressure at the turbine inlet is about 0.05 ppm (Straub, 1964). Styrikovich (1958), in agreeing with this estimate, concludes that the total dissolved solid content (salts plus silica) of steam used in normal turbine operations should not exceed 0.05 to 0.2 ppm. But steam of such purity is difficult to achieve, even in the best boiler systems.

An estimate of the silica concentration in steam for geothermal applications may be obtained from the work of Heitmann (1965), who equilibrated silica gel, silicic acid, and quartz with high-pressure steam. Results on silica gel are presented as smoothed data in Fig. 1. Silica gel and the various silicic acids will interconvert, depending on temperature and steam-pressure conditions. Hence, the figures given make no distinction among them. Quartz shows a lower solubility in water than does silica gel by a factor ranging from 10 at 210°F (370°K) to 2 at 620°F (600°K). In steam, volatilized-silica concentrations are lower for quartz than for silica gel, but the differences are usually small; the maximum difference amounts to a factor of 2 at about 620°F (600°K) and 30 to 50 atm steam pressure. The silica concentration in steam is given in Fig. 1 as 9 ppm at 662°F (623°K) and 80 atm steam pressure, and is therefore estimated to be in the range of 5 to 9 ppm for an underground nuclear chimney, since either quartz or the more active forms of silica may be present. It becomes apparent that the silica content of the 80-atm steam from the underground chimney is too high by about 2 orders of magnitude for normal turbine operations, and that severe deposition will occur on the turbine blades if the steam is fed directly into the turbine. The situation is further complicated by the presence of radioactivity in the steam, since radionuclides such as ^{125}Sb, ^{137}Cs, and ^{134}Cs may coprecipitate on the turbine blades with the silica and various salts.

Natural-steam fields operate with "dirty" steam but at lower temperatures and pressures, as shown in Fig. 1, so that scaling is only a marginal problem. The turbine cycles are correspondingly less efficient, but this inefficiency is counterbalanced by the low cost of natural steam.

If geothermal steam from a nuclear chimney is to be used directly in turbine operations at high pressures, it will likely need purification to prevent inefficient turbine operation and excessive radioactivity buildup. A method of purification suggested by Straub (1964) is to scrub

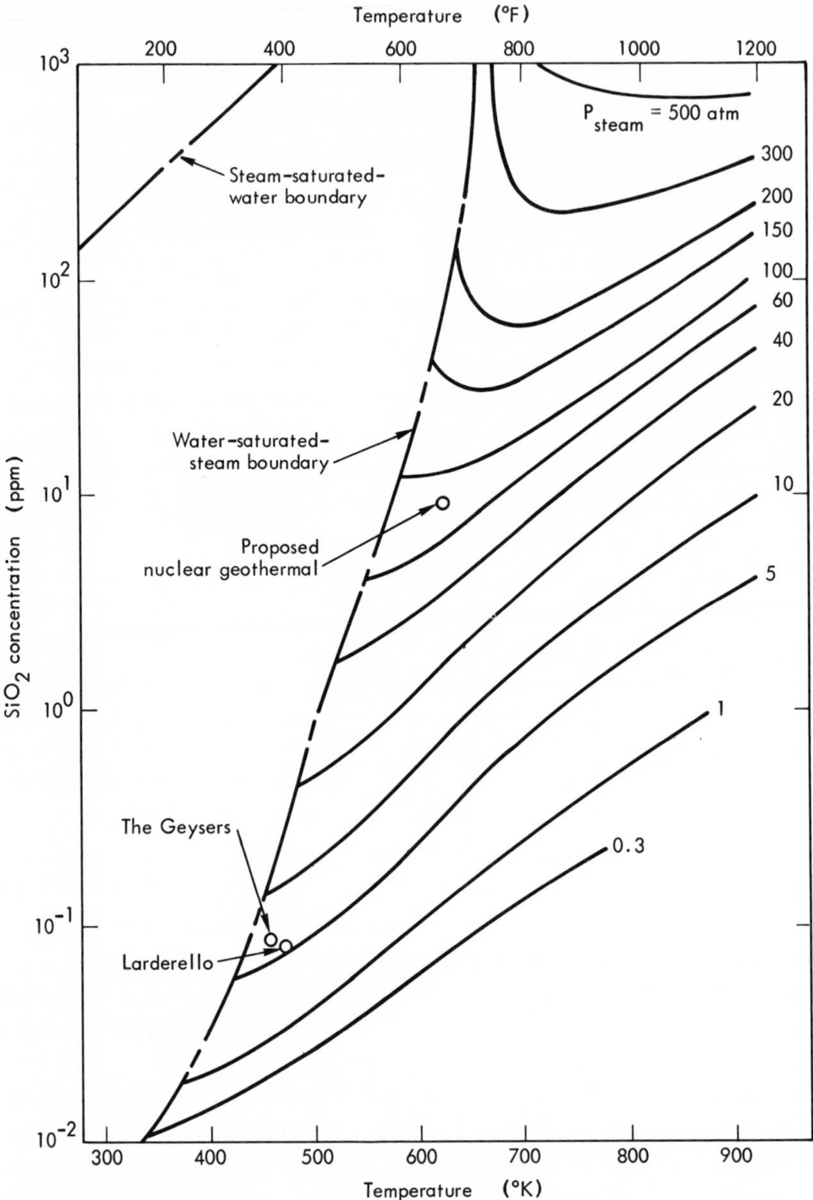

Fig. 1. Temperature and pressure dependence of silica concentrations in steam and steam-saturated water equilibrated with silica gel.

TABLE 1
Steam/Water Distribution Ratios for Silica and Silica Contents of Nuclear-Chimney Steam Before and After Scrubbing

Steam temperature °F(°K)	Saturation pressure, atm	C_{st}/C_w for SiO_2	SiO_2 (ppm)	
			In chimney steam	In scrubbed steam
484(524)	40	1×10^{-3}	1.7	0.0017
565(569)	80	5×10^{-3}	7	0.035
594(585)	100	9×10^{-3}	12	0.11
662(623)	163	1×10^{-2}	50	2.5

the steam with high-purity water before passing it through the turbine. Unfortunately, such a process cools the steam, degrades its energy, and lowers efficiency. Also, after scrubbing, the steam is at its saturation point, and its use in the turbine could lead to erosion of the turbine blades by the impinging of water droplets. Consequently, the steam would need to be reheated.

The effectiveness of such scrubbing treatments may be evaluated from Styrikovich's (1968) data on distribution ratios of various "solute" materials in saturated steam and in water. Table 1 gives distribution ratios for silica in steam vs. silica in water (C_{st}/C_w) for several temperatures. Also given are the concentrations of silica in steam, assuming equilibrium with condensed-phase silica in the nuclear chimney (from Fig. 1) and the predicted silica content of the steam after scrubbing.

From Table 1, it may be concluded that scrubbing is a feasible method of purifying steam up to temperatures in the vicinity of 565°–594°F (569°–585°K), corresponding to saturation pressures of 80 to 100 atm. Styrikovich also gives distribution ratios for NaCl, NaOH, and a variety of other salts and hydroxides. The C_{st}/C_w ratio is about 10^{-4} or less for these compounds; therefore the various salts and hydroxides should be removed even more effectively by scrubbing than SiO_2 is. If deposits are not allowed to form on the turbine, any radioactive materials that are not removed by the scrub water will depart the system in the condensate of the turbine discharge.

Other alternatives warrant consideration. A solid-phase scrubber such as limestone might prove effective in removing silica. Another approach is to add a heat exchanger in which water or other secondary fluid (e.g. isobutane, as described by Anderson, this volume) would be boiled and the steam from the nuclear chimney condensed. Though this additional equipment would decrease the thermal efficiency of the process, it would concentrate the scaling and radioactivity. The economic trade-off with this approach lies in its ability to use a less expensive turbine and more conventional equipment in the rest of the plant. If scaling in the heat exchangers proves severe, either a second heat exchanger could be provided or chemical or mechanical descalers could be added.

If the steam is cleaned, the contaminants will be present in the scrubbing systems. If a fluid scrubbing system is used, it might be possible to dispose of the radioactivity and contaminants present in the scrubbing system by reinjection into the nuclear chimney.

Radioactive Materials in Steam

Green and Lessler (1970) give the radioactive contaminants that might be volatile in steam after 180 days from either fission or thermonuclear explosives. The fission products are ^{85}Kr*, ^{103}Ru*, ^{106}Ru, ^{106}Rh*, ^{125}Sb*, ^{127m}Te*, and ^{137}Cs*; the induced activities are ^{3}H, ^{22}Na*, ^{32}P, ^{35}S, ^{37}Ar, and ^{134}Cs* (significant gamma emitters carry asterisks). For 1 Mt of yield, the amount of gamma radiation remaining after 180 days of decay is 3.0×10^{16} MeV/sec for an all-fission device and 1.1×10^{15} MeV/sec for a thermonuclear explosive employing 3 kt of fission (see Table 2). Although the thermonuclear explosive shows a lower level of gamma radiation for volatile radionuclides, it has a very high level of beta activity in the form of tritium.

It should be noted that the induced activities indicated in Table 2 are for an explosive without neutron shielding. Calculations show a factor of about 10 reduction in activation of the underground formation for every 15 cm of boric-acid neutron shielding (Lessler, 1970). The induced activities also depend on the composition of the geologic media (calculations in Table 2 are for Hardhat granite) and the details of the explosive design.

Noncondensable gases (including radioactive nuclides) that are present in the steam will be separated in the condenser. In natural-steam fields, noncondensable gases are exhausted from the condenser and vented to the atmosphere. For a nuclear-stimulated geothermal power plant, the gases could be collected, and the radioactive contaminants, such as ^{85}Kr, ^{37}Ar, and ^{3}H, could be separated or concentrated, then stored underground. Low levels of volatile chemical and radioactive contaminants could be vented to the atmosphere.

TABLE 2
Potentially Volatile Radioactive Nuclides Present 180 Days After a Total Yield of 1 Mt of Nuclear Explosives

		Fission		Thermonuclear (3-kt fission)[a]	
Nuclide	Half-life	kCi	MeV/sec	kCi	MeV/sec
		Fission Products			
^{85}Kr	10.7 y	19	1.41×10^{12}	0.057	4.2×10^{9}
^{103}Ru	40 d	1,150	2.1×10^{16}	3.45	6.3×10^{13}
^{106}Ru (^{106}Rh)	1 y	1,000	3.37×10^{15}	3.0	1×10^{13}
^{125}Sb	2.7 y	60	9.6×10^{14}	0.18	2.9×10^{12}
^{127m}Te	109 d	90	6.7×10^{11}	0.27	2.0×10^{9}
^{137}Cs	30 y	180	3.8×10^{15}	0.54	1.14×10^{13}
		Induced Activity in Soil[b]			
^{3}H	12.3 y	220	—	20,290[c]	—
^{22}Na	2.6 y	—	—	0.6	4.9×10^{13}
^{32}P	14.3 d	2	—	2.5	—
^{35}S	88 d	29	—	40.0	—
^{37}Ar	35 d	70	—	200	—
^{134}Cs	2 y	14	7.8×10^{14}	18.3	1.0×10^{15}

SOURCE: J. Green and R. Lessler, Lawrence Livermore Laboratory, private communication to H. B. Levy, 1970.

[a] Fission products have been listed for one fusion explosive; the amounts of fission radioactivity should be multiplied by the number of thermonuclear explosives used.

[b] The amounts of these activities are derived from direct linear scaling of amounts produced by 50-kt fission and thermonuclear explosives as calculated by Green and Lessler.

[c] To the 290 kCi of soil-induced activity has been added the residual tritium from the thermonuclear fuel; assuming 2 g of residual tritium per kt of fusion yeld, this is an additional 20,000 kCi (Lessler, 1970).

Thermonuclear explosives, if used, would yield several orders of magnitude less krypton but much more argon and much more tritium. Once the tritium has exchanged with the hydrogen in the water or steam it would be most difficult to remove, and adequate measures to ensure its safe containment would be needed.

Actual volatilities of the various gamma-emitting radionuclides, estimated with a thermodynamic model, can be used to either qualitatively or semiquantitatively estimate the gamma activity of the steam, since the assumption of complete volatility of these species appears to be unnecessarily extreme. To apply thermodynamic calculations to the problem, certain assumptions must be made about processes and reactions occurring during formation of the nuclear-explosion chimney. Soon after the nuclear event, virtually all of the vaporized rock and explosive materials will have condensed (except for steam and permanent gases) to form a molten glass. Refractory radioactive species and some portion of the more volatile radioactive species will be in this glass.

The glass will form both as droplets within the expanded cavity and as a surface layer on the walls, and will tend to settle into the bottom of the cavity, forming a puddle of glass. After collapse of the overburden, this puddle will cool rapidly at first, because of convective and radiative losses, more slowly as the temperature drops. A significant decrease in the cooling rate can be expected if minerals begin to crystallize out. The crystallization range for a typical basalt at 80 atm steam pressure is about 1,970° to 2,240°F (1,350° to 1,500°K); and for a typical granite, 1,700° to 2,060°F (1,200° to 1,400°K). Whether or not crystallization occurs, the viscosities of silicate melts in these temperature regimes are very high, and thus limit the rates with which the gas and liquid can interact.

From these considerations, it seems appropriate to assume that the late-time condensation, which is critical to the disposition of volatile radioactivity, occurs at temperatures below about 2,240°F (1,500°K). (The term "late-time condensation," as it is used here, refers to the condensation that occurs when molten rock surfaces have so cooled that diffusional and chemical interaction with subsequent vapor condensate is exceedingly slow.) A number of factors are expected to contribute to variations in the amount and composition of the late-time condensate. Perhaps the most important factor is the time of chimney collapse. With an early collapse, the melt would be quenched at high temperature and the cavity would release steam containing a high proportion of moderately volatile rock components (and also the moderately and highly volatile radionuclides) into the chimney. With late collapse, and consequently lower melt temperatures, a greater amount of the moderately volatile rock constituents and radionuclides would remain in the melt phase. Other factors, such as steam pressure, oxidizing or reducing conditions, and amount of nuclear yield, are also expected to affect the amount of late-time condensation.

An estimate of the amount of late-time condensate may be made on a basis of 80 atm steam pressure and 2,240°F (1,500°K) as the conditions during separation from the melt region. The steam pressure in the cavity will depend upon the amount of water in the rock and the depth of the explosion. A steam pressure of 80 atm at a cavity temperature of 2,240°F (1,500°K) is obtained if it is assumed that the rock contains either 3 weight percent water at 0.5 km depth or 1 weight percent at 3 km, and that 1.2×10^9 kg of rock release water as steam per

Mt of yield. The corresponding cavity void volumes are estimated at $3 \times 10^6 m^3/Mt$ and $1 \times 10^6 m^3/Mt$ at 0.5 and 3 km depths, respectively. The last of the vaporized rock constituents to condense in the presence of steam would be residual amounts of the alkali metals and silica. The principal alkali-vapor species are believed to be NaOH(g) and KOH(g) under the above conditions. The condensed forms are probably carbonates and silicates that can be assumed, for purposes of thermodynamic calculations, to be $Na_2Si_2O_5$, Na_2SiO_3 , K_2CO_3 , $K_2Si_4O_9$, and $K_2Si_2O_5$. The compound Na_2CO_3 is not stable under the conditions considered here. Whether an alkali condenses as a carbonate or reacts with a rock to form a silicate will depend upon the partial pressure of CO_2 in the gas and upon kinetic factors. Condensed NaOH and KOH will not form because of their high vapor pressures under these conditions. On the basis of available thermodynamic data (Kelley, 1962; Dow Chemical, 1970; Stern and Weise, 1969), and assuming a steam pressure of approximately 80 atm, the gaseous alkali-hydroxide vapor pressure is calculated to be approximately 10^{-3} atm at 2,240°F (1,500°K). Alkali chlorides are also believed to be very volatile, especially in the presence of steam. Data are not available, however, on the volatilities of these chlorides in rock melts at the low chlorine concentrations (~100 ppm) typically found in igneous rocks. The vapor pressure of volatile silicon-hydroxide species under these conditions exceeds 10^{-3} atm (Krikorian, 1970), so that direct gas-phase reactions can also lead to silicate formation. Thus, either gas-surface reactions or gas-gas reactions can occur, which will then lead to late-time condensation of alkali silicates and carbonates at temperatures of about 2,240°F (1,500°K) or less.

Assuming that the vapor phase contains a total of 10^{-3} atm of alkali-hydroxide molecules such as NaOH(g) and KOH(g) in a volume of $10^6 m^3$ means that about 2×10^4 g-atoms of alkali will condense at temperatures below about 2,240°F (1,500°K). This rough estimate of late-time condensate can be compared in Table 3 with the amounts of potentially available volatile gamma emitters from 1 Mt of all-fission explosives. It is apparent that the concentrations of radionuclides are low compared to the total amount of late-time condensate. We postulate that all of the potentially volatile species listed in Table 3, with the exception of permanent gases, are associated with the late-time condensation of alkali silicates and carbonates. It is further assumed that this

TABLE 3
Relative Volatilities in Steam of Potentially Volatile Gamma Emitters Produced by a Total of 1 Mt of Fission Explosives in Igneous Rock, After 180 Days

Nuclide	Nuclide, g-atoms	Volatility in steam, 80 atm at 662°F (623°K)
^{85}Kr	0.57	Permanent gas
^{103}Ru	0.35	Oxidizing, high; reducing, low
^{106}Ru (^{106}Rh)	2.8	Oxidizing, high; reducing, low
^{125}Sb	0.45	Probably high
^{127m}Te	0.075	High
^{137}Cs	15.1	Initially high, declining with time; CO_2 decreases volatility, Cl^- increases volatility
^{134}Cs	0.08	Similar to ^{137}Cs

condensation occurs over a time scale of minutes. Because of the late-time nature of this condensate, it will probably extend into the chimney region and be coated on rock surfaces.

Radioactive Volatilities in Power-Plant Operation

Some predictions can be made on the volatilities of the gamma emitters listed in Table 3 under power-plant conditions. The steam produced in the nuclear chimney is assumed to be at 80 atm and 662°F (623°K). The volatilities of NaCl, KCl, and NaOH in steam under the conditions of interest were studied by Spillner (1940); the volatilities of NaCl and NaOH were studied by Straub (1964). Results of the studies have been examined by Elliott (1952). The findings of Straub and Spillner are reasonably consistent for NaCl, and KCl appears to have an identical volatility. At 80 atm steam pressure and 662°F (623°K), the alkali-halide partial pressures in steam are approximately 1×10^{-4} atm above the pure-halide salts. For pure-liquid NaOH, Spillner found the partial pressure of gaseous alkali hydroxide to be about 1×10^{-3} atm under these steam conditions. Straub worked only with water solutions of NaOH but noted that the NaOH volatilities in steam were higher than for identical concentrations of NaCl in water. Elliott showed that the gaseous species accounting for the volatilities of alkali halides and hydroxides under these conditions are very complex. He tentatively proposed species such as $NaCl \cdot 7H_2O(g)$ and $NaOH \cdot 7H_2O(g)$. Although a positive identification of these species has not been made, they are useful in showing the marked dependence of alkali-halide and -hydroxide volatilities on steam pressure. An order of magnitude change in steam pressure was found to cause seven orders of magnitude change in alkali-halide or -hydroxide volatility. Such complex species would not be expected to be important at much higher temperatures, as, for example, at 2,240°F (1,500°K), where simpler molecules such as NaCl(g) or NaOH(g) would be expected to predominate.

In the nuclear chimney, alkali hydroxides are not stable as condensed phases, though alkali halides may be. If alkali halides are present as pure phases, they would contribute about 1×10^{-4} atm partial pressure to the steam. Any gaseous alkali hydroxides that are transported into the chimney as part of the late-time condensate are predicted to form either condensed silicates or condensed carbonates. Thus they would have a lower NaOH activity than the pure NaOH studied by Spillner.

The thermochemical NaOH activity in the rock can be estimated by comparing the calculated vapor pressure of NaOH(g) from a reaction such as

$$Na_2SiO_3\ (glass) + \tfrac{1}{2}\,H_2O(g) = \tfrac{1}{2}\,Na_2Si_2O_5\ (glass) + NaOH(g)$$

with the NaOH(g) vapor pressure over pure NaOH(l). The activity of NaOH(l) is thus found to be 1.3×10^{-2}. Applying this factor to Spillner's value for the volatility of pure NaOH in steam gives a value of 10^{-5} atm for the sodium volatility in steam above the sodium silicates.

It does not seem fruitful at this time to pursue a more complete consideration of the possible volatility reactions that can occur. The alkali metals will all be assumed to have about the same volatilities, either as hydrated-halide or -hydroxide species, with a partial pressure of 10^{-5} to 10^{-4} atm at 80 atm steam pressure and 662°F (623°K).

It should be noted that the alkali partial pressure at a constant 80 atm steam pressure decreases by only about a factor of 10 to 100 in going from 2,240° to 662°F (1,500° to 623°K). This means that about 2.5 to 25 percent of the alkali still remains in the gaseous state after cooling from 2,240° to 662°F (1,500° to 623°K), if concentrations are calculated according to the ideal gas law. There is quite a change in the types of gaseous species at the two temperatures. Whereas at 2,240°F (1,500°K) species such as NaOH(g) are important, at 662°F (623°K) species such as $NaOH \cdot 7H_2O(g)$ are believed to be important.

The radioactive cesium, as a first approximation, may be expected to form an ideal mixture with the principal alkali-metal constituents, either in the gas phase or in late-time condensate. Thus, as much as 25 percent of the radioactive cesium may be present in the gaseous state in a nuclear chimney at 80 atm steam pressure and 662°F (623°K). Any other radioactive-alkali species, such as ^{22}Na, should behave similarly.

Some complicating factors must be considered. First, it must be recognized that ^{137}Cs has volatile precursors that exist during part, if not all, of the early condensation period. Known precursors are ^{137}I with a 24-sec half-life, ^{138}I with a 6-sec half-life, and ^{137}Xe with a 3.9-min half-life. A more realistic explanation of the disposition of the ^{137}Cs might be to assume that a portion of the ^{137}I and ^{138}I isotopes become entrapped in the alkali-silicate or -carbonate matrix during the early stages of late-time condensation, but that most of the material migrates further up the chimney to convert to ^{137}Xe and finally condenses as it converts to ^{137}Cs. Thus, the final condensate will be farther from the puddle region and more highly enriched in ^{137}Cs relative to the alkali matrix. In the region in which the ^{137}Cs is enriched in the condensate, it will tend to form a carbonate rather than a silicate. Although thermodynamic data on cesium silicates are not available, the thermodynamic stabilities of alkali carbonates are known to increase relative to the silicates as the atomic number of the alkali increases (Kelley, 1962; Stern and Weise, 1969). In the carbonate form, the cesium volatility will be lowered by about two or three orders of magnitude relative to the silicate for $p(H_2O)/p(CO_2)$ ratios in the range of 1 to 1000.

As steam is extracted from the chimney, the radioactive-alkali content of the steam should vary qualitatively as follows. Initially, the radioactive content of the steam will be high, probably not as high as expected on the basis of ideal mixing, but rather of the order of 1 percent of the total radioactive alkali. During the initial extraction, the ^{137}Cs would be expected to be proportionately lower in steam (compared to ^{134}Cs and ^{22}Na) than its abundance would indicate, as a consequence of carbonate formation. As steam flow progresses, the ^{134}Cs and ^{22}Na, which are expected to be present as silicates, should volatilize first, leading to a gradual decline in the total radioactive alkali content of the steam and an enrichment of the proportion of ^{137}Cs in the gaseous radioactive alkali. At late times, the main species should be ^{137}Cs, which would originate from decomposition of a carbonate phase.

In the cases of ^{103}Ru and ^{106}Ru, the volatile species in the presence of steam are not known, so a calculation cannot be made. Radiochemical observations do indicate the presence of these isotopes in the chimney region in some nuclear events, although the behavior is erratic. The radionuclide ^{103}Ru has as its precursors 62-sec ^{103}Mo and 50-sec ^{103}Tc, while ^{106}Ru has 37-sec ^{106}Tc as its precursor. Molybdenum is known to be volatile in steam as $MoO_2(OH)_2(g)$ under oxidizing conditions, and it may be speculated that technetium is volatile as $TcO_3(OH)(g)$ by analogy with rhenium (Jackson, 1970, 1971). Ruthenium might also be expected to form volatile species, such as $RuO_2(OH)_2(g)$. According to calculations based on estimated data given by Jackson (1971), the lower-oxygen-content species, $RuOH(g)$ and $Ru(OH)_2(g)$, are unimportant. Also, the experience of Blasewitz and Mendel (1970) in calcining of nitrate solutions of radioactive wastes shows that ruthenium volatility increases as a function of hydrogen-ion concentration. Their results can also be interpreted as indicating that the ruthenium volatility is a function of oxygen partial pressure, since the HNO_3 present in the acid solutions would decompose to produce oxygen during calcining.

Qualitatively, it may be speculated that under neutral conditions, where the oxygen partial pressure is determined only by the dissociation of steam, species such as $MoO_2(OH)_2(g)$, $TcO_3(OH)(g)$, and $RuO_2(OH)_2(g)$ are probably volatile at 2,240°F (1,500°K). If significant amounts of hydrogen are produced by the explosion, the volatilities of these species will decrease. At 662°F (623°K) and 80 atm steam pressure, it is anticipated that the volatilities of ^{103}Ru and ^{106}Ru will be very low under reducing conditions and probably low under neutral conditions, but may be high under oxidizing conditions.

The precursors for ^{125}Sb are 9.4-day ^{125}Sn and 9.7 min ^{125m}Sn. Tin and antimony undoubtedly have volatile hydroxides, but their molecular species and thermodynamic data are not presently known. The radionuclide ^{127m}Te has 2.1-hr ^{127}Sn and 93-hr ^{127}Sb as its precursors. Tellurium is believed to form $TeO(OH)_2(g)$, which is quite volatile (Jackson, 1970, 1971). Fortunately, ^{127m}Te is much less important than the other fission products considered.

The order-of-magnitude nature of the volatility calculations used here should be emphasized. The primary objective has been to make reasonable estimates of the concentration of volatile gamma-emitting radionuclides in the steam. As experimental information becomes available, these conclusions may change. The radionuclides ^{137}Cs and ^{125}Sb may

TABLE 4
Chemical Analyses of Geothermal Waters from Several Sources

Rock type	Site	Maximum source temperature °F(°K)	Total solids (ppm)	Na^+	K^+	NH_4^+
Basaltic lava	Hengill, Iceland	446 (503)	913	174	10	0.1
Basaltic lava	Krysuvik, Iceland	446 (503)	2,030	500	68	0.0
Ignimbrites & pumiceous breccias	Wairakei, New Zealand	509 (538)	5,000	1,285	201	0.2
Ignimbrites & pumiceous breccias	Waiotapu, New Zealand	563 (568)	3,500	825	122	1.0
Volcanic origin	Tatio, Chile	104–194 (313–363) [a]	—	—	—	—
Volcanic origin	Reykjavik, Iceland	190 (361) [a]	285	159	1.4	0.0

[a] Surface waters.

be less volatile than anticipated, or at least less available in the chimney region for volatilization to occur. Other radioactivities that have been considered refractory here may actually be more significant. It is clear that these complex chemical interactions must be studied in the laboratory and the field before they can be defined accurately.

Assuming that the well is put into production 1½ to 2 years after the 180-day test, the chemical processes for carrying of radionuclides by steam should remain unchanged, but radioactivity levels will have been reduced. From the half-lives involved, it can be seen that ^{103}Ru and ^{127m}Te will have declined to very low levels, ^{106}Ru will have been reduced by a factor of $\sim$4, ^{125}Sb, ^{22}Na, and ^{134}Cs by a factor of 2 or less, and ^{137}Cs will have remained about the same. Considering the order-of-magnitude nature of the volatility calculations, and that ^{125}Sb and the alkali radionuclides are probably the most volatile radionuclides in steam, the general conclusions reached for the 180-day test are also applicable to the production period.

Noncondensable Gases in Steam

To evaluate the corrosion behavior of geothermal steam, an assessment must be made of the impurities present in the rocks and source waters of the wells. It may then be possible to derive some conclusions

TABLE 4 (*continued*)
Chemical Analyses of Geothermal Waters from Several Sources

Mg^{++}	Ca^{++}	H_3BO_3	Al^{+++}	Fe^{+++}	SiO_2	Cl^-	HCO_3^{-b}	$CO_3^{=b}$	$SO_4^=$	Total sulfide as H_2S (ppm)[b]
0.0	2.8	4.9	0.4	0.1	283	152	24	57	72	5.2
0.5	8.7	9.7	0.1	0.0	425	735	0	50	67	7.0
2.6	12	160	—	—	602	2,178	71	35	34	2.4
3.5	9	85	—	—	470	1,380	145	106	52	11
—	7–28	60–1,050	3–19	3–27	11–36	—	—	—	—	—
0.3	1.9	0.3	0.1	0.0	126	30	—	43	17	0.2

[b] Analyses for total ammonia, carbon dioxide, and H_2S will vary according to the method used to separate water from steam. The values determined by chemical analyses are lower than the actual concentrations in the underground water.

concerning the condensable and noncondensable impurities picked up by the steam and the chemical effects of these impurities on the steam-turbine system. Much of the information on corrosion and scaling is, of necessity, based on experience with natural geothermal wells. Caution must be exercised in extrapolating this information to the nuclear case, since different steam conditions and geology may prevail for the nuclear wells.

Chemical analyses of geothermal well waters from Iceland and New Zealand and surface waters from Chile are shown in Table 4. Geothermal-steam analyses from several sources are shown in Table 5. Except for the Showa-shinzan Volcano (Krauskopf, 1967), analyses were taken from various reports in the *Proceedings of the* [1961] *United Nations Conference on New Sources of Energy* (Elizondo, 1964; Bodvarsson, 1964; Burgassi, 1964; Ellis, 1964; Einarsson, 1964; Hansen, 1964). A number of minor constituents have not been included in the tables. The rock types represented are primarily igneous in origin.

In Table 4, analyses for solutes that can form gaseous products are not representative of the actual solute concentrations in the underground well waters because of losses incurred during separation of the geothermal waters from steam. During the separation, volatile solutes

TABLE 5

Chemical Analyses of Noncondensable Gases in Geothermal Steam

Site	Maximum source temperature, °F(°K)	Wellhead temperature, °F(°K)	Noncondensable gas content of total discharge (wt %)	Mole percent						
				CO_2	H_2	H_2S	CH_4	NH_3	N_2	Residuals
Hengill, Iceland	446 (503)	~320 (433)	0.3	84.6	2.1	4.9	0.0	—	—	8.4[a]
Hveragerdi, Iceland	446 (503)	~320 (433)	0.1	78.5	1.1	17.2	—	—	—	3.2[a]
Krysuvik, Iceland	446 (503)	~320 (433)	1.3	83.9	5.4	9.6	0.1	—	—	1.0[a]
Wairakei, New Zealand	509 (538)	~383 (468)	0.01–0.5	93.0	0.8	3.8	0.8	~0.2	1.4	—
Waiotapu, New Zealand	563 (568)	—	0.07–0.2	90.0	1.5	7.8	0.3	~0.2	0.2	—
Larderello, Italy (1870)	—	—	—	90.5	2.0	4.2	1.4	—	1.9	—
Larderello, Italy (1960)	473 (518)	~374 (463)	4.5	92.4	1.4	2.5	1.0	1.7	0.6	0.4[b]
The Geysers, California	>401 (478)	347 (448)	0.7	69.3	12.7	3.0	11.8	1.6	1.6	—
Showa-shinzan Volcano, Japan	—	381 (467)	0.6	69.4	12.4	3.9	0.1	0.0	9.1	5.1[c]
Showa-shinzan Volcano, Japan	—	622 (601)	2.2	84.9	6.6	1.0	0.1	0.0	5.2	2.2[c]

[a] Includes N_2.
[b] Primarily H_3BO_3.
[c] Primarily HCl, with lesser amounts of HF and SO_2.

tend to concentrate in the steam phase, whereas nonvolatile solutes tend to concentrate in the water. As an example, the actual CO_2 content of the well waters at Wairakei, New Zealand, is about 200 to 600 ppm in the underground condition, and of those at Waiotapu, New Zealand, about 800 to 2,500 ppm (Ellis, 1964), as compared with the much lower values for HCO_3^- and $CO_3^=$ listed in Table 4.

Tables 4 and 5 show that substances known to cause corrosion are present in significant amounts. Similar substances may be anticipated for the nuclear case. Whether or not corrosion and scaling reactions actually occur will depend on the specific conditions in each case. Chlorides, NH_3, CO_2, and H_2S are known to be corrosive to metals in certain instances. Steam and hot water are corrosive to all silicate materials, such as grouting in pipe joints, reinforced concrete pipes or tanks, and silicate filter beds. All such materials should either be avoided or protected with suitable coatings. Scaling (or deposition) may occur in pipes, turbines, or condensers as a consequence of volatile impurities in the steam, such as the chlorides, hydroxides, carbonates, sulfates, and silicates of the alkalies, calcium carbonate, and silica. Noncondensable impurities in the steam, such as CO_2, H_2S, H_2, and CH_4, may not necessarily be corrosive, but will need to be exhausted from the turbine discharge in order to have an efficient turbine operation.

Some information is available on the corrosion resistance of common alloys and other materials to geothermal fluids (Einarsson, 1964; Foster, Marshall, and Tombs, 1964; Bruce, 1964). The alloys most resistant to corrosion are found to be titanium, austenitic stainless steels, and chromium-plated steels. Epoxy resins are found to be resistant to corrosion at temperatures up to 212°F (373°K). Tests have been performed on alloy coupons with water at 122° to 464°F (323° to 513°K) and with steam at 212° to 392°F (373° to 473°K) for periods up to 150 days.

Although a wide variety of corrosion mechanisms has been identified, the most serious in alloys is probably stress corrosion. It is known that stress corrosion can be induced by trace amounts of either chlorides or sulfides, and can lead to formation of fissures and fractures in a variety of alloys that are used in the construction of turbines. A minimum concentration and temperature requirement for chloride-stress corrosion is believed to be 5 ppm of chlorides at 122°F (323°K). The presence of oxygen accelerates the rate of stress corrosion. There is no limiting stress below which cracking will not occur. Sulfide-induced stress cracking can occur in the presence of H_2S at temperatures up to at least 374°F (463°K). The gas H_2S also causes a phenomenon known as hydrogen infusion, which can lead to the embrittlement, blistering, and fracture of stainless steels.

Other corrosion mechanisms that should be mentioned are chemical corrosion and mechanical erosion. Chemical corrosion can occur from the presence of NH_3, H_2S, CO_2, and chlorides in geothermal steam. Generally, the corrosive action of these chemicals is enhanced by the presence of air. Chemical corrosion has been minimized or avoided in most natural geothermal power plants by the proper selection of alloys and by avoiding aeration. At the higher steam pressures and temperatures proposed for the nuclear application, however, chemical corrosion may be a more serious problem.

The problem of mechanical erosion is encountered when wet steam at high velocities impinges on the turbine blades. Such impingement can lead to a combined corrosion-erosion process that can seriously damage the turbine blades. A certain amount of success has been achieved in erosion resistance through the use of 13 Cr-stainless-steel alloys.

The experience gained with corrosion and scaling in natural geothermal power plants makes it apparent that similar problems are likely to be encountered in the nuclear case. These problems can be explored more fully by performing exposure tests on alloy coupons in geothermal steam and water under the proposed nuclear-application conditions. It is important in such tests that the compositions of the steam and the water be representative of the particular geothermal formation. Steam taken directly from a nuclear well would be most desirable; or, lacking an actual well, it may be possible to infer the impurities to be expected in the steam from analyses of core samples from the formation. It seems likely, as has been the case for natural geothermal wells (Ellis, 1964; Hansen, 1964; Foster, Marshall, and Tombs, 1964) that suitable alloys and other construction materials for turbine systems can be found through such screening tests.

ACKNOWLEDGMENTS

The author gratefully acknowledges the information provided him by H. B. Levy, and the encouragement and helpful suggestions of C. E. Chapin.

REFERENCES

Blasewitz, A. G., and J. Mendel. 1970. Battelle Memorial Institute Northwest, Richland, Washington. Private communication.

Bodvarsson, G. 1964. Physical characteristics of natural heat resources in Iceland. *In* Geothermal energy I: Proc. United Nations Conf. on New Sources of Energy, Rome, 1961, v. 2, pp. 82–90.

Bruce, A. W. 1964. Experience generating geothermal power at The Geysers power plant, Sonoma County, California. *In* Geothermal energy II: Proc. United Nations Conf. on New Sources of Energy, Rome, 1961, v. 3, pp. 284–95.

Burgassi, R. 1964. Prospecting of geothermal fields and exploration necessary for their adequate exploitation performed in various regions of Italy. *In* Geothermal energy I: Proc. United Nations Conf. on New Sources of Energy, Rome, 1961, v. 2, pp. 117–33.

Dow Chemical. 1970. JANAF thermochemical tables, with supplements through June 30, 1970. Dow Chemical Co., Midland, Mich.

Einarsson, S. S. 1964. Proposed 15-megawatt geothermal power station at Hveragerdi, Iceland. *In* Geothermal energy II: Proc. United Nations Conf. on New Sources of Energy, Rome, 1961, v. 3, pp. 354–62.

Elizondo, J. R. 1964. Rapporteur's summation. *In* Geothermal energy I: Proc. United Nations Conf. on New Sources of Energy, Rome, 1961, v. 2, pp. 48–50.

Elliott, G. R. B. 1952. Gaseous hydrated oxides, hydroxides, and other hydrated molecules. Ph.D. thesis, University of California Radiation Laboratory, Berkeley. Report UCRL-1831.

Ellis, A. J. 1964. Geothermal drillholes: Chemical investigations. *In* Geothermal energy I: Proc. United Nations Conf. on New Sources of Energy, Rome, 1961, v. 2, pp. 208–16.

Foster, P. K., T. Marshall, and A. Tombs. 1964. Corrosion investigations in hydrothermal media at Wairakei, New Zealand. *In* Geothermal energy II: Proc. United Nations Conf. on New Sources of Energy, Rome, 1961, v. 3, pp. 186–94.

Green, J., and R. Lessler. 1970. Lawrence Livermore Laboratory. Private communication to H. B. Levy.

Hansen, A. 1964. Thermal cycles for geothermal sites and turbine installation at The Geysers power plant, California. *In* Geothermal energy II: Proc. United Nations Conf. on New Sources of Energy, Rome, 1961, v. 3, pp. 365–77.

Heitmann, H. 1965. Glastechn. Ber., v. 38, p. 41.

Jackson, D. D. 1970, 1971. Thermodynamics of the gaseous hydroxides. Lawrence Livermore Laboratory, Livermore, California. Reports UCRL-71132 and UCRL-51137.

Kelley, K. K. 1962. Heats and free energies of formation of anhydrous silicates. U.S. Dept. of Interior, Bur. of Mines Report of Investigations 5901.

Krauskopf, K. B. 1967. Introduction to geochemistry. New York: McGraw-Hill, p. 461.

Krikorian, O. H. 1970. Thermodynamics of the silica-steam system. *In* Symposium on Engineering with Nuclear Explosives, January 14–16, 1970, Las

Vegas, Nevada. CONF-700101 (v. 1). Clearinghouse for Federal Scientific and Technical Information, U.S. Dept. of Commerce, Springfield, Va.
Lessler, R. M. 1970. Reduction of radioactivity produced by nuclear explosives. *In* Symposium on Engineering with Nuclear Explosives, January 14–16, 1970, Las Vegas, Nevada. CONF-700101 (v. 2). Clearinghouse for Federal Scientific and Technical Information, U.S. Dept. of Commerce, Springfield, Va.
Spillner, F. 1940. Chem. Fabrik, v. 13, no. 2, p. 405.
Stern, K. H., and E. L. Weise. 1969. High-temperature properties and decomposition of inorganic salts. Pt. 2, Carbonates. U.S. Dept. of Commerce, Nat. Bur. of Standards Report NSRDS-NBS 30.
Straub, F. G. 1964. Steam turbine blade deposits. Urbana: University of Illinois, Bull. 43, no. 59. Engineering Experiment Station Bull. ser. no. 364.
Styrikovich, M. A. 1958. Investigation of the solubility of low volatility substances in high pressure steam by radioisotopes. *In* Radioisotopes scientific research; Proc. International Conf., Paris, 1957. London: Pergamon Press, v. 1, pp. 411–25.

18. Geothermal Resources Research

JESSE C. DENTON AND DONALD D. DUNLOP

For several thousands of years man has known of the existence of geothermal energy: ancient man was familiar with volcanoes; the American Indian was aware of the geysers and warm springs in our Northwest. For more than a century now, wells have been drilled to produce petroleum. And for at least 40 years holes have been drilled at various angles, in groups from a single platform, in deep water, at great depths, etc. In spite of all this drilling expertise in the United States, geothermal energy has been tapped for power production only during the last 20 years. Why has this resource not become an important contributor to the nation's overall energy budget? To some extent, simply because not enough is known about geothermal resources, nor about generating economic power from geothermal fluids, except in such rare, special cases as The Geysers, in California. The purpose of this paper is to report on efforts being made by the Federal Government toward identifying the more pressing geothermal-resources research needs, and describing a research program that could yield a significant energy contribution from these resources.

For a number of years, various Government agencies have been conducting research directed toward an increased understanding of geothermal resources. In fiscal year 1973, funds for identified geothermal programs of Federal agencies, as requested in the President's budget, total less than $6.5 million; including all related work the total funding

Jesse C. Denton was with the National Science Foundation, Advanced Technology Department, Washington, D.C., when this paper was prepared. He is now at the University of Pennsylvania, National Center for Energy Management and Power, Philadelphia. Donald D. Dunlop is located in Fairfax, Virginia.

probably does not exceed $10 million. The Office of Science and Technology, in cooperation with the Federal Council on Science and Technology and the appropriate Federal agencies, is carrying out an extensive assessment of various energy technologies, as directed by the President in his message on energy of June 4, 1971. In December 1971, the Office of Science and Technology requested the Department of the Interior to assume the responsibility for assessing the technology of geothermal energy and recommending a program of research and development. The United States Geological Survey is conducting the assessment with help from two groups: the panels of a Geothermal Resources Research Conference held in September 1972; and an informal interagency geothermal coordinating committee, established in September 1971 to promote exchange between the interested Federal agencies, meeting at least monthly, with participants from the Bureau of Reclamation, Office of Saline Water, U.S. Geological Survey, Bureau of Mines, Bureau of Land Management, Atomic Energy Commission, National Aeronautics and Space Administration, Advanced Research Projects Agency, Environmental Protection Agency, and the National Science Foundation.

In response to a proposal submitted by the University of Alaska, with Walter J. Hickel as principal investigator, the National Science Foundation granted funds in support of a working conference on geothermal-resources research. The objective of the conference, held in September 1972, was to assess the state of the art of geothermal science and technology and to recommend a research program establishing the proper role of geothermal resources in providing (1) additional energy to alleviate the nation's impending shortage, (2) water to supplement present supplies, and (3) additional mineral resources. Though attendance was by invitation only, the entire geothermal community was invited to provide written recommendations to the conference. The conference was organized into six substantive panels with the following titles: (1) resource exploration, (2) resource evaluation, (3) reservoir development and production, (4) utilization technology and economics, (5) environmental effects, and (6) institutional considerations. A report of this meeting has been released (Hickel, 1972).

This paper identifies some topics that are appropriate for geothermal-resources research, drawing chiefly on the contributions to the Seattle conference and using its six-part classification.

Resource Exploration

The objectives of geothermal-resource exploration are to locate areas underlaid by hot rock; to estimate their volume, temperature, and permeability; and to determine the nature of any producible fluids. For convenience, geothermal resources may be divided into four types: convective hydrothermal systems; geopressured systems; hot, impermeable rock; and magma systems. Each type has its own physical properties and poses its own problems. Some of the more important problems are concerned with determining the age, size, and magmatic type of igneous occurrences related to convective hydrothermal systems; the nature and cause of structural features controlling the location of convective hydrothermal and hot, impermeable rock systems; and the relationship of convective hydrothermal systems to broad regions of elevated heat flow.

Geochemical data collected during exploration are useful in all phases of evaluation, development, and utilization. Several hydrochemical indicators have been used successfully in geothermal exploration, but others are difficult to interpret. Research is needed in the chemical, physical, and thermodynamic properties of aqueous solutions at temperatures between 100° and 400°C (212° and 750°F); the relationship between the chemical composition of geothermal fluids and the temperature of the host rock; and the isotopic variation of geothermal waters.

Geothermal systems exhibit variations in electrical resistivity. Research should be directed toward understanding the effect of porosity, water salinity, and temperature upon electrical resistivity in geothermal reservoirs; improving electric-field techniques and procedures for extracting true resistivity values from field data; and developing complementary exploration techniques that will improve the interpretation of resistivity data. Electrical-exploration research should include further research on dc-resistivity, self-potential, electromagnetic, telluric, and magnetotelluric techniques. Airborne methods may be of particular interest in large-scale reconnaissance.

Both active and passive seismic techniques are useful in geothermal exploration, as noted by Combs and Muffler (this volume). Active seismic methods should be studied to characterize energy absorption and attenuation, as well as frequency shifts in known high-temperature systems. Accurate determination of earthquake patterns can help delineate faults that may channel hot fluids to drillable depths. A greater understanding of the effects of fluids and high temperatures on fault strength and slip characteristics is also needed. Finally, seismic-noise studies should be undertaken to evaluate temporal and spatial noise variations; the characteristics of recognized noise sources, noise spectra, noise coherency, location of source, and cause of noise; and the direction of noise propagation and apparent noise velocity.

The gravitational and magnetic characteristics of geothermal systems vary greatly from one geologic province to another. Further research is needed to determine the source of the gravitational and magnetic anomalies associated with known geothermal areas and to decide whether these anomalies can be used as indicators of the internal temperature of the system.

Thermal-exploration techniques provide a direct method for assessing the size and potential of a geothermal system. More regional heat-flow determinations are needed to refine the estimates of available resources. And research is needed to determine the relationships between temperature gradients, subsurface isotherm patterns, and the geometry of geothermal systems. Laboratory temperature experimentation on model geothermal systems should be useful. Hydrologic studies are required to understand more fully the effect of ground-water movement on local geothermal gradients.

The development of new, less expensive techniques for drilling to depths in excess of 8 km would benefit geothermal, oil, and gas exploration greatly, though drilling research is needed at shallower depths as well. The major difficulties with present technology are in its application at high temperatures. Problems are encountered with rubber seals, valves, cements, drilling muds, heat shields, sound mufflers, etc. Formation testing in uncased holes can be useful in the exploration for reservoirs. Problems of isolating tested intervals in unconsolidated sands and in fractured reservoirs have not been solved. Inexpensive core recovery with reservoir fluid in place would be useful in defining a model of a geothermal cell. Instrumentation for logging devices operating at temperatures above 180°C (350°F) is urgently needed, as is research on methods of transmission of information from the bore-hole face to the surface. Development of cheap, low-density, low-viscosity, thermally nonsensitive, thermally conductive, and high-surface-tension drilling fluids would lower drilling costs and leave the bore-hole face in a more nearly undisturbed state. Drill cuttings, cores, and logs should be preserved. The rotary drill must be improved and new techniques such as erosion drills, electric-melting drills, and turbine drills should be explored.

In addition to the more applied research, basic research should be pursued on physical properties of geothermal water-steam mixtures, thermal geophysics and hydrology, and computer modeling.

Resource Evaluation

We all recognize that geothermal resources include not only energy but water and, in many areas, valuable minerals as well. Thus the research program envisioned is one that attempts to evaluate all three potential contributions. It can be anticipated that the initial focus of the research program will be on the energy component. We know that the center of the Earth consists of magma, a molten mass at several thousand degrees temperature. There is a steady decline in temperature as distance from the magma increases. Theoretically, this energy source can be tapped from any point on Earth simply by drilling sufficiently deep holes, providing passage for some heat-transfer fluid, and extracting the heat. Practically speaking, this hot mass is much too deep to tap in many large areas of the world. Yet in other areas the resource is much closer to the surface, occasionally so close as to be manifested by such surface phenomena as geysers and hot springs.

The need for evaluation of the resource can readily be seen by noting the disparity among the estimates of persons knowledgeable in the field. For example, White (1965) estimated that 5,000 to 10,000 Mw of power could be generated for at least 50 years under present economic conditions and technology. The National Petroleum Council (1972) estimates that by 1985 the U.S. could be generating 7,000 to 19,000 Mw from geothermal resources. And R. W. Rex (1972), in his testimony before the Senate Interior Committee on June 15, 1972, estimated that the U.S. could develop 400,000 Mw capacity within 20 years in the West alone, and that this capacity could be maintained for 100 years. His latest estimates are given in this volume.

Resource-evaluation research should be directed at three classes of geothermal resources: (1) hydrothermal-convection systems; (2) geopressured systems; and (3) hot, dry rock resources. During the progress of the resource-evaluation research program a national inventory of geothermal resources, including energy, water, and minerals, should be established.

Two major types of hydrothermal-convection systems are recognized. The relatively rare vapor-dominated systems account for most of the present geothermal power production worldwide. But the more abundant liquid-dominated systems are expected to account for most increases in power production in the future. To achieve increased utilization of hydrothermal-convection systems, we must study (1) the size, life, origin, and dynamics of natural hot-water geothermal systems; and (2) the geological, geochemical, hydrologic, and geophysical characteristics of geothermal systems, as a basis for engineering development of geothermal resources.

Geopressured aquifers occur in zones bounded by faults that generally parallel the Gulf coastline. Three forms of energy are available in the produced water, the thermal energy, the mechanical energy available from the high pressure, and the methane contained in the fluid. A regional resource study should be conducted to compile data on temperature, structure, salinity, sedimentary facies, pressure-potential distribution, and possible waste-water disposal methods in the Gulf Coast region extending well into the coastal plain (~1,000 km) and out onto the adjacent continental shelf. This study should identify three suitable sites for subsequent field research leading to a determination of the life, size, and total energy content of typical geopressured reservoirs. It will be necessary to design, drill, and test the formations selected and to instrument and observe the performance of disposal wells.

No adequate means has yet been demonstrated to deliver the thermal energy of hot, dry rock as usable energy at the Earth's surface. Present concepts involve the creation of artificial hot-water circulation systems by drilling a bore hole and then fracturing the rock. Various fracturing schemes may be employed—hydraulic fracturing, thermal and chemical fracturing, or nuclear explosions. A second well would be drilled into the fractured zone and a hot-water circulation system set up using the deep well as an injection well and the other as a production well, with a heat exchanger between the two wells on the Earth's surface. In order to establish the viability of hot, dry rock as an energy resource it will be necessary to select a test site; fracture the formation; study and model the rock mechanics and hydrology of the holes, site, and fracture system; drill and instrument monitoring wells; and design, fabricate, and install heat exchangers between the extraction and injection wells. Obviously, new wells must be drilled on new sites to test alternative fracturing methods.

Reservoir Development and Production

A fair amount is known about flow of fluids in porous media from the production of oil and gas. Fracturing techniques for oil and gas reservoirs are available to increase the permeability in reservoirs. Knowledge of heat flow in porous media is available from studies of in-situ combustion. An important first step is to adapt and apply these techniques to the development of geothermal resources.

For the generation of power, a hot fluid must be produced at the surface. The flow may be derived from the reservoir fluid itself or it may be necessary to inject a fluid at the surface that would be heated by the reservoir and subsequently produced at the surface. Whether the natural reservoir fluids or injected fluids are produced, the production of dissolved minerals may well be important. Hot water is an excellent solvent for many minerals. This mineral production may create problems with respect to energy production but at the same time may make available the mineral resource itself.

A great deal of research will be needed to form a sound basis for reservoir development and production methods. Research is required in drilling technology, geological formation evaluation, reservoir engineering, production engineering and production management, and well-stimulation technology and techniques. Research in drilling technology and formation evaluation has been discussed.

Reservoir engineering involves assessment of the size and deliverability of the geothermal resource and planning for the optimum development of the resource. Fundamental research into the physical, chemical, and thermal behavior of multiphase aqueous systems in porous media is needed to establish physical principles leading to viable physical and mathematical models of geothermal-fluid systems. These studies would include thorough transport phenomena studies. There is an urgent need for field-reservoir engineering studies to construct mathematical models of sufficient detail to permit optimization studies of the effects of well spacing, liquid recharge from aquifers, reinjection of condensate and cooled liquids, cooling owing to in-situ boiling of liquids, changes in pore space and fluid conductivity owing to precipitation of salts and solids, and other important parameters. To this end field geothermal research sites are necessary for acquisition of information and for validation of reservoir engineering and hydrologic theory.

Production engineering and management problems concern well workovers, well stimulation, corrosion and scaling, geothermal-fluid separation and metering, critical velocity of flow in pipes and valves, and deep-well pumps.

Well-stimulation technology and techniques are concerned with improving the permeability of the formation, and thus are important in both production and reinjection. Hydrofracturing, acid treatments, thermal fracturing, and chemical- and nuclear-explosion fracturing should all be considered. In particular, the large void volumes produced by nuclear explosives could provide space for mineral deposition to collect from flashing hot water. Nuclear explosives might also dry up a wet flashing system by providing space for the flashing to occur underground.

Utilization Technology and Economics

Another question for research is the choice of technology best suited to each type of geothermal resource. For the vapor-dominated systems and the higher-temperature (above 200°C/390°F) liquid-dominated systems, the technology for economic production of power appears well developed. In the case of vapor-dominated systems, such as that at The Geysers field, the steam is available at the wellhead. After filtering, it is fed to relatively low-pressure turbines that drive generators. In the higher-temperature liquid-dominated systems, the liquid in the reservoir flashes into a mixture of steam and liquid upon reduction of pressure by the well. After the liquid is separated from the steam, the steam is fed to a turbine. It appears that there is no pronounced need for research in these two cases. If it proves feasible to recover the thermal energy from hot, dry rock systems, the same may be said if the temperature of the formation is high enough to heat introduced fluids sufficiently.

In all cases it would be helpful to know more about gas content, particulate removal, noise control, removal of dissolved salts such as silica and boron, and disposal of brine, condensate, and solids.

It is believed that there are many geothermal reservoirs (perhaps as many as 80 percent of all reservoirs) in which the temperature is not high enough to provide fluids that may be used with existing technology to produce economic power. Successful demonstration that these marginal reservoirs can be utilized economically for generating electric power and for other uses, such as commercial recovery of minerals, would expand the recoverable geothermal resources many times. For the generation of power from such reservoirs, U.S. industry has made some progress in developing the technology of vapor-turbine (binary-fluid) systems in which heat is transferred to a low-boiling point fluid (such as freon or isobutane), which is then used as the working fluid in a closed-cycle system (see Anderson, this volume). Further research on a variety of such systems, as well as other power cycles deriving energy from hot concentrated brines, is needed, and demonstration plants will need to be built. Desalination pilot plants using geothermal energy are needed, and mineral production may also require pilot plants.

Immiscible heat-transfer systems for extracting geothermal energy should be studied. These systems involve injecting a water-immiscible fluid into the formation and then producing the combined fluids. Mechanical separation at the surface would provide a hot, clean working fluid. Corrosion rates of materials exposed to geothermal brines are required. In the case of geopressured resources, research is needed to remove methane from the formation fluid. Direct conversion of geothermal heat to electricity, as by thermoelectric techniques, should be studied.

Environmental Effects

Geothermal developments are unique in the sense that all activities related to the power-production cycle are localized to the immediate vicinity of the power plant. Support operations such as mining, fuel processing, transportation, and other handling facilities do not exist. For this reason the environmental effects are site-dependent in origin. Certain undesirable effects, such as the production of hydrogen sulfide gas, can extend for several miles from the geothermal field itself and thus introduce environmental problems into the surrounding region.

Potential environmental effects that need to be examined are gaseous and particulate emissions; land modification; subsidence; seismic hazards; surface- and ground-water pollution; biological effects; noise effects; and social effects. Research should be directed toward an accurate determination and evaluation of the character and magnitude of these effects. It will be necessary to set up a careful program of monitoring as geothermal developments progress, and it will probably be necessary to develop techniques for decreasing some environmental impacts. Reinjection of geothermal fluids into the formation, for example, cannot be done without proper regard for ground-water quality.

Institutional Considerations

An important objective is to define the organizational, legal, and regulatory considerations that will permit geothermal development in a manner deemed safe, economical, environmentally sound, and equitable for all concerned parties. These parties are chiefly the general public as users, private industry as developers, and local, state, and Federal governmental agencies as regulators. The geothermal industry is an infant with respect to the total energy industry, and institutional problems will have to be faced squarely and solved at the highest level if geothermal resources are to become a major supplier of energy. There are questions of law to address as well as questions of the proper role of the Federal Government vis-à-vis industrial research and development. The adequacy of the existing body of law should be assessed. Moreover, the usual practice of leaving potential profitmaking ventures to industry is desirable, but how does one recognize beforehand that the profit potential is sufficient to induce industrial development? What regulations will be necessary to ensure the optimal development of the resource for the good of the country? The entire regulatory process should be reexamined.

How, then, should Government research be structured? At present, a number of Federal agencies are conducting small projects in geothermal areas. Should a new agency be created to coordinate new research programs? Should the Government create a national energy council of which geothermal energy would be a part? Clearly some critical questions concerning the institutional impact of the geothermal-resources research program need to be addressed and solved. We recommend that a Presidential study commission weigh the merits of different organizational structures.

At another level the international political potential of geothermal resources should not be overlooked. For example, if geothermal energy is indeed available all over the world, many political benefits could accrue to the U.S. by exploiting the technology for the development of geothermal energy in India or elsewhere in Asia or other parts of the world.

From a purely selfish point of view, another opportunity to export U.S. technology would present itself. And the exploitation of these resources would stand to raise the living standards of many countries and perhaps improve their economic stability. All of these benefits would accrue indirectly to the rest of the world.

A great deal of study must precede solutions to these institutional problems. Such efforts should include the compilation of state, Federal, and local laws and regulations governing geothermal developments and an analysis of their functioning, administration, and interactions; the preparation of a model code; an analysis of legal ownership of geothermal resources and the problems of utilization pending final determination of ownership; the establishment of effective liaison between all sectors of the field and exploration of mechanisms for cooperation; the establishment of a Federal information-gathering function to promote systematic gathering and rapid distribution of geothermal-resource information; and the establishment of a Presidential study commission to recommend organizational and mission changes.

Summary and Conclusions

From the outset, a program of research should include both short-range and long-range goals. One of the first tasks is to define a program commensurate with the magnitude of the resource and adequate to its accelerated development. Such a program should emphasize the development of (1) vapor-turbine power-generating systems; (2) improved exploration methods; (3) desalination methods for geothermal fluids; (4) cheaper methods for deep drilling into hot formations; and (5) better models of geothermal reservoirs.

All of this research activity, however, will become irrelevant if the economic parameter is not exhaustively researched. If comparisons are to be valid, the first stage of analysis must use the economic parameters that other primary-energy producers use. But because present uses are related to utility economics, perhaps the only legitimate approach is to focus on the cost to the ultimate consumer. Utilities are a regulated industry and the regulation itself may be brought into question.

It is apparent that the energy, the water, and perhaps the mineral potentials of geothermal resources are enormous. It is also clear that the considerable experience of other fields suggests that this resource may be tapped initially by extending the existing technology. But methods of defining and locating the resource per se, as well as evaluating the economics of the development, require additional research. The environmental aspects and the institutional factors also need study but appear tractable. The real question is whether the bulk of this reserve is available under economically feasible conditions.

REFERENCES

Hickel, W. J., et al. 1972. "Geothermal energy: A national proposal for geothermal resources research" (report of working conference held September 1972, Seattle, Washington). A limited number of single copies of this 95-page report are available from Kenneth M. Rae, Vice President for Research, University of Alaska, College, Alaska 99701.

National Petroleum Council. 1972. "U.S. energy outlook, an interim report," an initial appraisal by the New Energy Forms Task Group, 1971–85.

Rex, R. W. 1972. Testimony before the Senate Interior Committee.

White, D. E. 1965. Geothermal energy. U.S. Geological Survey Circular 519.

Index

Index